An Omics Perspective on Cancer Research

William C.S. Cho
Editor

An Omics Perspective on Cancer Research

Editor
William C.S. Cho
Department of Clinical Oncology
Queen Elizabeth Hospital
Hong Kong SAR
PR China

ISBN 978-90-481-2674-3 e-ISBN 978-90-481-2675-0
DOI 10.1007/978-90-481-2675-0
Springer Dordrecht Heidelberg London New York

Library of Congress Control Number: 2009929347

Printed on acid-free paper

Springer is part of Springer Science+Business Media (www.springer.com)

Preface

Postgenome science is characterized by omics data related to genome, transcriptome, epigenome, proteome, metabolome and interactome. In the omics era, it is a revolution in cancer research which fundamentally shifts the strategy from piece-by-piece to global analysis and from hypothesis-driven to discovery-based research.

This book attempts to take a comprehensive overview on different areas of omics technologies for cancer research. It expounds important omics technologies which are multidimensional tools that may translate into clinical applications serving as the basis for personalized medicine of the 21st century.

This book not only serves as an introduction to novices to the area and a useful reference for those already involved, but also serves as a stimulus to these and others to develop new approaches to cancer research.

November 2009 William C.S. Cho

Contents

Chapter 1
Omics Approaches in Cancer Research

William C.S. Cho

Abstract Cancer is a complex genetic, proteomic, and cellular disease caused by multiple factors via genetic mutations (hereditary or somatic) or environmental factors. The emerging omics technologies are being increasingly used for cancer research and personalized drug discovery, including genomics, epigenomics, proteomics, cytomics, metabolomics, interactomics, and bioinformatics. Recent advances in high-throughput omics technologies have provided new opportunity in the molecular analysis of human cancer in an unprecedented speed and details.

The detection and treatment of cancer is greatly facilitated by the omics technologies. For example, genomics analysis provides clue for gene regulation and gene knockdown for cancer management. The approval of Mammaprint and Oncotype DX indicates that multiplex diagnostic marker sets are becoming feasible. Discovery of the involvement of microRNAs in human cancers has opened a new page for cancer researchers. Some therapeutic drugs targeting on DNA methylation and histone deacetylation are currently undergoing keen studies. Proteomics also plays an important role in cancer biomarker discovery and quantitative proteome-disease relationships provide a mean for connectivity analysis. Fluorescent dye enables a more reliable analysis and it facilitates the progress of biochip and cytomics. The huge amount of information collected by multiparameter single cell flow- or slide-based cytometry measurements serves to investigate the molecular behavior of cancer cell populations. Cancer is also an ideal field of application for metabolite profiling owing to its unique biochemical properties.

It is envisioned that omics technologies will enhance our understanding of molecular signatures of cancer on both qualitative and quantitative patterns. The novel omics technologies have brought powerful abilities to screen cancer cells at the gene, transcript, protein, metabolite, and their interaction network level in searching of novel drug targets, expounding the drug mechanism-of-action, identifying adverse effects in unexpected interaction, validating current drug targets, speeding up the discovery of new targets, exploring potential applications for novel drugs,

W.C.S. Cho (✉)
Department of Clinical Oncology, Queen Elizabeth Hospital, Hong Kong SAR, PR China
e-mail: chocs@ha.org.hk

W.C.S. Cho (ed.), *An Omics Perspective on Cancer Research*,
DOI 10.1007/978-90-481-2675-0_1, © Springer Science+Business Media B.V. 2010

and enabling the translation from bench to bedside. The field is moving fast, specialized techniques are being developed to integrate omics information and to enable new research avenues that can take advantage of and apply this information to new therapies. In this chapter, different omics technologies are briefly introduced.

1.1 Introduction

After the Human Genome Project, the scientific era of omics has emerged to revolutionize our way of studying and learning about cancer (Keusch 2006; Nicholson 2006; Finn 2007; Hamacher et al. 2008). The Greek suffix "*ome*" means collection or body, the term omics represents the rigorous study of various collections of molecules, biological processes, physiologic functions and structures as systems. It deciphers the dynamic interactions between the numerous components of a biological system to analyze networks, pathways, and interactive relations that exist among them, such as genes, transcripts, proteins, metabolites, and cells (Keusch 2006).

Recent studies use a combination of high-throughput omics technologies, including genomics, transcriptomics, epigenomics, proteomics, metabolomics, interactomics, and bioinformatics. Omics research has launched the era of cancer molecular medicine. Application of omics in cancer research provides multi-dimensional analytical approach that reveals cancer molecular portraits. It provides a good deal of biological information and new insights into the gene, protein, and metabolite profiles during various stages of cancer. The recent developments in screening omics technologies have allowed the discovery of combinatorial cancer biomarkers (Cho 2010a). Omics analysis may be translated into practice for risk stratification, early detection, diagnosis, biomarker identification, treatment selection, prognostication, and the monitoring for recurrence (Cho 2007a). Several commonly used omics technologies in cancer research are overviewed in this chapter.

1.2 Genomics

Genomics is the study of the genomes of organisms. The application of genomics may lead to the discovery of a host of novel oncogenes and tumor suppressors, which will have a significant impact in our understanding of tumorigenesis and in the clinical management of cancer patients (Shih and Wang 2005). Genomics can also be used to identify molecular pathways that are deregulated in cancer which will not only elucidate underlying tumorigenic mechanisms, but may also help to determine the classes of drugs that are used for cancer treatment (Furge et al. 2007).

It is a challenge for genomics studies to identify unique biomarkers in complex biological mixtures that can be unambiguously corrected to biological events so as to validate novel drug targets and predict drug response. Clinically useful biomarkers are informative for regulatory and therapeutic decision making regarding candidate drugs and their indications which may help bring new medicines to the right patients.

With the next generation sequencing technologies, a wide range of applications are now affordable and within reach. Systems can be used for applications ranging from megagenome and genome sequencing (*de novo* or re-sequencing) to transcritpome analysis (e.g. cDNA, serial analysis of gene expression, cap analysis gene expression), to regulome studies (e.g. chromatin immunoprecipitation, microRNA). Genetic profiling associating with chemotherapeutic outcome contributes to the understanding of oncogenesis, tumor growth, and therapeutic response which may indicate new targets for cancer treatment.

1.3 Epigenomics

Epigenomics is a relatively new omics technology that can be useful for cancer management. Aberrant gene function and altered patterns of gene expression are key features of cancer. Increasing evidences show that acquired epigenetic abnormalities participate with genetic alterations to cause this dysregulation (Jones and Baylin 2007). Epigenetic silencing of tumor suppressor genes plays an important role in the pathogenesis of most cancers (Hatada 2006). Recent technological advances are now allowing cancer epigenomics to be studied genome-wide, an approach that may provide both biological insight and new avenues for translational research (Esteller 2007). Since epigenetic modifications contribute to carcinogenesis evolution, it may help to optimize the potency of epigenome targeting agents (Karamouzis et al. 2007).

1.4 Transcriptomics

Global mRNA transcript expression profiling is a very powerful tool in modern research because it encompasses the cell's transcription of activated genes. Transcriptomics plays several roles in advanced management of cancer in the post-genome era. Its main applications involve cancer diagnostics and prognostics based on tumor gene expression profiling of mRNA, as well as biomarker applications in drug discovery and development (He 2006). Even though less than 2% of the mammalian genome encodes proteins, a significant fraction can be transcribed into non-coding RNAs. MicroRNAs are single-stranded small non-coding RNA molecules which main function acts as epigenetic regulator of their corresponding target genes at the post-transcriptional level. Numerous microRNAs are deregulated in human cancers, and growing evidences indicate that they can play roles as oncogenes or tumor suppressor genes (Negrini et al. 2007). In recent years, microRNA is becoming important for the understanding of tumorigenesis and some of them are believed to have diagnostic and prognostic roles (Cho 2007b, 2009, 2010b,c; Cho et al. 2009; Rosenfeld et al. 2008).

1.5 Proteomics

During the past few years, the emphasis on genomics has shifted via transcriptomics to proteomics, the science of understanding how the whole set of proteins are expressed and function at the cellular level. Proteins are the physiological/pathological active key players, the relationship between gene expression measured at the mRNA level and the corresponding protein level is complex in cancer. The mRNA/protein correlation coefficient varied among proteins with multiple isoforms, indicating potentially separate isoform-specific mechanisms for the regulation of protein abundance (Reymond and Lippert 2008). Proteomics enable the quantitative investigation of both cellular protein expression levels and protein–protein interactions involved in signaling networks. Monitoring the protein expression pattern in tumor cells by high-throughput proteomics technologies offers opportunities to discover potential cancer biomarkers (Cho 2007c). There is an intense interest in applying proteomics to foster an improved understanding of cancer pathogenesis, develop new cancer biomarkers for diagnosis, and early detection using functional proteomic signatures (Sanchez-Carbayo 2006).

Recent progress in clinical proteomics is mostly contributed by sophisticated new methodologies for proteome analyses (Rosenblatt et al. 2004). Different proteomics tools such as two-dimensional difference gel electrophoresis, protein microarray, mass spectrometry (MS) platforms including matrix-assisted laser desorption/ionization, electrospray ionization, surface-enhanced laser desorption/ionization, isotope-coded affinity tag, isobaric tags for relative and absolute quantification, as well as multidimensional protein identification technology have been used for differential analysis of biological samples (Cho and Cheng 2007; van der Merwe et al. 2007). Oncoproteomics has the potential to revolutionize clinical practice, including early cancer diagnosis and screening based on proteomic portraits as a complement to histopathology, individualized selection of therapeutic combinations that target the entire cancer-specific protein network, real-time assessment of therapeutic efficacy and toxicity, and rational modulation of therapy based on changes in the cancer protein network associated with prognosis and drug resistance (Cho 2007d). With the application of mathematical tools in oncoproteomics, scientists can even describe the connectivity of chemical and/or biological systems using networks (González-Díaz et al. 2008).

1.6 Metabolomics

Metabolome is the complete complement of all small molecule (<1.5 kDa) metabolites found in a specific cell, organ or organism. Metabolomics is a dynamic portrait of the metabolic status of living systems. This new omics technology purports to give us a cross-section of the small molecular weight components in cells, tissues, organ, body fluids or the whole body at any moment in time, such that the constellation of molecules and their relative proportions can provide us with information about the functional state (or the dysfunctional state) at that time. From reading such a profile, information may be gleaned which indicates some activity state that is meaningful in the present circumstances. Elucidation of cellular responses to molecular damage, including evolutionarily conserved inducible molecular defense systems, can be achieved with metabolomics and can lead to the discovery of new biomarkers of molecular responses to functional perturbations. Metabolomics is useful for cancer management which permits simultaneous monitoring of many small molecules, as well as functional monitoring of multiple pivotal cellular pathways (Claudino et al. 2007). Animal study has shown that metabolomics may have a greater chance of success in toxicology and biomarker assessment than genomics and proteomics (van Ravenzwaay et al. 2007). It is well-known that cancer cells typically consume glucose and glutamine voraciously. Many cancer cells use glucose inefficiently, through glycolysis rather than oxidative phosphorylation (Warburg effect) (Hsu and Sabatini 2008). For human studies, the Human Metabolome Project attempts to identify and catalog all of the metabolites found in the human bodies, aiming to complete a metabolite inventory for human beings, to generate resources that can facilitate metabolomics research across many different disciplines, as well as to provide detailed information about the linkage between human metabolites and the genes, proteins, and pathways with which they are involved (Wishart et al. 2006; Wishart 2007).

Metabolomics is the study of metabolism at the global level. Metabolomic studies capture global biochemical events by assaying thousands of small molecules in cells, tissues, organs, or biological fluids-followed by the application of informatic techniques to define metabolomic signatures (Kaddurah-Daouk et al. 2008). Technical developments in ultra-high pressure liquid chromatography, Fourier transform MS, orbitrap MS, higher field and cold-probe nuclear magnetic resonance magnets have already had a significant impact in metabolomics. Continuing developments in lab-on-a-chip technologies can be expected to make metabolomics much cheaper. However, the analysis of the metabolome is particularly challenging due to the diverse chemical nature of metabolites (Gowda et al. 2008). We need to understand more about the basic concepts of physiology and metabolism in the body before we can seriously use metabolomics to great advantage.

1.7 Interactomics

The behavior, morphology, and response to stimuli in biological systems are dictated by the interactions between their components. These interactions are shaped by genetic variations and selective pressure (Cesareni et al. 2005). The interactomics architectural map may represent the first step toward the attempt to decipher the carcinogenesis at the systems level (Hsu et al. 2007). Large-scale mappings of protein–protein interactions give new insights of the complex molecular mechanisms inside a cell (Futschik et al. 2007). High-throughput technologies are employed to chart dynamic interactions between the components of a biological system, as well as to simulate and analyze pathways. The significant advances in microarray and proteomics analyses have resulted in an exponential increase in potential new targets and have promised to shed light on the identification of cancer biomarkers and cellular pathways (Cesareni et al. 2005; Heck 2008).

1.8 Cytomics

Cytomics is an omics technology that applying various bioinformatic techniques to investigate the functions and molecular architecture of the cytome. It has received great attention in recent years as it allows the qualitative and quantitative analyses of individual cells, cell constituents, as well as their intracellular and functional interactions in a cellular system. High-content and high-throughput single-cell analysis may lead to systems biology and cytomics. The application of cytomics in cancer research is very broad, ranging from the better understanding of the tumor cell biology to the identification of residual tumor cells after treatment. The ultimate goal is to pinpoint these processes on the molecular, cellular, and tissue level. A comprehensive knowledge of cytomics requires multiplex and functional cellular and tissue analyses (Tárnok et al. 2006).

1.9 Phenomics

Human phenome can be viewed as a landscape of interrelated diseases, reflecting overlapping molecular causation. Cancer is a highly complex and heterogeneous disease involving a succession of genetic changes which resulting in a molecular phenotype with malignant specification. It does not have a clear inheritance pattern which involves multiple genes with modest effects acting independently or interacting (Mei et al. 2007). Cancer phenomics uses objective and systematic acquisition of phenotypic data at many levels which may help to evaluate the genetic influences of cancer (Zbuk and Eng 2007). Oncological research using systematic analysis of phenotype relationships to study human biology is still in its infancy. The major challenges for the identification of genetic effects are genetic heterogeneity and difficulty in analyzing high-order interactions (Lussier and Liu 2007).

1.10 Bioinformatics

Bioinformatics is the application of information technology to the field of molecular biology. Computational analysis (e.g. data mining and machine learning algorithms) has become an essential element of cancer research with a main role of bioinformatics being the management and analysis of huge data. Cancer bioinformatics is a branch of bioinformatics. The bridge between information and modeling in cancer can be achieved by the bioinformatics strategies (Stransky et al. 2007). Many databases on cancer research have been set and are useful for further bioinformatics manipulation. Comparative and structural omics can be applied for understanding the underlying tumorigenesis process. In the post-genome era, omics challenges us with the necessity of using and communicating huge information outside the existing paradigm of bench and bedside services.

1.11 From Omics to Personalized Medicine

Several omics technologies are being increasingly used for personalized drug discovery and their efforts are in progress in major therapeutic areas. Antibody drugs (such as trastuzumab, cetuximab, and bevacizumab), small molecule inhibitors for tyrosine kinases (such as gefitinib, erlotinib, and imatinib), conventional cytotoxic drugs, and antihormonal drugs are used for cancer chemotherapy. Biomarker monitoring may contribute to therapeutic optional choice and drug dosage determination. Biomodal targeting of single oncoproteins may become latter-day combination therapy, e.g. retinoic acid and arsenic trioxide for PML-RAR, Herceptin and lepatinib for HER2. Predictive genetic tests may allow individuals to learn their own susceptibilities and to reduce the risks for interventions. Although the present omics technologies are not ready for immediate clinical use as diagnostic tools, it can be envisaged that simple, fast, robust, portable, and cost-effective clinical diagnosis systems may be available in the future for home and bedside use (Zhang et al. 2007).

1.12 Challenges and Prospective

Advances in the large-scale omics technologies have led to a proliferation of putative cancer biomarkers. However, before the results can be implemented in the management of cancer patients, thorough validation and the issues of sensitivity, specificity, reproducibility, and accuracy need to be addressed. On the other hand, the validation of biomarker has a complicated interaction with known oncogenes or oncoproteins that has established further links with molecular pathways implicated in malignant transformation. The major challenge is how to bring the best results from the omics research into clinical use as accurate and reliable standardized tests that integrate into the clinical work-up. Each omics approach has its strengths and drawbacks. The integration of various omics data and their functional interpretation in conjunction with clinical results is another challenge.

The wealth of new information in omics databases provides unlimited possibilities for designing new therapeutic agents for cancer. For example, the multiplex diagnostic biomarker is becoming favorable as indicated by the approval of MammaPrint and Oncotype DX. This will ultimately beneficial to the patients, whose detection of cancer and treatment thereof will be greatly facilitated by the work. It is likely that omics-based cancer research will take a central place in the understanding, diagnosis, treatment, and monitoring of cancer in the near future.

References

Cesareni G, Ceol A, Gavrila C et al (2005) Comparative interactomics. FEBS Lett 579:1828–1833

Cho WC (2007a) Cancer biomarker discovery: the contribution of "omics". BIOforum Eur 11:35–37

Cho WC (2007b) OncomiRs: the discovery and progress of microRNAs in cancers. Mol Cancer 6:60

Cho WC (2007c) Contribution of oncoproteomics to cancer biomarker discovery. Mol Cancer 6:25

Cho WC (2007d) Proteomic approaches to cancer target identification. Drug Discov Today Ther Strateg 4:245–250

Cho WC (2009) Role of miRNAs in lung cancer. Expert Rev Mol Diagn 9:773–776

Cho WC (2010a) Cancer biomarkers (an overview). In: Hayat EM (ed) Methods of cancer diagnosis, therapy and prognosis. Springer, Netherlands

Cho WC (2010b) MicroRNAs as potential biomarkers for cancer diagnosis, prognosis and targets for therapy. Int J Biochem Cell Biol in press

Cho WC (2010c) MicroRNAs in cancer - from research to therapy. BBA - Rev Cancer in press

Cho WC, Cheng CH (2007) Oncoproteomics: current trends and future perspectives. Expert Rev Proteomics 4:401–410

Cho WC, Chow AS, Au JS (2009) Restoration of tumour suppressor *hsa-miR-145* inhibits cancer cell growth in lung adenocarcinoma patients with epidermal growth factor receptor mutation. Eur J Cancer 45:2197–2206

Claudino WM, Quattrone A, Biganzoli L et al (2007) Metabolomics: available results, current research projects in breast cancer, and future applications. J Clin Oncol 25:2840–2846

Esteller M (2007) Cancer epigenomics: DNA methylomes and histone-modification maps. Nat Rev Genet 8:286–298

Finn WG (2007) Diagnostic pathology and laboratory medicine in the age of "omics": a paper from the 2006 William Beaumont Hospital Symposium on Molecular Pathology. J Mol Diagn 9:431–436

Furge KA, Tan MH, Dykema K et al (2007) Identification of deregulated oncogenic pathways in renal cell carcinoma: an integrated oncogenomic approach based on gene expression profiling. Oncogene 26:1346–1350

Futschik ME, Chaurasia G, Herzel H (2007) Comparison of human protein–protein interaction maps. Bioinformatics 23:605–611

González-Díaz H, González-Díaz Y, Santana L et al (2008) Proteomics, networks and connectivity indices. Proteomics 8:750–778

Gowda GA, Zhang S, Gu H et al (2008) Metabolomics-based methods for early disease diagnostics. Expert Rev Mol Diagn 8:617–633

Hamacher M, Herberg F, Ueffing M et al (2008) Seven successful years of Omics research: the Human Brain Proteome Project within the National German Research Network (NGFN). Proteomics 8:1116–1117

Hatada I (2006) Emerging technologies for genome-wide DNA methylation profiling in cancer. Crit Rev Oncol 12:205–223

He YD (2006) Genomic approach to biomarker identification and its recent applications. Cancer Biomark 2:103–133

Heck AJ (2008) Native mass spectrometry: a bridge between interactomics and structural biology. Nat Methods 5:927–933

Hsu CN, Lai JM, Liu CH et al (2007) Detection of the inferred interaction network in hepatocellular carcinoma from EHCO (Encyclopedia of Hepatocellular Carcinoma Genes Online). BMC Bioinform 8:66

Hsu PP, Sabatini DM (2008) Cancer cell metabolism: Warburg and beyond. Cell 134:703–707

Jones PA, Baylin SB (2007) The epigenomics of cancer. Cell 128:683–692

Kaddurah-Daouk R, Kristal BS, Weinshilboum RM (2008) Metabolomics: a global biochemical approach to drug response and disease. Annu Rev Pharmacol Toxicol 48:653–683

Karamouzis MV, Konstantinopoulos PA, Papavassiliou AG (2007) Epigenomics in respiratory epithelium carcinogenesis: prevention and therapeutic challenges. Cancer Treat Rev 33:284–288

Keusch GT (2006) What do -omics mean for the science and policy of the nutritional sciences? Am J Clin Nutr 83:S520–522

Lussier YA, Liu Y (2007) Computational approaches to phenotyping: high-throughput phenomics. Proc Am Thorac Soc 4:18–25

Mei H, Cuccaro ML, Martin ER (2007) Multifactor dimensionality reduction-phenomics: a novel method to capture genetic heterogeneity with use of phenotypic variables. Am J Hum Genet 81:1251–1261

Negrini M, Ferracin M, Sabbioni S et al (2007) MicroRNAs in human cancer: from research to therapy. J Cell Sci 120:1833–1840

Nicholson JK (2006) Reviewers peering from under a pile of omics data. Nature 440:992

Reymond MA, Lippert H (2008) Proteomics in lung cancer. In: Daoud SS (ed) Cancer proteomics: from bench to bedside. Human Press, Totowa, NJ, pp 139–159

Rosenblatt KP, Bryant-Greenwood P, Killian JK et al (2004) Serum proteomics in cancer diagnosis and management. Annu Rev Med 55:97–112

Rosenfeld N, Aharonov R, Meiri E et al (2008) MicroRNAs accurately identify cancer tissue origin. Nat Biotechnol 26:462–469

Sanchez-Carbayo M (2006) Antibody arrays: technical considerations and clinical applications in cancer. Clin Chem 52:1651–1659

Shih IeM, Wang TL (2005) Apply innovative technologies to explore cancer genome. Curr Opin Oncol 17:33–38

Stransky B, Barrera J, Ohno-Machado L et al (2007) Modeling cancer: integration of "omics" information in dynamic systems. J Bioinform Comput Biol 5:977–986

Tárnok A, Bocsi J, Brockhoff G (2006) Cytomics – importance of multimodal analysis of cell function and proliferation in oncology. Cell Prolif 39:495–505

van der Merwe DE, Oikonomopoulou K, Marshall J et al (2007) Mass spectrometry: uncovering the cancer proteome for diagnostics. Adv Cancer Res 96:23–50

van Ravenzwaay B, Cunha GC, Leibold E et al (2007) The use of metabolomics for the discovery of new biomarkers of effect. Toxicol Lett 172:21–28

Wishart DS (2007) Proteomics and the human metabolome project. Expert Rev Proteomics 4:333–335

Wishart DS, Knox C, Guo A et al (2006) DrugBank: a comprehensive resource for *in silico* drug discovery and exploration. Nucleic Acids Res 34:D668–672

Zbuk KM, Eng C (2007) Cancer phenomics: RET and PTEN as illustrative models. Nat Rev Cancer 7:35–45

Zhang X, Li L, Wei D et al (2007) Moving cancer diagnostics from bench to bedside. Trends Biotechnol 25:166–173

Chapter 2
Recent Advances in Cancer Genomics and Cancer-Associated Genes Discovery

Bin Guan, Tian-Li Wang, and Ie-Ming Shih

Abstract Human cancer is a personalized disease characterized by complex molecular genetic abnormalities unique to individual patients. Studying cancer genome has defined much of the molecular pathogenesis of neoplasia we have understood so far and has supported the view that cancer is a genetic disease caused by sequential accumulation of genetic alterations. Recent advances in genome-wide technologies have provided unprecedented tools to reveal the genomic landscape of cancer in great detail, and thus have offered new opportunity in deciphering the specific genomic changes participated in tumor initiation and progression. Here, we review these emergent array- or sequencing-based technologies and provide examples of how they can be applied in discovering molecular genetic changes in cancer and in facilitating mining of important cancer genes. From a clinical perspective, it appears a daunting challenge in translating those molecular genetic findings from cancer cells to cancer patients. Therefore, we will also briefly discuss the potential problems in translational cancer genomic research and propose the possible solutions.

2.1 Introduction

Cancer is a complex genetic disease caused by mutations that can be hereditary but most of time are somatic (Kinzler and Vogelstein 2002). It has become accepted that cancer develops as a result of accumulated genetic alterations which serve as the driving forces in initiating tumor development and propelling tumor progression. The Darwinian evolution theory of cancer predicts that clinically detectable tumors harbor the clonal molecular genetic changes that are causally related to uncontrolled tumor growth, survival in dynamic tumor microenvironment, invasion into surrounding normal tissues and metastasis to distant organs (Merlo et al. 2006).

B. Guan, T.L. Wang, and I.M. Shih (✉)
Departments of Pathology, Oncology and Gynecology/Obstetrics, Johns Hopkins Medical Institutions, Baltimore, MD 21231, USA
e-mail: ishih@jhmi.edu, shihie@yahoo.com

W.C.S. Cho (ed.), *An Omics Perspective on Cancer Research*,
DOI 10.1007/978-90-481-2675-0_2,

Various forms of genomic abnormalities have been documented to occur in cancers, such as point mutations (silent, missense, nonsense and frameshift mutations), DNA copy number alterations (CNAs, including gene duplication, amplification and deletion), and chromosomal rearrangements (insertion, inversion, intra- and inter-chromosomal translocations) (Fig. 2.1). These mutations can be grouped into two broad categories, microscopic changes involving large segments of DNA (typically larger than 3 Mb) that can be detected using traditional G-banded karyotyping and fluorescence *in situ* hybridization (FISH), and submicroscopic alterations less than 3 Mb that have been identified using an array of molecular and cellular biology techniques (Feuk et al. 2006). Proto-oncogenes are typically activated by gene amplifications, gene translocations, and activating intragenic mutations whereas tumor suppressors are inactivated by gene deletions (loss of heterozygosity or homologous deletion), inactivating intragenic mutations, and epigenetic silencing (Haber and Settleman 2007). Therefore, decoding the genetic history present in tumor DNA, the identification and characterization of these molecular changes involving cancer-associated genes and the pathways they controlled, have not only shed new light on the molecular etiology of cancer, but also promised for the development of new diagnostic markers and novel therapeutic targets (Kinzler and Vogelstein 2002).

Recent advances in genome-wide technologies and bioinformatics have provided new opportunity in genomic analysis of human cancer in an unprecedented speed and details (Fig. 2.2). These high-throughput and high-resolution techniques have produced a long list of exciting candidate cancer-associated genes. In this chapter, we will focus on reviewing these emergent technologies and provide examples of how they can be applied in discovering cancer associated genetic changes and facilitate new cancer gene discovery. The advantage and disadvantages inherent to each method will be briefly reviewed. We will also discuss the challenges in transforming these findings into biologically interesting and clinically relevant knowledge, and the challenges faced in translating these findings into clinical applications that could directly benefit cancer patients.

2.2 Array-Based Technologies

2.2.1 Array Comparative Genomic Hybridization (aCGH)

First published in 1992, comparative genomic hybridization (CGH) is the first genome-wide method in detecting DNA copy number alterations (CNAs). In the original method, total genomic DNA is isolated from test and reference samples, differentially labeled and hybridized to metaphase chromosomes from normal individuals (Kallioniemi et al. 1992). Measuring the fluorescence intensity ratio along each chromosome reveals the gain or loss in the test sample relative to reference sample at a genome-wide scale. However, the resolution afforded by metaphase chromosome CGH is typically only 5–10 Mb (Carter 2007). Substitution of metaphase chromosomes with DNA arrays for CGH theoretically greatly increases its resolution in CNA

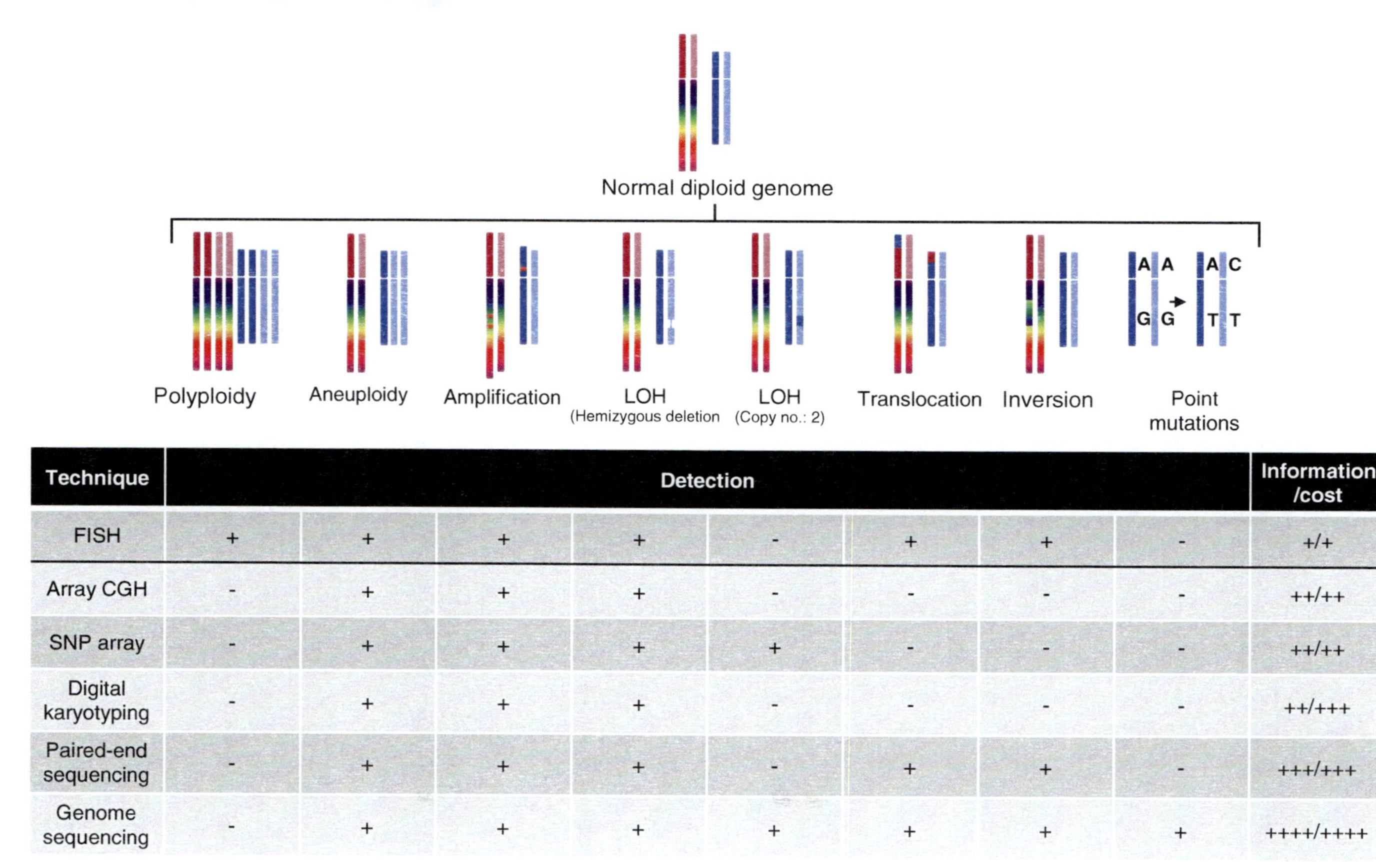

Technique	Detection								Information /cost
	Polyploidy	Aneuploidy	Amplification	LOH (Hemizygous deletion	LOH (Copy no.: 2)	Translocation	Inversion	Point mutations	
FISH	+	+	+	+	-	+	+	-	+/+
Array CGH	-	+	+	+	-	-	-	-	++/++
SNP array	-	+	+	+	+	-	-	-	++/++
Digital karyotyping	-	+	+	+	-	-	-	-	++/+++
Paired-end sequencing	-	+	+	+	-	+	+	-	+++/+++
Genome sequencing	-	+	+	+	+	+	+	+	++++/++++

Fig. 2.1 Common genetic alterations found in cancer and their detection techniques. Cellular cytogenetic techniques (such as FISH) are capable of detecting gene copy number changes occurred in individual cells, while other methods examine the average DNA content from a group of cells. Detection of copy number alterations using sequencing-based techniques relies on state-of-art DNA preparation methods and requires sophisticated bioinformatics analyses. Refer to text for details

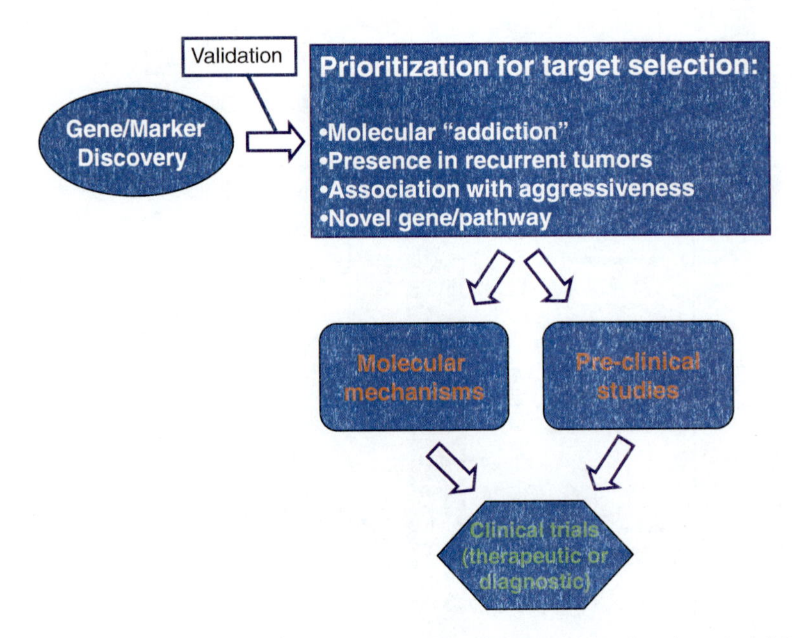

Fig. 2.2 Strategy to study cancer-associated genes. The flowchart summarizes the sequential steps for marker discovery and prioritization aiming to understand the biology of the genes in cancer development and to provide new molecular marker for diagnosis and targets for novel therapeutics

detection, since the resolution of the array CGH (aCGH) is determined by the size and density of the probes (DNA sequences) present on the arrays.

The DNA sequences spotted for aCGH have included large-insert BAC clones, cDNA clones (80–200 kb in length), fosmid and cosmid clones (40 kb), genomic PCR products (100 bp to 1.5 kb), and oligonucleotides (25–80 bp) (Carter 2007). Among these DNA arrays, BAC clones have been thought to offer the most complete genome coverage and the highest signal-to-noise ratio. Assuming each BAC clone is 100 kb in size, only 30,000 clones are needed to obtain complete chromosome coverage. Indeed, a tiling array consisting of 32,433 overlapping BAC clones covering the entire human genome has been constructed to permit a sub-megabase resolution (Ishkanian et al. 2004). However, the large size of BACs limits its ability to detect single-copy alterations smaller than 50 kb (Carter 2007). Microarrays with smaller elements, such as fosmid and genomic PCR products array can potentially provide higher genomic resolution, with the expense of increasing production cost and higher noise-to-signal ratio due to hybridization kinetics (Carter 2007; Pinkel and Albertson 2005). It is noteworthy that more probes are apparently needed to cover the whole genome when smaller DNA sequences are spotted. For example, assuming the PCR-products of genomic DNA average 1 kb, then 3 million 1-kb elements will be needed to cover the 3-billion-base human genome. Unfortunately, it is difficult to spot more than 60,000 DNAs onto a glass slide using current printing devices (Carter 2007). As a result, 50 slides would be needed to obtain full-genome coverage at a 1-kb resolution. Alternatively, hybridization using one slide would only provide a resolution of 50 kb. It should be noted that above calculations did not count for the highly repetitive regions of the genome, which could not be effectively analyzed using hybridization technologies.

Recently, CGH using commercial oligonucleotide array platform has gained popularities for their high-resolution, ready-for-use availability and relatively low price (Ylstra et al. 2006). Oligonucleotide array can now be constructed using synthesis-on-slide technologies at a 1–2 million probes per slide (Carter 2007). For example, NimbleGen HD2 tiling array contains 2.1 million probes of 60 mer, with a median probe spacing of 1,169 bp (http://www.nimblegen.com/products/cgh/index.html#hd2hg18). Practically, these high-density arrays may present a ~5-kb resolution for CNAs studies.

2.2.2 *Representational Oligonucleotide Microarray Analysis*

A major disadvantage of oligonucleotide arrays is the poor signal-to-noise ratio of hybridization due to the short probes on the slides (Carter 2007). The representational oligonucleotide microarray analysis (ROMA) improves the signal-to-noise ratio by reducing the complexity of the input genomic DNA (Lucito et al. 2003). With ROMA, genomic DNA samples are digested using a restriction enzyme (such as BglII), and the fragments are ligated to adapters followed by PCR amplification using universal primers. It was estimated that up to 97.5% reduction in complexity of input DNA can be achieved using the digestion-amplification process (Lucito et al. 2003). Although restriction digestion and PCR process may lead to differential representation between test and reference samples and thus introduce additional artifacts (Carter 2007), ROMA has been widely used for both oligonucleotide-based CGH and SNP array analyses.

2.2.3 *SNP Arrays*

Single nucleotide polymorphisms (SNPs) are normal variations of nucleotide sequences and are frequently present in the genome. In fact, SNP is the most common and well-catalogued genetic variation. Of the estimated over 10 million common human SNPs, 3.9 million (1 SNP/700 bp) have now been identified (HapMap release 23a, http://www.hapmap.org/), and have served as the basis for designing oligonucleotide high-resolution SNP genotyping arrays by companies including Affymetrix and Illumina. SNP genotyping array has been proven as a powerful tool for the genome-wide genetic association studies, in which thousands or millions of SNPs were compared between cases and controls to identify loci associated with disease phenotypes, including cancers.

This platform has successfully identified novel cancer susceptibility loci in breast cancer (Easton et al. 2007) and prostate cancer, such as the locus harboring the *kallikrein-3* (*KLK3*) gene encoding the prostate specific antigen (PSA) (Eeles et al. 2008; Gudmundsson et al. 2008; Thomas et al. 2008).

Furthermore, SNP arrays offer an effective high-resolution method in detecting chromosomal regions undergoing loss of heterozygosity (LOH), which may harbor tumor suppressor genes. In addition, as the hybridization signal on SNP arrays is relatively proportional to the copy number of input DNA sample, SNP array can be used to measure DNA copy number change. Numerous studies have demonstrated the power of high-density SNP arrays in detecting LOH, large-scale and minute CNAs, and uniparental disomies that retain a normal copy number. Due to its high-content information and relatively low cost, SNP array has become one of the most popular tools for cancer genomic studies in recent years (Mao et al. 2007). For example, using the Affymetrix 250 K Sty SNP array, an analysis of 371 lung tumors reveals 57 significantly recurrent amplifications/deletions including 31 focal events (Weir et al. 2007).

The Affymetrix SNP Array contains 25-base long matched and mismatched probes for each SNP allele designed for better sensitivity and wider linear range. As with ROMA, the input genomic DNA is digested with a restriction enzyme (StyI and/or NspI) and is PCR-amplified to improve the signal-to-noise ratio. In contrast to aCGH, only one labeled DNA sample is hybridized onto the chip. The intensity of hybridization is then compared to that predetermined from normal DNA from a group of individuals. Therefore, highly standardized slide handling and processing procedures and precision in array fabrication are essential to obtain reliable results (Carter 2007). Illumina has developed another SNP array platform using 50-bp oligonucleotides attached to indexed beads, in which allele-specific primer extension generates signal amplification. The longer probe and non-PCR amplification might result in better signal-to-noise ratio (Wang et al. 2007). However, to our best knowledge, no side-by-side comparison using different SNP array platforms have been published. As SNPs are not uniformly distributed in the genome, the resolution of copy number detection has a limit of 10 kb (Carter 2007). Newer versions of Illumina and Affymetrix SNP arrays are now available and they contain non-SNP probes to increase the resolution. For example, the Affymetrix Genome-Wide Human SNP Array 6.0 claims to have an averaged probe spacing of 700 bp including more than 906,600 SNPs and more than 946,000 non-SNPs probes for the detection of copy number change (http://www.affymetrix.com/support/technical/datasheets/genomewide_snp6_datasheet.pdf).

2.3 Sequencing-Based Technologies

2.3.1 Mutational Analyses of Cancer Genome

The completion of the human genome project has revolutionized the biological and medical research. It is the foundation for technology innovations to detect genomic alterations in cancer and provided the database for designing oligonucleotides probes in DNA microarrays as we have discussed above. Recently, with the advance in sequencing technologies accompanied by much reduced cost, it becomes a reality to sequence individual human genome, either using the traditional Sanger dideoxy sequencing method (Levy et al. 2007), or next-generation sequencing methods (Wheeler et al. 2008). The whole genome sequencing in cancer can provide unbiased information on genomic alterations in individual cancer specimen, and thus would likely produce a complete genomic map of individual cancers.

The first large-scale sequencing effort in cancer was reported by a Johns Hopkins University group, which initially sequenced protein coding regions of 13,023 genes in 11 breast and 11 colorectal cancers (Sjoblom et al. 2006), and later extended to include all of the 18,191 genes in the Reference Sequence database in the same set of samples (Wood et al. 2007). Another study from the Wellcome Trust Sanger Institute focused on a small set of genes encoding 518 protein kinases, in a larger set of tumor samples including 210 cases of various tumor types (Greenman et al. 2007). Approximately 200 new cancer genes with somatic mutations were discovered by the Hopkins group from 22 tumor samples, and ~120 kinase gene mutations were found by the Sanger group. In 2008, the Johns Hopkins group further extended their mutational analyses to 24 pancreatic cancers (Jones et al. 2008) and 20 glioblastoma multiformes, a highly malignant brain tumors (Parsons et al. 2008). These studies have helped shaping the current view of human cancer as a molecularly personalized disease of extreme complexity and heterogeneity, which is attributed to individual's genetic background and environmental exposure. For example, the near genome-wide sequencing analyses conducted by the Hopkins group showed that the genomic landscape of breast and colorectal cancers are characterized by a few commonly mutated gene "mountains" and a much larger number of gene "hills" that are mutated at a much lower frequency (Wood et al. 2007). In addition, these mutational analyses suggest that tumor-type-specific mutations exist, but the vast majority of mutations occur in less than 10% of tumors. Interestingly, mutations enriched in a subset of signaling pathways or interactomes were identified, suggesting that strategies targeting multiple pathways may be a promising option that might be more effective and benefit a wider pool of patients (Jones et al. 2008; Wood et al. 2007).

Recently, the introduction of the "next-generation" sequencing instruments has poised to revolutionize genetic/genomic research in many facets (Schuster 2008; Wold and Myers 2008). Currently, there are three such systems that are commercially available and vying for the spotlights, i.e. 454 technology from Roche Diagnostics, Solexa technology from Illumina, and SOLiD technology from Applied Biosystems (Chi 2008). While these instruments use various sample preparation methods and sequencing strategies (Rusk and Kiermer 2008), they all achieved the goal of massively parallel sequencing with the advantages of bacterial-cloning free, high-throughput (over 1 Gb per run) and relatively low cost. Solexa and SOLiD instruments read about 30–35 bp, and the 454 technology can sequence up to 400 bp but with relatively higher operating cost (Rusk and Kiermer 2008; Schuster 2008). Currently these platforms are still relatively error-prone and generate shorter reading than ~750 bp by the traditional Sanger dideoxy method

(Dohm et al. 2008; Hillier et al. 2008). However, it is likely that accuracy and reading length would be improved in the near future. For example, a learning-phase using a standard sample can improve the number of accurate reads to 78 bases using the Solexa Genome Analyzer (Erlich et al. 2008). The strength of the next-generation sequencing platform has been demonstrated by the completion of the diploid whole-genome sequencing of individuals in a period of months (Wang et al. 2008; Wheeler et al. 2008). Recently, the first whole-genome sequencing of a primary human cancer genome has been reported using the Illumina/Solexa technology (Ley et al. 2008). In this study, the DNA contents of both tumor and normal skin cells obtained from a leukemia patient were sequenced and analyzed. Interestingly, by comparing the tumor genome to normal genome, only ten genes with acquired somatic non-synonymous mutation were found in leukemia cells (Ley et al. 2008).

Large-scale and high-throughput nucleotide sequencing accompanied by the state-of-art sample preparation methods can also be applied to study several biological questions including CNAs, protein-DNA interaction and identification of potential fusion genes. In the following, we will discuss applications in cancer genomics using the sequencing-based methods.

2.3.2 Digital Karyotyping

The principle of digital karyotyping (Wang et al. 2002) is similar to the serial analysis of gene expression (SAGE) method. The sequence tags in digital karyotyping are obtained from genomic DNA via restriction enzyme digests, and are then linked into ditags, concatenated, cloned, and sequenced. These 21 bp tags are mapped back to their corresponding genomic loci from which they are derived. With sufficient number of tags (usually 160,000 tags), the tag densities along the chromosomes can then be transformed to copy numbers, yielding a digital readout of genome-wide copy numbers in cancer samples in a high-resolution and high-throughput manner. Typically, SacI is used as the mapping enzyme in digital karyotyping, which cuts DNA once per 4,096 bp on average ($4^6 = 4{,}096$). Therefore, ~4 kb resolution can be theoretically achieved assuming SacI recognition sites are evenly distributed in genome and a sufficient number of tags can be sequenced. Indeed, digital karyotyping was shown to be superior to the then-standard aCGH and SNP arrays in detecting copy number changes (Shih and Wang 2005).

Digital karyotyping has been used in profiling CNAs in colorectal (Wang et al. 2004) and ovarian cancers (Shih et al. 2005). In two of four 5-fluorouracil-resistant colorectal tumors analyzed, an amplicon spanning an approximately 100-kb region on 18p11.32 was identified that contains a gene encoding thymidylate synthase (TYMS), a molecular target of 5-fluorouracil (Wang et al. 2004). Using FISH, TYMS gene amplification was shown to occur in 23% of 31 patients who received 5-fluorouracil treatment, whereas no amplification was observed in metastases of patients who did not receive 5-fluorouracil treatment, suggesting that genetic amplification of TYMS is one of the mechanisms of 5-fluorouracil resistance *in vivo* (Wang et al. 2004). In an analysis of seven high-grade ovarian serous carcinoma samples using digital karyotyping , an amplification at 11q13.5 was found in three of seven cases, and amplicon mapping delineated a 1.8-Mb core of amplification that contained 14 genes, of which the chromatin-remodeling factor *Rsf-1* (or called *HBXAP*) (Shih Ie et al. 2005), p21/cdc42/Rad-activated kinase *PAK1* (Schraml et al. 2003), and adaptor protein *GAB2* (Bentires-Alj et al. 2006), have been found to be important oncogenes in several types of human cancer.

As a sequencing-based technology, digital karyotyping overcomes many disadvantages associated with array-based technologies in detecting genomic structural changes. For example, array-based technologies are limited by the sequences present on chips and also cannot reliably detect regions with low-complexity. However, while very powerful, digital karyotyping has not gained similar popularity as compared to array-based technologies. This is mainly attributed to the relatively higher cost and being more technically-demanding for digital karyotyping than for array-based platforms. In contrast, microarray service is now available in both commercial and academic sectors with affordable cost. Nonetheless, we expect that the continuously decreasing cost of sequencing, the coupling of the next-generation sequencing in particular, could make digital karyotyping more accessible for cancer researchers. In addition, adaptation of digital karyotyping method might also improve its utility. For example, similar to superSAGE (Matsumura et al. 2003), the type III-endonuclease EcoP15I of phage P1 could be used to produce 26 bp tags instead of 21 bp tags. The longer tag can significantly improve the digital readout by enhancing mapping accuracy to reference genome.

2.3.3 Genomic DNA End-Sequencing: BAC, Fosmid and Paired-End

The main disadvantages of array-based technologies and digital karyotyping are that they are incapable of detecting balanced genomic rearrangements, either translocations or inversions. Besides, both techniques provide limited resolution in detecting breakpoints that are associated with copy number changes. DNA end-sequencing is an emerging tool in detecting point mutations, genomic rearrangement, as well as copy number alterations. Earlier studies employed the construction of a bacterial artificial chromosome (BAC) library from cancer samples, followed by BAC-ends sequencing and subsequent mapping end-sequence pairs onto the human reference genome (Volik et al. 2003). As in digital karyotyping, the density of the end-readings corresponds to their copy numbers; in addition, the paired ends that map in an opposite orientation, far apart, or too close on the reference genome suggest that the BAC contains genomic rearrangements, such as inversion, deletion/translocation, or insertion, respectively. The subsequent full-insert sequencing of these BAC clones produces the breakpoints information at a base-pair level. However, the large insert size of BAC library limits the resolution of BAC-end-sequencing to be over 100 kb. Although the resolution of end-sequencing can be improved using smaller-sized fosmid library (Tuzun et al. 2005), constructing a BAC or fosmid library is a laborious process.

High-throughput next-generation sequencing technologies have also been applied to identify possible genomic structural alterations and assess DNA copy number changes omitting the bacteria cloning steps. By 454 sequencing of paired-ends produced from ~3 kb genomic DNA fragments, over 1,000 structural variations spanning over 3 kb were found from two normal individuals (Korbel et al. 2007). It is very likely that the same methodology could be adapted to sequence human cancer genome. In another effort to analyze genomic structural alterations, investigators have used 200 or 400 bp genomic DNA fragments from the two lung cancer cell lines, NCI-H1770 and NCI-H2171, and generated millions of paired reads of 29–36 bases at each end of these fragments using the Solexa sequencing platform (Campbell et al. 2008). Computational mapping of these paired reads to the human reference genome found over 1 million aberrantly mapping reads that could be result of chromosomal rearrangements. As the Solexa sequencing is theoretically error-prone, only paired-ends mapped with high uniqueness scores were prioritized for PCR validation of the potential breakpoint regions using cancer and normal cell line derived from the same patients. Subsequent sequencing of these PCR products using the conventional Sanger method identified a total of 103 somatic rearrangements and 306 germline variations, including deletions, tandem duplications and inversions. Interestingly, while most of the germline rearrangements involve deletions of AluY elements and LINE repeats, most of the somatically acquired rearrangements are from amplicons. To identify rearrangements that may lead to fusion genes, aberrantly paired reads were examined in annotated gene databases, and those with both ends fell within the coding footprint of two different genes in the correct orientation were further studied for in-frame fusion products. This strategy identified two fusion transcripts from internal tandem duplication and two created by inter-chromosomal rearrangements (Campbell et al. 2008). For the DNA copy number analyses, the comparison of the frequency of DNA regions sequenced to that of the theoretical frequency from *in silico* simulation, has revealed that its performance is comparable to Affymetrix SNP 6.0 array with 1.85 million loci (Campbell et al. 2008). Interestingly, paired-end sequencing produced markedly higher estimates of copy number than SNP array for some highly-amplified regions, which might be result of signal saturation on the SNP array (Campbell et al. 2008).

2.3.4 Paired-End diTags (PETs) cDNA Sequencing

Gene fusions play important roles in the development of both hematological disorders and malignant solid tumors (Mitelman et al. 2007). While paired-end sequencing of genomic DNA can detect gene fusions, it requires additional steps to determine whether a fusion transcript is produced from rearranged genomic regions as discussed above. The Paired-End diTags (PETs) sequencing method appeared to be extremely powerful in detecting gene fusions, as it directly sequence diTag library produced from cancer cDNA sample (Ruan et al. 2007). The Paired-End diTags method involves making full-length cDNA library from test sample, extracting 18 base-pair from each of the 5′ and 3′ end of the full length cDNA, ligating the 5′ and 3′ together to form Paired-End diTags, and sequencing these diTags (Ng et al. 2005). Initially employed the conventional sequencing method, the coupling of PETs with 454 sequencing dramatically reduced the cost and improved the throughput (Ng et al. 2006). Using this technique, 70 candidate gene fusions were identified from the breast cancer cell line MCF7 and the colon cancer cell line HCT116 (Ruan et al. 2007).

2.4 Identification of Cancer-Associated Genes

The genome-wide cancer genomics studies, conducted by individual research group and by consortium efforts such as the NIH's Cancer Genome Atlas (http://cgap.nci.nih.gov/) and the Wellcome Trust Sanger Institute's Cancer Genome Project (http://www.sanger.ac.uk/genetics/CGP/), have generated an enormous amount of data, and hopefully in the near future, these efforts can lead to the completion of genomic maps of various cancers. However, challenges remain to fully understand and translate genomic alterations documented in these studies. Cancer-associated gene discovery can sometimes be assisted by the known function of the candidate genes or that of their homologue in other organisms. Nevertheless, given the multi-functionality nature of proteins and the presence of non-protein-coding genomic regions (such as microRNA), the full functional annotation of human genome appears to be a daunting task. Therefore, each candidate cancer-associated genes needs to be characterized using experimental approaches to validate their roles in cancer biology. In addition, the examination of cancer genome is performed after a cancer is surgically removed, which provides a snapshot of DNA abnormalities reflecting the genetic history of cancer cells during their journey of tumor evolution. Some mutations are thought to initiate the malignant transformation or promote progression, by conferring cancer cells with growth advantages or the ability escaping from cell death, and are therefore called drivers. The remaining passenger mutations, in contrast, are result of genome instability or just located next to driver mutations. While passenger mutations could serve as biomarkers, driver mutations are considered the key to understanding the mechanisms of tumorigenesis and the key to the selection of novel targets for therapeutic interventions. Various strategies have been successfully employed for the quest of important cancer-associated genes, and an example is shown in a latter section.

In mutational analyses, driver mutations can be distinguished from passenger mutations by comparing the observed-to-expected ratio of synonymous mutations with that of non-synonymous mutations (Greenman et al. 2007). This is based on the assumption that synonymous (silent) mutations do not change amino-acid composition and thus do not have an effect in tumorigenesis. Using this strategy, 158 predicted driver mutations were identified in 120 kinase genes (Greenman et al. 2007). However, it should be noted that there are increasing number of reports on the link between silent mutations with cancer risk, and synonymous mutation representing rare codons may alter the mRNA stability and protein synthesis (Duan et al. 2003; Nackley et al. 2006) or protein conformation (Kimchi-Sarfaty et al. 2007).

Functional screenings can be also useful in the identification of cancer-associated genes. Using transformation screen in NIH-3T3 cells with cDNA library prepared from a lung tumor, a gene fusion was identified involving a tyrosine kinase gene (*ALK*) that affects 7% of non-small-cell lung cancer (Soda et al. 2007). Alternatively, cDNA libraries or RNAi libraries can be used to identify potential cancer-related genes, by examination of cancer-related phenotypes, such as proliferation (Schlabach et al. 2008), anchorage-independent growth, migration (Witt et al. 2006), cell-cycle progression (Kittler et al. 2007; Mukherji et al. 2006), and drug resistance (Bosma et al. 2003). The major advantage of functional screening is that it offers an unbiased genome-wide over-expression/downregulation experimental approach that directly demonstrates the potential biological function of the genes in tumor development. Obviously, clinical relevance needs to be further established for those candidate genes identified from libraries not produced from patients.

2.5 Strategy to Prioritize Candidate Genes for Validation and Functional Characterization

Application of established and emerging genomic technologies is expected to identify numerous genes that are associated with human cancer. The real challenge is how to select and focus on the biologically most important and clinically most relevant cancer genes from a list of candidates for further studies. Fig. 2.2 represents our preferred strategies in facilitating the target gene selection and characterization. After validation of a repertoire of candidate genes using independent methods, genes will be prioritized for further characterization based on the following criteria: (1) molecular addition to the specific gene expression or genetic alteration, (2) expression and alteration of the gene in recurrent tumors, (3) association with aggressive clinical behavior, and (4) novel genes and pathways. These criteria can be applied to look for driver genes within an amplicon or a deleted region, and from a large-scale sequence mutational analysis. For example, based on digital karyotyping, we have previously detected frequent amplification at the chr11q13.5 locus in several types of human cancer including ovarian cancer (Schwab 1998). Fine mapping by FISH in 11q13.5 amplified tumors further defines the minimal amplicon where potential driver gene(s) may reside (Fig. 2.3). Prior to identification and characterization of the driver gene(s) in the 11q13.5 amplicon, we assessed whether 11q13.5 amplification is associated with disease aggressiveness in ovarian

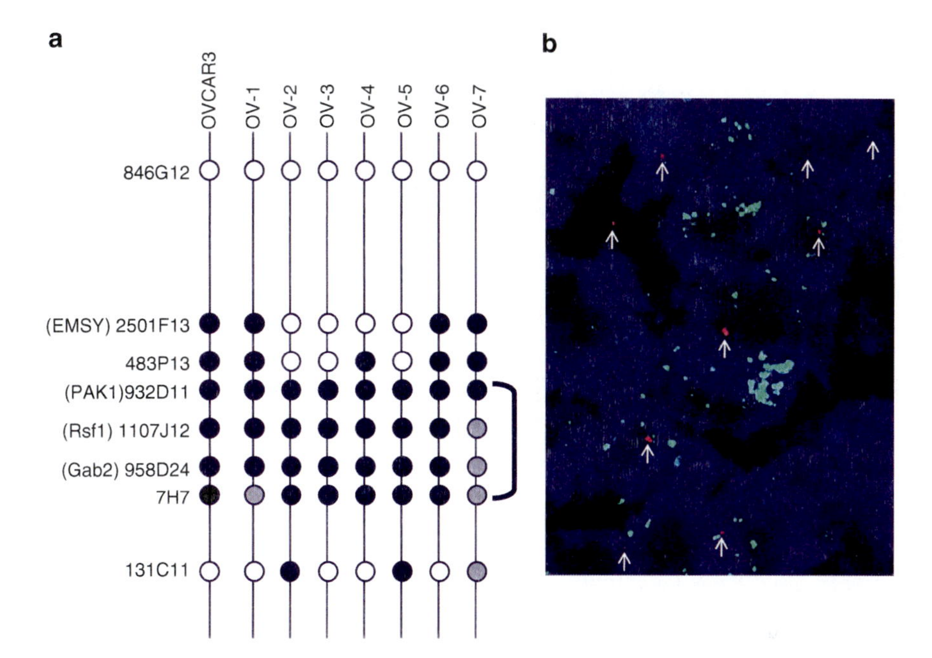

Fig. 2.3 Minimum mapping to facilitate the discovery of important cancer-driving genes. After identification of an amplicon (or a deletion region) from the discovery phase, fine mapping of a minimum overlapping chromosome region present in multiple tumor samples can help identify driver genes. (**a**) Using multiple probes hybridizing to the regions flanking the 11q13.5 amplicon, a minimum amplicon (*bracket*) was deduced by FISH analyses on one ovarian cancer cell line (OVCAR3) and seven high-grade ovarian serous carcinoma samples (OV-1 to OV-7) with 11q13.5 amplification. *Each dot* represents FISH signal of BAC probes. *White*, no gain; *black*, high gain; *grey*, low gain. (**b**) Example of FISH using an 11q13.5 probe (*green*) on a high-grade ovarian serous carcinoma specimen. *Blue*, DAPI. *Red*, a signal from the reference probe

cancer. Using FISH, we were able to demonstrate 11q13.5 amplification in 13–15% of high-grade ovarian serous carcinomas (Nakayama et al. 2007; Shih et al. 2005). 11q13.5 amplification was significantly associated with worse clinical outcome in patients with high-grade serous carcinomas in independent retrospective cohorts (Brown et al. 2008; Shih et al. 2005). These findings suggest that gene(s) within this amplicon may contribute to clinical aggressiveness in neoplasm with 11q13.5 amplification. With this "green light", we proceeded to identify the potential gene(s) within the 11q13.5 amplicon which may play a mechanistic role in maintaining the survival of cancer cells and in developing drug resistance in high-grade ovarian serous carcinomas. We first selected the top six genes within this amplicon that demonstrate the most significant correlation between DNA and mRNA copy number from a total of 14 genes within the amplicon. After knockdown of each gene using RNA interference, we determined their sensitivity to paclitaxel and carboplatin which are routinely used in treating ovarian cancer patients after cytoreduction surgery. We found that *Rsf-1* (*HBXAP*) knockdown significantly inhibited cellular proliferation and decreased the IC_{50} of paclitaxel in ovarian cancer cells (Choi et al. 2008). Therefore, we selected *Rsf-1* (*HBXAP*) to study its molecular mechanisms in contributing to tumor progression and assess its potential in *Rsf-1* (*HBXAP*)-targeted cancer therapy in the future.

2.6 Cancer Genomics from Bench to Bedside

Since the discovery of the Philadelphia chromosome in chronic myelocytic leukemia (CML) in 1960, knowledge gained from cancer genomics studies has been steadily applied in practice of cancer patient care, including the screening and prevention of familiar and hereditary forms of cancer, classification, diagnosis and prognosis prediction, and the therapeutics of cancers (Dalton and Friend 2006; Ward and Hawkins 2001). While only accounts for 5–10% of all cancer cases, the hereditary cancer represents the most cogent example of successful application of genetic knowledge in clinic settings, as these patients often benefit the most from early detection by genetic screening, counseling and preventive surgery and treatment. For example, mutations in the *BRCA1/2* genes account for approximately 90% of the hereditary ovarian carcinomas, and the lifetime risk of ovarian cancer in a *BRCA1* mutation carriers approaches 40–50%, whereas that of *BRCA2* mutation carriers risk is 20–30% depending on the population studied (Prat et al. 2005). The prophylactic bilateral salpingo-oophorectomy in *BRCA* mutation carriers after childbearing has shown to reduce the risk of ovarian, fallopian tube, or primary peritoneal cancers by 98% (Prat et al. 2005). The other examples of successful application of cancer genomics studies to clinics are Novartis' imatinib (Gleevec, a tyrosine kinase inhibitor against BCR-ABL) for treating chronic myeloid leukemia with *BCR-ABL* fusion, and Genentech's Herceptin (trastuzumab, a humanized monoclonal antibody that binds to HER2) for treating breast cancers with *HER2/neu* (*ErbB2*) over-expression and/or amplification (Baselga 2006).

Gene expression analyses using microarray and quantitative RT-PCR have being widely used to identify genes whose expression is altered in cancer tissues compared to their normal counterparts. The expression of these differentially expressed genes can be used for molecular cancer classification, diagnosis and prediction of treatment outcome. Recently, two commercially available gene expression assays, Oncotype DX (Genomic Health) and MammaPrint (Agendia, Amsterdam, the Netherlands) have been applied in the assessment of the risk of tumor recurrence in patients with stage I or II node-negative breast cancer (Dobbe et al. 2008). The Oncotype DX 21-gene assay employs quantitative RT-PCR to analyze the expression levels of 16 cancer-related genes and 5 reference genes from formalin-fixed and paraffin-embedded specimens. These 16 genes were selected from a panel of 250 candidate genes that might be associated with breast cancer behavior based on literature, expression status and other information. A recurrence score can then be calculated from the normalized expression levels of these 16 genes and be used for prediction of recurrence (Paik et al. 2004). Patients with low recurrence score may receive hormonal therapy (tamoxifen) alone without adjuvant chemotherapy, while patients with high recurrence score may require adjuvant chemotherapy in addition to tamoxifen administration (Dobbe et al. 2008; Paik et al. 2004). The MammaPrint is an oligonucleotide-array assay performed on fresh-frozen tumor samples to analyze the expression levels of 70 genes that have been shown to have the highest association with cancer metastasis among a total of 25,000 genes (van de Vijver et al. 2002). The potential advantages of recurrence-assessment for node-negative breast cancer patient lie on the facts that 85% of these women are expected to be free from distant metastasis in 10 years and most of these patients are still considered candidates for systematic chemotherapy (Dobbe et al. 2008). Therefore, the introduction of these prognosis prediction tests to clinical oncology is expected to have an impact on the decision making of treatment plans that are not possible without molecular analysis of cancer transcriptome (Dobbe et al. 2008).

Although in the next coming years, we will likely embrace the new era of genomic medicine, the long-awaited personalized cancer care has yet to come of age. This is mainly because of tremendous efforts needed to be taken in order to translate the clinical significance of the enormous body of genomic data and to project human genomic alterations to specific types of therapeutic regimen. Furthermore, the extremely complex and unique nature of cancer appear to be a result of the Darwinian evolution of tumor species. Each cancer specimen is presented as unique products of accelerated evolution process that maximizes the cancer cells' survival in a dynamic micro-environment. This is clearly illustrated by the findings from aforementioned genome-wide mutational studies of cancer, showing that the genomic landscape of breast and colorectal cancers are characterized by a few commonly mutated gene "mountains" and a much larger number of gene "hills" that are mutated at a much lower frequency (Wood et al. 2007). The other recent whole-genome sequencing of an acute myeloid leukemia (AML) genome further demonstrated the uniqueness of each tumor sample, i.e. none of the eight novel genes with somatic mutations found in the patient was detected in 187 additional cases of AML (Ley et al. 2008). It also should be noted that the sequencing of normal genome of several individuals have showed that a large quantities of DNA content differ from individual to individual (Levy et al. 2007; Ley et al. 2008; Wang et al. 2008; Wheeler et al. 2008). This polymorphism among individuals would pose an additional hurdle in the interpretation of the tumor genome. On the other hand, these variations in such large quantity also provide a useful source in identifying cancer susceptibility genes by genome-wide association study. These studies suggest that much more mutations are yet to be identified, and much more number of cancer genomes need to be studied before we can identify the most important mutations or common pathways affected in cancers. In addition to case-to-case variation, the heterogeneity of cancer in individual patients and cells with normal DNA in the specimen would further complicate the genomics analysis of cancer. Until we obtain deeper understanding of the abnormalities occurred in cancer, the clinical translation of cancer genetics would be of limited use.

2.7 Concluding Remarks

In summary, we have witnessed the advances in applying genomic tools to further understand the molecular etiology of cancer in recent years. Cancer is a personalized disease characterized by a medley of genomic and molecular genetic abnormalities. While no single technology can adequately detect all kinds of genomic and molecular genetic changes (Feuk et al. 2006), the whole-genome sequencing is thought to likely provide the most comprehensive genomic landscape of cancer. However, currently the relatively high cost associated with whole-genome sequencing appears as a daunting limitation for large-scale cancer genetics studies, especially when a larger sample size is needed. Therefore, other array- or sequencing-based technologies will likely continue making substantial contribution in cancer genomics and cancer-associated gene studies. Since cancer contains balanced and unbalanced, large-scale and minute small mutations, until unbiased screening methods at several levels of resolution are applied, our quest for important genomic changes within cancer cells will be just like the story of blind men and an elephant (Heim and Mitelman 2008). In addition, the inquisition of cancer-associated genes that are causative genes or druggable targets needs to be accompanied by transcriptome and proteome analyses, both in human samples and in animal models. Lastly, it is expected that knowledge stemming from cancer genomics studies will be translated to clinically useful tools for diagnostics and therapeutics in cancer patients, following the successful lead of Novartis' Gleevec (imatinib) for treating chronic myeloid leukemia with *BCR-ABL* fusion, and Genentech's Herceptin (trastuzumab) for treating breast cancers with *HER-2/neu* (*ErbB2*) amplification.

Acknowledgement Part of this study was supported by an NIH/NCI grant (RO1 CA129080).

References

Baselga J (2006) Targeting tyrosine kinases in cancer: the second wave. Science 312:1175–1178

Bentires-Alj M, Gil SG, Chan R et al (2006) A role for the scaffolding adapter GAB2 in breast cancer. Nat Med 12:114–121

Bosma PT, van Eert SJ, Jaspers NG et al (2003) Functional cloning of drug resistance genes from retroviral cDNA libraries. Biochem Biophys Res Commun 309:605–611

Brown LA, Kalloger SE, Miller MA et al (2008) Amplification of 11q13 in ovarian carcinoma. Genes Chromosomes Cancer 47:481–489

Campbell PJ, Stephens PJ, Pleasance ED et al (2008) Identification of somatically acquired rearrangements in cancer using genome-wide massively parallel paired-end sequencing. Nat Genet 40:722–729

Carter NP (2007) Methods and strategies for analyzing copy number variation using DNA microarrays. Nat Genet 39:S16–21

Chi KR (2008) The year of sequencing. Nat Methods 5:11–14

Choi JH, Sheu J, Guan B et al (2009) Functional analysis of 11q13.5 amplicon identifies *Rsf-1 (HBXAP)* as a gene involved in paclitaxel resistance in ovarian cancer. Cancer Res 69:1407–1415

Dalton WS, Friend SH (2006) Cancer biomarkers – an invitation to the table. Science 312:1165–1168
Dobbe E, Gurney K, Kiekow S et al (2008) Gene-expression assays: new tools to individualize treatment of early-stage breast cancer. Am J Health Syst Pharm 65:23–28
Dohm JC, Lottaz C, Borodina T et al (2008) Substantial biases in ultra-short read data sets from high-throughput DNA sequencing. Nucleic Acids Res 36:e105
Duan J, Wainwright MS, Comeron JM et al (2003) Synonymous mutations in the human dopamine receptor D2 (DRD2) affect mRNA stability and synthesis of the receptor. Hum Mol Genet 12:205–216
Easton DF, Pooley KA, Dunning AM et al (2007) Genome-wide association study identifies novel breast cancer susceptibility loci. Nature 447:1087–1093
Eeles RA, Kote-Jarai Z, Giles GG et al (2008) Multiple newly identified loci associated with prostate cancer susceptibility. Nat Genet 40:316–321
Erlich Y, Mitra PP, DelaBastide M et al (2008) Alta-Cyclic: a self-optimizing base caller for next-generation sequencing. Nat Methods 5:679–682
Feuk L, Carson AR, Scherer SW (2006) Structural variation in the human genome. Nat Rev Genet 7:85–97
Greenman C, Stephens P, Smith R et al (2007) Patterns of somatic mutation in human cancer genomes. Nature 446:153–158
Gudmundsson J, Sulem P, Rafnar T et al (2008) Common sequence variants on 2p15 and Xp11.22 confer susceptibility to prostate cancer. Nat Genet 40:281–283
Haber DA, Settleman J (2007) Cancer: drivers and passengers. Nature 446:145–146
Heim S, Mitelman F (2008) Molecular screening for new fusion genes in cancer. Nat Genet 40:685–686
Hillier LW, Marth GT, Quinlan AR et al (2008) Whole-genome sequencing and variant discovery in *C. elegans*. Nat Methods 5:183–188
Ishkanian AS, Malloff CA, Watson SK et al (2004) A tiling resolution DNA microarray with complete coverage of the human genome. Nat Genet 36:299–303
Jones S, Zhang X, Parsons DW et al (2008) Core signaling pathways in human pancreatic cancers revealed by global genomic analyses. Science 1164368 321:1801–1806
Kallioniemi A, Kallioniemi OP, Sudar D et al (1992) Comparative genomic hybridization for molecular cytogenetic analysis of solid tumors. Science 258:818–821
Kimchi-Sarfaty C, Oh JM, Kim IW et al (2007) A “silent” polymorphism in the MDR1 gene changes substrate specificity. Science 315:525–528
Kinzler KW, Vogelstein B (2002) The genetic basis of human cancer, 2nd edn. McGraw-Hill, Toronto
Kittler R, Pelletier L, Heninger AK et al (2007) Genome-scale RNAi profiling of cell division in human tissue culture cells. Nat Cell Biol 9:1401–1412
Korbel JO, Urban AE, Affourtit JP et al (2007) Paired-end mapping reveals extensive structural variation in the human genome. Science 318:420–426
Levy S, Sutton G, Ng PC et al (2007) The diploid genome sequence of an individual human. PLoS Biol 5:e254
Ley TJ, Mardis ER, Ding L et al (2008) DNA sequencing of a cytogenetically normal acute myeloid leukemia genome. Nature 456:66–72
Lucito R, Healy J, Alexander J et al (2003) Representational oligonucleotide microarray analysis: a high-resolution method to detect genome copy number variation. Genome Res 13:2291–2305
Mao X, Young BD, Lu YJ (2007) The application of single nucleotide polymorphism microarrays in cancer research. Curr Genomics 8:219–228
Matsumura H, Reich S, Ito A et al (2003) Gene expression analysis of plant host–pathogen interactions by SuperSAGE. Proc Natl Acad Sci USA 100:15718–15723
Merlo LM, Pepper JW, Reid BJ et al (2006) Cancer as an evolutionary and ecological process. Nat Rev Cancer 6:924–935
Mitelman F, Johansson B, Mertens F (2007) The impact of translocations and gene fusions on cancer causation. Nat Rev Cancer 7:233–245

Mukherji M, Bell R, Supekova L et al (2006) Genome-wide functional analysis of human cell-cycle regulators. Proc Natl Acad Sci USA 103:14819–14824

Nackley AG, Shabalina SA, Tchivileva IE et al (2006) Human catechol-O-methyltransferase haplotypes modulate protein expression by altering mRNA secondary structure. Science 314:1930–1933

Nakayama K, Nakayama N, Jinawath N et al (2007) Amplicon profiles in ovarian serous carcinomas. Int J Cancer 120:2613–2617

Ng P, Tan JJ, Ooi HS et al (2006) Multiplex sequencing of paired-end ditags (MS-PET): a strategy for the ultra-high-throughput analysis of transcriptomes and genomes. Nucleic Acids Res 34:e84

Ng P, Wei CL, Sung WK et al (2005) Gene identification signature (GIS) analysis for transcriptome characterization and genome annotation. Nat Methods 2:105–111

Paik S, Shak S, Tang G et al (2004) A multigene assay to predict recurrence of tamoxifen-treated, node-negative breast cancer. N Engl J Med 351:2817–2826

Parsons DW, Jones S, Zhang X et al (2008) An Integrated Genomic Analysis of Human Glioblastoma Multiforme. Science 1164382

Pinkel D, Albertson DG (2005) Array comparative genomic hybridization and its applications in cancer. Nat Genet 37:S11–S17

Prat J, Ribe A, Gallardo A (2005) Hereditary ovarian cancer. Hum Pathol 36:861–870

Ruan Y, Ooi HS, Choo SW et al (2007) Fusion transcripts and transcribed retrotransposed loci discovered through comprehensive transcriptome analysis using Paired-End diTags (PETs). Genome Res 17:828–838

Rusk N, Kiermer V (2008) Primer: Sequencing–the next generation. Nat Methods 5:15

Schlabach MR, Luo J, Solimini NL et al (2008) Cancer proliferation gene discovery through functional genomics. Science 319:620–624

Schraml P, Schwerdtfeger G, Burkhalter F et al (2003) Combined array comparative genomic hybridization and tissue microarray analysis suggest PAK1 at 11q13.5–q14 as a critical oncogene target in ovarian carcinoma. Am J Pathol 163:985–992

Schuster SC (2008) Next-generation sequencing transforms today's biology. Nat Methods 5:16–18

Schwab M (1998) Amplification of oncogenes in human cancer cells. Bioessays 20:473–479

Shih Ie M, Sheu JJ, Santillan A et al (2005) Amplification of a chromatin remodeling gene, *Rsf-1/HBXAP*, in ovarian carcinoma. Proc Natl Acad Sci USA 102:14004–14009

Shih IM, Wang TL (2005) Apply innovative technologies to explore cancer genome. Curr Opin Oncol 17:33–38

Sjoblom T, Jones S, Wood LD et al (2006) The consensus coding sequences of human breast and colorectal cancers. Science 314:268–274

Soda M, Choi YL, Enomoto M et al (2007) Identification of the transforming EML4-ALK fusion gene in non-small-cell lung cancer. Nature 448:561–566

Thomas G, Jacobs KB, Yeager M et al (2008) Multiple loci identified in a genome-wide association study of prostate cancer. Nat Genet 40:310–315

Tuzun E, Sharp AJ, Bailey JA et al (2005) Fine-scale structural variation of the human genome. Nat Genet 37:727–732

van de Vijver MJ, He YD, van't Veer LJ et al (2002) A gene-expression signature as a predictor of survival in breast cancer. N Engl J Med 347:1999–2009

Volik S, Zhao S, Chin K et al (2003) End-sequence profiling: sequence-based analysis of aberrant genomes. Proc Natl Acad Sci USA 100:7696–7701

Wang J, Wang W, Li R et al (2008) The diploid genome sequence of an Asian individual. Nature 456:60–65

Wang K, Li M, Hadley D et al (2007) PennCNV: an integrated hidden Markov model designed for high-resolution copy number variation detection in whole-genome SNP genotyping data. Genome Res 17:1665–1674

Wang TL, Diaz LA Jr, Romans K et al (2004) Digital karyotyping identifies thymidylate synthase amplification as a mechanism of resistance to 5-fluorouracil in metastatic colorectal cancer patients. Proc Natl Acad Sci USA 101:3089–3094

Wang TL, Maierhofer C, Speicher MR et al (2002) Digital karyotyping. Proc Natl Acad Sci USA 99:16156–16161

Ward RL, Hawkins NJ (2001) Checking the scoreboard: the impact of cancer genetics on clinical practice. Intern Med J 31:249–253

Weir BA, Woo MS, Getz G et al (2007) Characterizing the cancer genome in lung adenocarcinoma. Nature 450:893–898

Wheeler DA, Srinivasan M, Egholm M et al (2008) The complete genome of an individual by massively parallel DNA sequencing. Nature 452:872–876

Witt AE, Hines LM, Collins NL et al (2006) Functional proteomics approach to investigate the biological activities of cDNAs implicated in breast cancer. J Proteome Res 5:599–610

Wold B, Myers RM (2008) Sequence census methods for functional genomics. Nat Methods 5:19–21

Wood LD, Parsons DW, Jones S et al (2007) The genomic landscapes of human breast and colorectal cancers. Science 318:1108–1113

Ylstra B, van den Ijssel P, Carvalho B et al (2006) BAC to the future! or oligonucleotides: a perspective for micro array comparative genomic hybridization (array CGH). Nucleic Acids Res 34:445–450

Chapter 3
An Integrated Oncogenomic Approach: From Genes to Pathway Analyses

Jeff A. Klomp, Bin T. Teh, and Kyle A. Furge

Abstract Several technologies now exist that allow the simultaneous evaluation of the amount of RNA produced by each cellular gene. Application of these technologies to measure transcriptional activity in cancer cells has provided a rich source of information that is being used to understand tumor biology. Analysis of the resulting gene expression data has evolved from the identification of individual gene expression differences between tumor and non-diseased cells to model-based evaluation of complex signal transduction pathways. Pathway-based models that utilize gene expression data have yielded new insights into tumor cell biology by more accurately describing both pleiotropic and polygenic cell processes. Further description and integration of gene expression-based models will be critical to fully exploit the information contained in gene expression data and to develop a more in-depth understanding of tumor cell development and progression.

3.1 Introduction to Gene Expression Profiling

Gene expression profiling commonly refers to the simultaneous measurement of the population of RNA within a biological sample of interest (Quackenbush 2006). The procedure begins with the isolation of RNA from the biological sample, which is often tumor cells growing in tissue culture or a small section of tumor tissue. Following RNA extraction, a fluorescent chemical moiety is enzymatically attached to each individual molecule of RNA. The attachment of this chemical moiety to the RNA is often termed RNA labeling, as this modification is used to quantify the amount of RNA molecules in later experimental steps. To quantify the RNA, the entire population of labeled RNA is hybridized to short, complementary

J.A. Klomp and K.A. Furge (✉)
Laboratory of Computational Biology, Van Andel Research Institute,
Grand Rapids, MI, 49503, USA
e-mail: kyle.furge@vai.org

B.T. Teh
Laboratory of Cancer Genetics, Van Andel Research Institute, Grand Rapids, MI, 49503, USA

W.C.S. Cho (ed.), *An Omics Perspective on Cancer Research*,
DOI 10.1007/978-90-481-2675-0_3,

oligonucleotide sequences (probes) that have been arrayed onto a solid substrate (the gene expression "chip"). Complementary oligonucleotide probes are designed to hybridize with high affinity to the RNA produced by individual genes. During chip construction, photolithographic or fine-liquid handling techniques are used to deposit these gene-specific oligonucleotide probes in precise locations on the solid surface. Since the surface area required by the probes is very small, thousands to millions of unique probes can be arrayed on a single chip. The hybridization process involves washing the labeled RNAs over the chip so that they bind to their complementary probes. Poorly hybridized labeled RNA is washed away to prevent background fluorescence when the chip is scanned with a laser. The intensity of fluorescence at each probe is converted into a numerical value under the assumption that regions of the chip with more labeled RNA will fluoresce more brightly than regions with less RNA. Continual improvements are being made to the process to enhance both the accuracy and precision in which gene expression measurements are made. Probes can be rigorously interrogated to ensure each probe sequence maps to a unique gene sequence based on the latest genome assemblies (Dai et al. 2005). To more reliably estimate intensity values, differences in fluorescent dye incorporation rates and intensity-dependent fluorescence effects are corrected during data normalization or data preprocessing (Choe et al. 2005; Dabney and Storey 2006; Irizarry et al. 2003). These and other advances in gene expression profiling methodologies have led to high levels of concordance between the RNA measurements made using chip-based experiments and measurements made using more traditional molecular biology techniques.

Given the numerous processing steps that are required to generate a gene expression profile – ranging from the selection of appropriate chips, to the sample preparation, to the varied experience of laboratories in generating high-throughput data – there is a potential for data to be generated that lacks validity and reproducibility (Draghici et al. 2006; Simon et al. 2003). Moreover, due to the significant amount of information that is present in a gene expression profile and the continuing development of novel analysis methods, it is unlikely that a single research laboratory would be able to fully scrutinize the resulting data. To facilitate the critical evaluation of gene expression profiling data by a larger researcher community, a database infrastructure has been developed to gather, archive, and distribute transcriptional profiling data. As researchers publish papers involving gene expression microarray experiments, the basic components of the gene expression data are uploaded into a public database. Enough data needs to be made publically available so that another researcher can reproduce, and if desired, extend the original analysis. For this reason a standard has been established, termed the minimum information about a microarray experiment (MIAME), that outlines the information authors are expected to make publically available (Brazma et al. 2001). According to the MIAME guidelines, authors must provide access to pertinent information based on the following categories: (1) experimental design; (2) array design, including the sequences of the probes and the position of the probes on the chip; (3) samples and preparation used; (4) hybridization methods; (5) gene expression measurements; and (6) any normalization controls. MIAME-conforming data is typically submitted to a central public repository such as the Gene Expression Omnibus (GEO) operated by the National Center for Biotechnology Information or the Array Express operated by the European Bioinformatics Institute. The construction of these gene expression databases has facilitated the critical evaluation of the results of gene expression profiling studies. In addition, these databases provide the basis for a flourishing analysis community that has the benefit of a large base of experimental data in which to develop and test new analytical methods. As a consequence, analysis of gene expression data has evolved from the identification of individual gene expression differences to the identification of complex genomic, biochemical, and signal transduction pathway perturbations.

3.2 Identification of Discriminative Genes

Many of the initial studies that applied gene expression profiling to cancer biology focused on integrating the expression data with clinical information to identify genes that could be used to discriminate between distinct tissues (Alon et al. 1999; Bittner et al. 2000; Golub et al. 1999; Takahashi et al. 2003). One common study design was to search for individual genes that showed differences in expression between tumor and adjacent non-diseased tissue. Dramatic differences in transcription were found between tumor and corresponding normal tissue, demonstrating that gene expression data could be used to uniquely identify transcriptional patterns in the tumor tissue. A variation of this analysis was to identify gene expression differences between tumors of different histological subtypes. An example of this type of analysis was the use of gene expression information, rather than histologic information, to decide if leukemia was acute myeloid leukemia (AML) or acute lymphoblastic leukemia (ALL) (Golub et al. 1999). While some of the early gene expression profiling studies of cancer focused on classification of samples with clearly different histopathologies, an important conceptual advancement was the application of gene expression profiling data to classify tumors that lacked established histological markers. Examination of gene expression data derived from B-cell lymphomas, renal cell carcinomas, breast tumors, and hepatocellular carcinomas (Alizadeh et al. 2000; Perou et al. 2000; Takahashi et al. 2001; van't Veer et al. 2002; Ye et al. 2003) revealed that tumors with similar histological characteristics could be partitioned into unique groups with different transcriptional characteristics. The implication of these studies was that transcriptional data could provide a more detailed view into the molecular genetics of the tumor. As such, it became clear that a major emphasis should be placed on the identification of individual genes that are uniquely expressed between different tumor subtypes to aid in tumor classification. Two general types of gene identification approaches are often used for the selection of these discriminative genes. The first of these approaches is sometimes termed "supervised analysis", as it requires a researcher to assign *a priori* classifications to each tumor sample. For example, samples with AML histology could be placed in one group while samples with ALL histology would be placed in a different group. Alternatively, samples obtained from patients that died due to disease progression within 2 years could be placed in one group while tumor samples from patients showing no disease progression after 2 years would be placed in another group. Following partitioning of the samples into groups, a variety of statistical methods can be used to find differentially expressed genes, such as signal-to-noise statistics, variations of Student's *t*-tests, or rank-based approaches (Fig. 3.1a) (Troyanskaya et al. 2002). Recently, incorporation of robust estimates of measurement variance and corrections for multiple testing have been used to assign per-gene confidence levels, providing a robust framework for the identification of individual gene expression differences between two sets of samples (Dudoit et al. 2002; Rhodes et al. 2002; Smyth 2004; Tusher et al. 2001).

It was immediately clear that the set of differentially expressed genes isolated from supervised gene expression analysis could have significant utility in the classification of tumors. An effective and now established use of gene expression data is to identify markers that can be used for classical immunohistochemical analysis of tumor samples. This technique has revealed many potential diagnostic markers for many tumor types. For example in renal cell carcinoma (RCC), some of these markers include the expression of α-methyacyl-CoA racemase for papillary RCC,

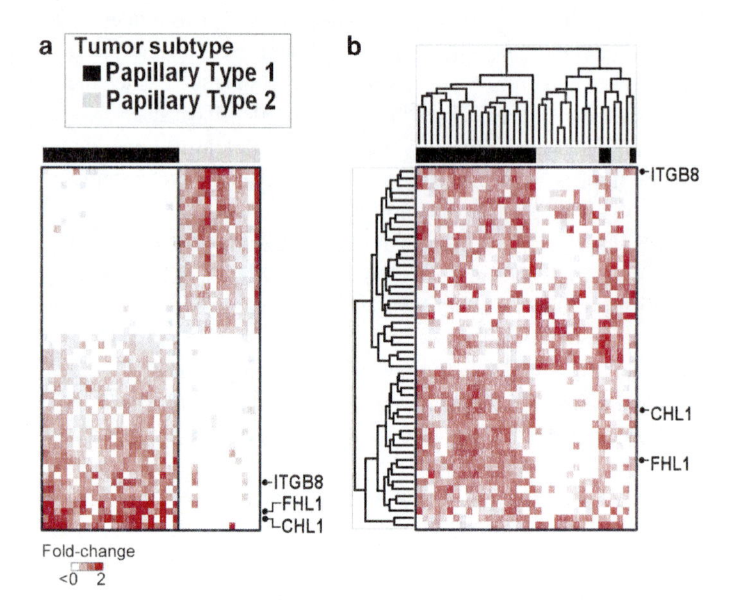

Fig. 3.1 Identification of gene expression differences between two histologically distinct classes of papillary renal cell carcinoma (RCC). Gene expression profiling was performed on a set of papillary RCC samples ($n = 35$). Using expert histopathological analysis, papillary RCC samples were designated as type 1 or type 2 based on large or small cell size and single layer or stratified layer tissue arrangement. (**a**) The 50 genes that were most differentially expressed between the subtypes were identified using a moderated *t*-statistic. (**b**) The same gene expression data used in A was filtered to identify the top 50 most variable genes. The samples and genes were organized by hierarchical clustering. At least three genes were identified in common in both analyses

glutathione S-transferase for clear cell RCC, and more recently, the *S100A1* gene to discriminate between renal oncocytoma and chromophobe RCC. Several other genes such as *vimentin*, *TIMP2*, *survivin*, and adipose differentiation-related genes have been identified as potential prognostic indicators (Kosari et al. 2005; Li et al. 2007; Lin et al. 2006; Moch et al. 1999; Rocca et al. 2007; Takahashi et al. 2001; Yao et al. 2007; Zhao et al. 2006). The use of markers identified from supervised gene expression analysis can complement and extend traditional patient stratification approaches such as TNM staging, tumor grade, and functional status (Gettman et al. 2001; Han et al. 2003; Tsui et al. 2000; Zisman et al. 2001).

Rather than use gene expression profiling data to identify potential immunohistochemical markers, a more powerful application of the data might be to use the gene expression profiles themselves to assist in sample classification. Using combinations of discriminative genes, it is possible to create unique signatures of gene expression for tumor subtypes. Thus, new samples might then be classified into one of the established groups for diagnostic purposes (Golub et al. 1999). A cogent example of this approach is a prognostic 70-gene signature that has been developed for breast cancer (van't Veer et al. 2002). Known commercially as MammaPrint, this test predicts patient outcome based on a gene expression signature and has been shown to outperform tests that use traditional clinical criteria (Buyse et al. 2006; Glas et al. 2006; van't Veer et al. 2002). This test has recently been granted FDA approval. Another test, known as Oncotype DX, uses 21 differentially expressed genes, identified in part, from microarray analysis, as a signature to predict recurrence in breast cancer patients (Habel et al. 2006; Paik et al. 2004). This test is now commonly performed on breast tumors. The advantage of a gene expression signature-based approach is that it provides a general framework for distinguishing between different tumor subtypes or different prognostic groups. As such, many tumor types are amenable to partitioning based on gene expression characteristics (Alizadeh et al. 2000; Perou et al. 2000; Takahashi et al. 2001; van't Veer et al. 2002; Ye et al. 2003). A disadvantage of gene expression-based classification is that in some cases, a gene expression signature that could differentiate between tumor subtypes in one study could not differentiate between tumor subtypes in other, independent studies (Draghici et al. 2006; Simon et al. 2003). Thus, controlled, statistically sound analytical approaches and robust validation are required before transitioning a gene signature to the clinical setting. However, as gene expression array technology matures, the performance of gene expression based classification, and hence clinical potential of gene expression signatures, is likely to mature as well (Dumur et al. 2008).

While supervised analysis is a useful method for finding discriminative genes between two pre-defined groups, an alternative method for identifying discriminative genes is to organize the genes into groups based on similar transcriptional characteristics (Alizadeh et al. 2000; Alon et al. 1999; Eisen et al. 1998). In this method, commonly termed "cluster analysis," genes are organized based on similarities in expression patterns using unbiased computational approaches. This approach is unsupervised since it does not require *a priori* classification of samples into groups by the researcher (Fig. 3.1b). Prior to the analysis, the gene expression data is typically filtered to remove genes that have low intensity or low variability across the samples. Following the removal of these uninformative genes, the remaining genes can be organized into clusters based on similarities in expression patterns. A variety of approaches can be used to measure similarities in expression, including various distance- and correlation-based metrics (Eisen et al. 1998). Once similarity measures between genes are computed, the relationship between these genes and samples can be visualized using a hierarchical dendrogram. While both unsupervised and supervised approaches contain very different mathematical frameworks, it is common for these approaches to identify similar sets of genes when there are large transcriptional differences between groups (Fig. 3.1). Interestingly, it was the visual interpretation of hierarchical clustering data that gave some of the first clues that gene expression data contained information that transcended individual genes. However, supervised analysis using advanced statistical models is often preferred for the identification of subtle transcriptional differences between samples.

3.3 Dissecting the Components of a Gene Expression Profile

The insights from hierarchical clustering analysis demonstrated that gene expression profiling data could expose underlying biological pathways. A simple example is based on the observation that relatively large groups of genes involved in cell proliferation and the cell division cycle were up-regulated in several cancers (Alizadeh et al. 2000; Chen et al. 2002; Perou et al. 2000). Specifically, genes that are periodically expressed in cells (indicating an association with the cell cycle) were often found to be up-regulated in the tumor tissue when compared to corresponding non-cancerous tissue (Whitfield et al. 2002). The identification of proliferation-related genes through gene expression profiling demonstrated the value of transcription information to investigate a variety of biological questions. While genes associated with cellular proliferation are expected to appear in analyses of cancer cells when compared to adjacent, non-cancerous tissue, other groups of genes related to inflammation, vascularization, and other general cellular physiology were also identified from gene expression profiling data. From these observations researchers first suggested that a gene expression profile could be subset into different components with more tangible biological underpinnings. For example, the set of genes associated with proliferation and the cell cycle can be partitioned into a proliferation component of the gene expression profile. Likewise, the set of genes associated with growth and development of new blood vessels make up a vascular component. The appearance of proliferation or vascularization components in the gene expression profile gave clues into the underlying biological state of the tumor samples. As more gene expression data was analyzed, it became clearer that insights into the cellular state of the tumor could be extended to other cellular phenomena ranging from oxygen deprivation to tissue repair (Chang et al. 2005; Chi et al. 2006).

While these insights shed light on the global cellular state of tumor cells, there are limitations to determining the cell state based on gene expression analysis. Perhaps the largest component of a gene expression profile can be attributed to the effect of cell lineage (Ross et al. 2000). When cells from different tissues are compared, the majority of gene expression variation can be directly attributed to differences between tissue types. For example, cells derived from ovarian tissue have much different expression characteristics than cells derived from kidney tissue. This can be a particularly important effect when designing a gene expression profiling study where tumor samples are compared to corresponding non-diseased tissue. While the tumor tissue may be dominated by the bulk tumor cell population, other cells are also present, such as stromal cells, endothelial cells, necrotic cells, and cells of the immune response. These cells are usually of a much different lineage and not representative of the population of cells found in the non-diseased tissue. This complexity adds to the challenge of placing a differentially expressed gene into a biological context. It is reasonable to speculate that this tissue-specific "noise" found in the gene expression data would make it difficult to obtain meaningful information from gene expression profiling experiments. The identification of a limited set of underlying mechanisms that directly account for the myriad of changes identified in gene expression profiling experiments would provide the easiest, most straightforward way to interpret the gene expression data. While the identification of cell-cycle genes provided initial support that specific biological processes could be interpreted from a gene expression profile, recent work has demonstrated that a gene expression profile can be divided into biologically meaningful components despite the presences of tissue-specific effects.

3.3.1 *Effects of Genomic Structure*

DNA amplifications and deletions were shown to have significant effects on gene expression in the budding yeast, *Saccharomyces cerevisiae* (Hughes et al. 2000a). In that study, mutant yeast cells containing a small region of amplification on one chromosome were compared to wild-type yeast cells that lacked the chromosomal amplification. When the gene expression data was arranged based on the genes' relative positions along the chromosome, it was noticed that the genes located within the region of amplification produced much more RNA than the genes located outside the region of amplification. Since chromosomal abnormalities are a common occurrence in many tumors (Struski et al. 2002), it was postulated that analogous effects were occurring in human tumors. Although studies in human cells lagged behind the yeast data due to the lack of an assembled human genome (Lander et al. 2001; Venter et al. 2001), several groups showed that chromosomal gains and losses have similar effects on gene expression in human tumors and human tumor-derived cell lines (Fig. 3.2). More RNA than normal is typically produced by genes located in regions of chromosomal amplification and less RNA is

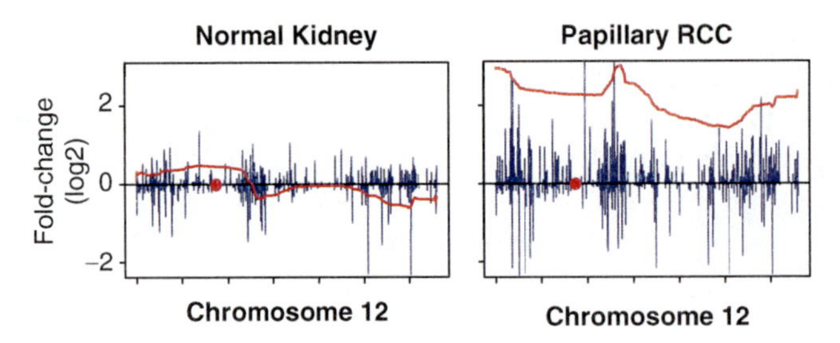

Fig. 3.2 Effects of aneuploidy on gene expression. Expression data from a normal kidney sample and a papillary renal cell carcinoma (RCC) were compared to expression data generated from a pool of normal kidney tissue. Gene expression ratios between the indicated sample and normal reference were log-transformed (base 2) such that a four-fold increase in expression has a transformed value of 2. Likewise, a four-fold decrease in expression has a transformed value of –2. Shown is the expression data from genes mapping to chromosome 12. The centromere is highlighted with a red circle. Trisomy of chromosome 12 is a common abnormality in papillary RCC and a clear enrichment of up-regulated genes is shown in the tumor sample. To highlight the increase in up-regulated genes, a data smoothing curve is also shown

typically produced by genes located in regions of chromosomal deletion (Crawley and Furge 2002; Mukasa et al. 2002; Phillips et al. 2001; Virtaneva et al. 2001; Xu et al. 2001). This gene expression effect has been observed in many tumor subtypes and these changes in gene expression nearly always indicate the presence of an underlying chromosomal abnormality (Furge et al. 2004; Hertzberg et al. 2006; Lindvall et al. 2004; Pollack et al. 2002).

Interestingly, not all transcripts that map to a region of deletion or amplification show a large upward or downward expression change (Fig. 3.2). This may be attributable to technical errors such as improper clone annotation, cross-hybridization, or genome assembly issues. It is also possible that an underlying biological mechanism, such as activation of signal transduction pathways, local genome architecture, or more complicated transcriptional regulation (e.g. gene silencing) control gene expression within a region of cytogenetic change (Platzer et al. 2002). Still, consistent with the observation that cell status and tissue type can be dissected from gene expression profiles, it is now clear that perturbations in genomic structure are reflected in gene expression data. Further, not only can the effects of large chromosomal gains and losses be found in the gene expression profile, but changes in transcription that result from certain other classes of chromosomal abnormalities can be identified. In some cases, chromosome translocations are reflected in the gene expression data. A prominent example of the discovery of translocations using gene expression data is the identification of *ETS* family fusions in prostate cancer (Tomlins et al. 2005). By searching for genes that showed abnormally high gene expression in prostate cancers, researchers discovered gene fusions between two members of the *ETS* family of transcription factors, *ERG* and *ETV1*, and the *TMPRSS2* gene. Moreover, analysis of gene expression profiling data has uncovered new types of chromosomal abnormalities. Somatic chromosome pairing is a phenomenon in which the two chromosome homologues become closely associated or joined in interphase. Analysis of transcriptional defects in a subtype of renal cancer revealed a somatic pairing abnormality (Koeman et al. 2008). Since nearly all solid tumors have associated chromosomal abnormalities, these data show that effects of genome structure represent an additional, valuable component of gene expression profiling data.

3.3.2 Effects of Cellular Pathways

While the aforementioned work showed that general cellular states and genomic perturbations are reflected in gene expression data, additional studies of yeast showed that modulation of specific cellular signal transduction pathways was also reflected in transcriptional data. For example, a panel of yeast strains, each of which had a unique mutation in a gene associated with cell wall synthesis, showed very similar transcriptional signatures (Hughes et al. 2000b). These results demonstrated that, at least in yeast, gene expression data can be exploited to detect perturbations in more subtle cellular processes. However, early gene expression studies attempting to identify gene expression signatures for modulation of signal transduction pathways in mammalian cells had more limited success. Particularly, initial attempts to characterize unique gene expression events downstream of activated growth factor receptors or transcription factors suggested that this type of transcriptional analysis would be difficult (Fambrough et al. 1999; Yu et al. 1999). These early reports generated a cloud of skepticism over gene expression signatures and their effectiveness as a method to determine pathway activation status of cancer cells. Moreover, since signal transduction events are often heavily influenced by post-transcriptional mechanisms such as protein modifications and changes in protein localization, isolation of specific transcriptional responses may not be possible in complex cells and tissues. Gradually, additional gene expression studies showed that at least a subset of signal transduction events could be measured using transcriptional data.

It is not clear whether advances in gene chip construction, advances in data processing methods, or simply the examination of more amenable signal transduction pathways lead to successes in pathway analysis. Regardless of the mechanism, accumulating evidence supports the observation that activation of oncogenic pathways (or inactivation of tumor suppressor pathways) can have unique and detectable gene expression effects. When a number of oncogenes, including *MYC*, *KRAS*, and *SRC*, were over-expressed in primary breast epithelial cells and compared to mock-transfected cells, a unique set of proliferation-independent gene signatures reflecting the activation of these genes were identified (Desai et al. 2002; Huang et al. 2003). It was also revealed that transcription factors in the *EF3* family and *HOX* family could have identifiable transcriptional signatures (Ferrando et al. 2002; Ishida et al. 2001). In these experiments, it was critical to demonstrate that these signatures were reflective of pathway activation and did not simply represent an aspect of cell cycle progression or apoptosis. Convincingly, the oncogenic gene signatures derived from several cell line studies were specific for monitoring activation of the corresponding molecular pathways in both *in vitro* cell culture and *in vivo* tumors. For example, genes up-regulated via over-expression of the *RAS* oncogene in breast cancer cell lines are also up-regulated in lung cancers known to contain activating *RAS* mutations, but not up-regulated in other lung cancers (Bild et al. 2005; Huang et al. 2003; Sweet-Cordero et al. 2005). Further, when cell lines with inactivating mutations in the *VHL* tumor suppressor gene are compared to cell lines with wild-type *VHL*, the identified genes are also deregulated in renal cell carcinomas that contain *VHL* mutations but not other kidney cancers (Furge et al. 2007b; Staller et al. 2003). Similar studies now recognize the possibility of detecting activation or inactivation of a variety of signal transduction pathways using gene expression data.

The dissection of a gene expression profile into pathway components has direct and important implications for guiding the use of molecular-targeted therapy in cancer. The accurate and objective identification of oncogenic pathways that are deregulated in a tumor sample is critical for the selection of molecular-targeted drugs that can lead to dramatic effects on tumor growth. For instance, activation of mammalian target of rapamycin (mTOR), due to activation of the phosphoinositide 3-kinase (PI3K) signaling pathway, is a frequent event in human cancers (Samuels and Ericson 2006). The application of specific mTOR inhibitors, such as rapamycin-derivatives, has been shown to be highly effective in inhibiting the growth of these tumor cells (Majumder et al. 2004). Aberrant PI3K pathway activity can be identified based on examination of gene expression profiling data (Creighton 2008; Majumder et al. 2004; Tiwari et al. 2003). As such, a PI3K activation signature can be used to rank patients based on their likelihood of response to rapamycin-based treatment regimens. An advantage of using gene expression-based models to identify pathway activation is that, like tumor sample classification, this approach is generalizable and therefore gene expression data can be used to monitor the activity of many different oncogenic pathways simultaneously (Rhodes and Chinnaiyan 2005). Thus, from a single gene expression profile, the activity of many pathways in addition to PI3K can be evaluated. Other important examples of molecular-targeted therapy involve the prominent class of anti-cancer drugs that inhibit the activity of receptor tyrosine kinases (RTK). Growth factors such as epidermal growth factor (EGF) or hepatocyte growth factor (HGF) bind to RTKs, EGFR and MET respectively, and stimulate cell proliferation, survival, and differentiation. Inappropriate activation of RTKs is associated with the development of many types of cancers. For many RTKs, small molecule inhibitors that prevent activation of the tyrosine kinase or antibody-based therapies that interfere with growth factor/receptor interactions are either currently clinically available or being evaluated in clinical trials (Matar et al. 2004; Toschi and Janne 2008). Like activation of the PI3K signaling pathway, activation of RTKs can also be monitored through evaluation of gene expression profiling data (Choi et al. 2007; Kaposi-Novak et al. 2006). The identification of RTK signal transduction defects using gene expression data is a compelling model for the screening of patients most likely to benefit from RTK inhibitors. While much work remains to validate gene expression-based models of pathway activation in a clinical environment, the outlook of this general approach is bright.

3.3.3 Pathway Analysis Methodology

Detection of pathway activation from gene expression data builds upon the statistical methods used to identify individual gene expression differences. Activation of an oncogene (or inactivation of a tumor suppressor gene) regulates the transcription of many downstream genes. A straightforward way to quantify the genes up- or down-regulated following oncogene activation is to perform a well-controlled genome-wide expression profiling experiment (Fig. 3.3). For example, primary breast epithelial cells transfected with *MYC* can be compared to mock-transfected primary breast epithelial cells and the set of genes up-regulated following *MYC* over-expression can be identified. Lists of genes that are up- or down-regulated following modulation of a given pathway are commonly reported in the literature either as tables in the manuscript, tables in the supplemental materials, or as data stored in gene expression databases. For clarity, we will use the term "empirically-derived" to describe a set of genes that are obtained from genome-wide expression

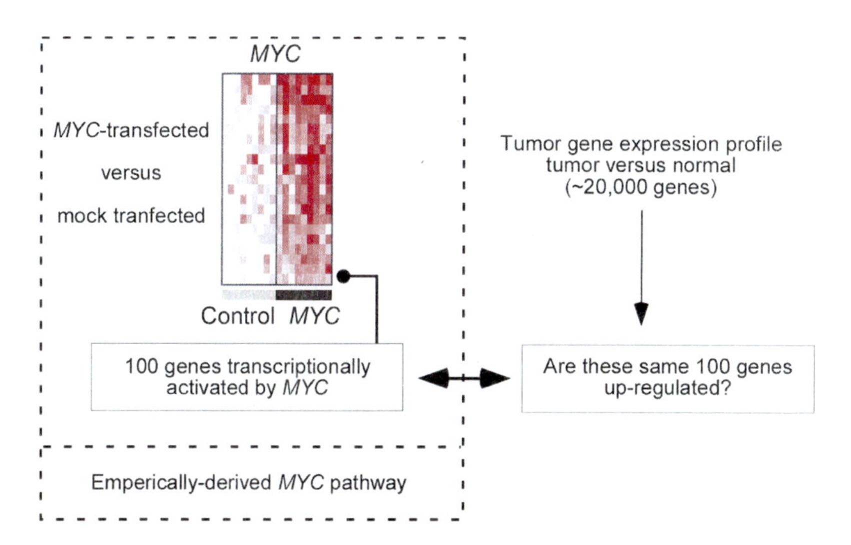

Fig. 3.3 Schematic of pathway analysis methodology. To generate an empirically-derived gene signature for *MYC* activation, gene expression profiling is performed on tissue culture cells in which *MYC* is over-expressed. Genes that are transcriptionally activated following *MYC* pathway activation are identified. To test if the *MYC* pathway is active in a particular tumor sample, these same transcriptionally responsive genes are examined for up-regulation in the tumor sample

profiling studies such as the *MYC* experiment mentioned previously. These empirically-derived gene sets can be obtained by a variety of cellular manipulations such as stimulation by protein growth factors, activation of oncogenes, exposure to limiting oxygen (hypoxia), etc. In each of these cases, gene sets associated with these perturbations are identified by direct examination of the gene expression data obtained from comparison of treated and control cells. Identification of sets of genes using this methodology is distinct from identification of sets of genes that are grouped based on functional similarities, such as the Gene Ontology, or are organized into sets based on descriptions of classical biochemical pathways, such as genes associated with the citric acid (Krebs) cycle (Ashburner et al. 2000; Joshi-Tope et al. 2005). We will refer to the latter sets of genes as "theoretically-derived" gene sets. In these cases, genes are grouped into sets based not on direct examination of gene expression data, but rather based on an accumulation and distillation of existing biochemical and molecular biological knowledge. As such, empirically-derived and theoretically-derived gene sets provide complementary information, but the distinction between the two is important for the interpretation of transcriptional data.

Similar to the identification of individual gene expression differences, there are several ways to determine if an empirically-derived or theoretically-derived set of genes is enriched in up-regulated genes (indicating possible pathway activation), down-regulated genes (indicating possible pathway inactivation), or that the set of genes does not contain any significant enrichment in up- or down-regulated genes in tumor samples (indicating no evidence for pathway deregulation). For example, to test if genes that are up-regulated by activation of the *MYC* oncogene are also up-regulated in tumor samples, gene expression levels in the tumor samples are compared to expression levels in non-diseased samples (Fig. 3.3). The genes are then ranked from most up-regulated in the tumor sample to most down-regulated in the tumor sample. One of the most common ways to test for enrichment (over-representation) is to use Fisher's exact test (Beissbart and Speed 2004; Bouton and Pevsner 2002; Hosack et al. 2003; Khatri et al. 2002). For this method, the top 100 genes up-regulated in the tumor sample are compared to the list of genes up-regulated by *MYC* over-expression. The higher the number of genes that occur in both lists, the more significant the prediction of pathway activation is. While this method is straightforward, effective, and easy to interpret, it requires that a somewhat arbitrary threshold be set to perform the intersection. Several other methods have been developed to avoid setting such thresholds and to take advantage of more subtle gene expression changes. Modifications to two classical statistical approaches, the Kolomogorov–Smirnov running sum statistic and the test statistic from the Student's *t*-test (Kim and Volsky 2005; Subramanian et al. 2005), can be used to test for gene set enrichments. Both of these approaches attempt to include genes that may have marginal statistical significance individually, but have high significance when analyzed as a set. While an exhaustive comparison of these and other statistical methodologies for comparing enrichment of one set of genes to a ranked list of genes is beyond the scope of this text, these approaches all share the same conceptual framework.

3.4 Integration of Gene Expression Components for Discovery

The benefit of dissecting gene expression profiles into different components is that it becomes possible to look for relationships between the components. If a tumor sample is predicted to have a genomic abnormality, such as a chromosomal amplification, the amplified region can be more closely scrutinized to look for deregulation of genes in the region. For example, if chromosome amplification causes over-expression of an oncogene, it is reasonable to expect that either an empirical pathway or a theoretical pathway that is associated with oncogene activation would also show evidence of transcriptional deregulation. If the expression data shows evidence for a genomic amplification, evidence for over-expression of an oncogene in the region, and evidence for activation of the oncogenic pathway, an integrative oncogenomic model can be built (Fig. 3.4). In this way, transcriptional data can be used to build cause-effect relationships between the different components of a gene expression profile. Moreover, based on recent work in chemical genomics, it is possible to link over-expression of an oncogene with activation of an oncogenic pathway and with identities of drugs that may modulate the activity of the oncogenic pathway (Hieronymus et al. 2006; Lamb et al. 2006). The advantage of this integrative approach is that these gene expression-based models can be rapidly built and used as screening tools to identify previously unappreciated mechanisms associated with tumor development. To extract similar information using molecular genetic

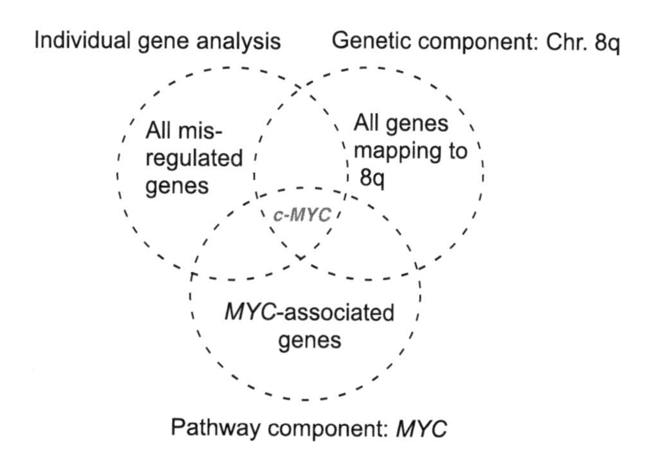

Fig. 3.4 Intersection of gene expression components. Gene expression analysis can be used to identify regions of genomic change. In this schematic, chromosome 8q was identified as being amplified in papillary renal cell carcinoma (RCC). Empirically-derived pathways can be analyzed to identify activated or repressed signal transduction components. In this schematic, the *MYC* pathway was identified as being active. Discriminative gene analysis can be used to identify individual genes that are deregulated. The *c-MYC* gene which maps to chromosome 8q was found to be over-expressed and showed evidence of pathway activation. Integration of these components suggests that the *c-MYC* gene that maps to chromosome 8q is a gene that is likely to be involved with the development of this tumor

approaches, a number of individual assays would be required, including comparative genomic hybridization studies to identify copy number abnormalities, a series of RT-PCR reactions to measure transcript abundance, and a series of Western blots to supply evidence of pathway activation. As these approaches often involve significant time and resource commitments, exploiting gene expression data can be an efficient approach to identify molecular genetic defects.

By integrating gene expression components, it is possible to identify a limited set of underlying mechanisms that can account for some of the vast differences in gene expression between tumor and normal tissue. However, a more complete understanding of the advantages and disadvantages of these pathway-based approaches remains to be worked out. A current limitation of pathway-based approaches is the potential for a large number of false-positives (Furge et al. 2007a). Effects of cellular proliferation can be found in many empirically-derived pathways. Oncogene activation and tumor suppressor gene inactivation often lead to increased cellular proliferation. If the effects of cellular proliferation dominate the transcriptional signature, then the signature will be predicted to be up-regulated in many tumors. This creates a false-positive result, as "activation" of the pathway simply represents proliferation of tumor cells in general, and can be a major limitation of pathway-based approaches. However, the tendency to over-predict oncogenic pathway activation based on proliferation effects should be tempered with the fact that gene expression predictions can still reveal true biological mechanisms. For example, samples from Burkitt's lymphoma patients often contain translocations involving the *MYC* gene and an immunoglobulin (*IG*) locus. Consistent with this translocation, a transcriptional signature indicative of *MYC* activation was strongly up-regulated in lymphomas that contained the *IG-MYC* translocation, but not in other lymphomas. However, there was a smaller, significant, subset of the non *IG-MYC* lymphomas that also contained strong transcriptional indications of *MYC* pathway activation (Hummel et al. 2006). In this subset of non *IG-MYC* lymphomas, it is possible that other translocation-independent abnormalities led to *MYC* activation and subsequent changes in the gene expression profiles. In this case, pathway activation would not be considered a false-positive. Rather, this transcriptional signature might reveal something important about the role of *MYC* in a subset of the non *IG-MYC* lymphomas.

3.5 The Evolution of Pathway Analysis

Full exploitation of gene expression profiling data in the future will require continued integration of the transcriptional data with the data produced by other "omics"-based approaches. For instance, it is now common to integrate gene expression data with genome-wide measurements of changes in chromatin structure, such as with the location of transcription factor binding sites, as identified by genome-wide chromatin immunoprecipitation experiments. In addition, pathway-based analysis of gene expression data requires a detailed knowledge of protein interaction networks.

The further elucidation of protein-protein interaction networks (interactomics) will be crucial for understanding how pathway modulations are reflected in gene expression profiling data. There are several public (Joshi-Tope et al. 2005; Okuda et al. 2008) and private (www.ingenuity.com) protein interaction databases that allow the integration of gene expression data with protein interaction networks. Integration of the gene expression and protein interaction networks has the potential to not only make predictions of pathway modulations, but also to shed light on changes in intracellular biochemistry and metabolite profiles (metabolomics). Integration of gene expression data with the metabolic pathway information can be used to highlight potential changes in metabolic flux. Given a kinetic model of a biochemical pathway, examination of the transcriptional changes that are occurring with key enzymes of a biochemical pathway can be used to infer changes in the rate at which various metabolites in a pathway are produced or consumed (Fig. 3.5). As more metabolic and signal transduction interaction networks are better described with mathematical models, integration of gene expression data has the potential

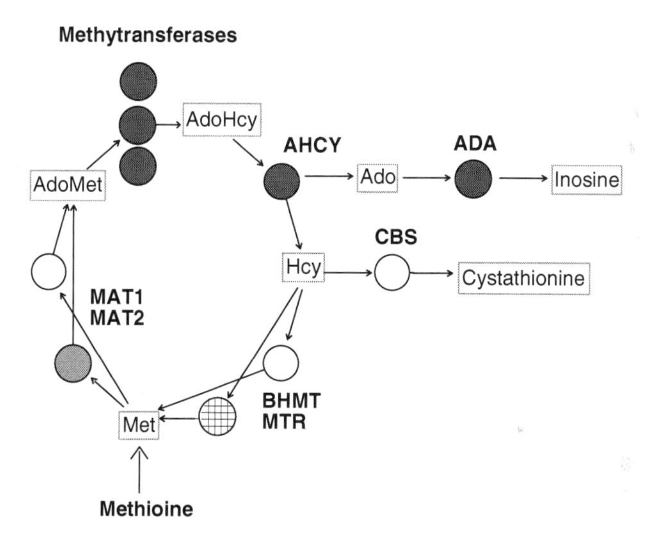

Fig. 3.5 Integration of gene expression data with protein-protein interaction data to infer metabolic changes in the methionine cycle. Methionine (Met) enters the pathway and is converted to S-adenosyl methionine (AdoMet) by a family of methionine adenosyltransferases (MAT1, MAT2). AdoMet is used as a substrate for all DNA, protein, and lipid methytransferases. S-adenosyl homocysteine (AdoHcy) is a product of the methytransferase reaction and is broken into adenosine (Ado) and homocystine (Hcy) by S-adenosyl homocysteine hydrolase (AHCY). Ado and Hcy are either fluxed out of the cycle by adenosine deaminase (ADA) and cystathionine beta-synthase (CBS), or cycled back into the pathway by the activity of betaine-homocysteine methyltransferase (BHMT) and 5-methyltetrahydrofolate-homocysteine methyltransferase (MTR). Overlaid on this pathway is gene expression data derived from comparisons of papillary renal cell carcinoma (RCC) and non-diseased tissue. *Dark gray* indicates increased gene expression and *Light gray* indicates decreased gene expression in the tumors. *Hatched* indicates the gene expression was not measured. The increased expression of key enzymatic components of this pathway suggests an increased methionine flux is occurring in the papillary RCCs

to yield significant insights into perturbations in the cellular networks and their associated biochemistries.

In the coming years, a goal of the computational community should be the continued refinement of pathway analysis to the point at which gene expression predictions of a given pathway modulation have as much robust historical validation as more traditional molecular biological approaches. While this is a lofty goal, limiting confounding effects and false-positives will be a key to unlocking the rich information present in a gene expression profile. In addition, more extensive work will be required to encourage acceptance of gene expression analyses by the biological community at large. It is likely that more subtle pathway perturbations will be difficult, if not impossible, to recreate accurately in the laboratory. Experimental model systems such as tissue culture cells, tumor xerographs, or even animal models may lack the more subtle characteristics of human tumor cell biology. The biological community requires more confidence in gene expression-based prediction methods before they can be fully accepted in translational or clinical settings. This is particularly important, since it is likely that additional clinically relevant biological components can be found in gene expression profiles. Effects of regulatory short RNAs (microRNAs), cytoskeleton modulations, and organelle dynamics all have the potential to be reflected in gene expression space. As additional gene expression components are identified from gene expression profiles, it will also be important to determine the extent by which a gene expression profile can be reduced to its component parts. Idealistically, all variability in gene expression of a tumor sample could be associated with defined biological components. While the complications of technical or experimental error coupled with the stochastic nature of gene expression may limit the understanding of all of the experimental variation, it will be crucial to categorize as much gene expression variability as possible.

References

Alizadeh AA, Eisen MB, Davis RE et al (2000) Distinct types of diffuse large B-cell lymphoma identified by gene expression profiling. Nature 403:503–511

Alon U, Barkai N, Notterman DA et al (1999) Broad patterns of gene expression revealed by clustering analysis of tumor and normal colon tissues probed by oligonucleotide arrays. Proc Natl Acad Sci USA 96:6745–6750

Ashburner M, Ball CA, Blake JA et al (2000) Gene ontology: tool for the unification of biology. The Gene Ontology Consortium. Nat Genet 25:25–29

Beissbart T, Speed TP (2004) GOstat: find statistically overrepresented gene ontologies within a group of genes. Bioinformatics 20:1464–1465

Bild AH, Yao G, Chang JT et al (2005) Oncogenic pathway signatures in human cancers as a guide to targeted therapies. Nature 439:353–357

Bittner M, Meltzer P, Chen Y et al (2000) Molecular classification of cutaneous malignant melanoma by gene expression profiling. Nature 406:536–540

Bouton CM, Pevsner J (2002) DRAGON View: information visualization for annotated microarray data. Bioinformatics 18:323–324

Brazma A, Hingamp P, Quackenbush J et al (2001) Minimum information about a microarray experiment (MIAME)-toward standards for microarray data. Nat Genet 29:365–371
Buyse M, Loi S, van't Veer L et al (2006) Validation and clinical utility of a 70-gene prognostic signature for women with node-negative breast cancer. J Natl Cancer Inst 98:1183–1192
Chang HY, Nuyten DS, Sneddon JB et al (2005) Robustness, scalability, and integration of a wound-response gene expression signature in predicting breast cancer survival. Proc Natl Acad Sci USA 102:3738–3743
Chen X, Cheung ST, So S et al (2002) Gene expression patterns in human liver cancer. Mol Bio Cell 13:1929–1939
Chi JT, Wang Z, Nuyten DS et al (2006) Gene expression programs in response to hypoxia: cell type specificity and prognostic significance in human cancers. PLoS Med 3:e47
Choe SE, Boutros M, Michelson AM et al (2005) Preferred analysis methods for Affymetrix GeneChips revealed by a wholly defined control dataset. Genome Biol 6:R16
Choi K, Creighton CJ, Stivers D et al (2007) Transcriptional profiling of non-small cell lung cancer cells with activating *EGFR* somatic mutations. PLoS ONE 2:e1226
Crawley JJ, Furge KA (2002) Identification of frequent cytogenetic aberrations in hepatocellular carcinoma using gene expression data. Genome Biol 3:RESEARCH0075
Creighton CJ (2008) Multiple oncogenic pathway signatures show coordinate expression patterns in human prostate tumors. PLoS ONE 3:e1816
Dabney AR, Storey JD (2006) A reanalysis of a published Affymetrix GeneChip control dataset. Genome Biol 6:R16
Dai M, Wang P, Boyd AD et al (2005) Evolving gene/transcript definitions significantly alter the interpretation of GeneChip data. Nucleic Acids Res 33:e175
Desai KV, Xiao N, Wang W et al (2002) Initiating oncogenic event determines gene-expression patterns of human breast cancer models. Proc Natl Acad Sci USA 99:6967–6972
Draghici S, Khatri P, Eklund AC et al (2006) Reliability and reproducibility issues in DNA microarray measurements. Trends Genet 22:101–109
Dudoit S, Yang YH, Callow MJ et al (2002) Statistical methods of identifying differentially expressed genes in replicated cDNA microarray experiments. Stat Sinica 12:111–129
Dumur CI, Lyons-Weiler M, Sciulli C et al (2008) Interlaboratory performance of a microarray-based gene expression test to determine tissue of origin in poorly differentiated and undifferentiated cancers. J Mol Diagn 10:67–77
Eisen MB, Spellman PT, Brown PO et al (1998) Cluster analysis and display of genome-wide expression patterns. Proc Natl Acad Sci USA 95:14863–14868
Fambrough D, McClure K, Kazlauskas A et al (1999) Diverse signaling pathways activated by growth factor receptors induce broadly overlapping rather than independent, sets of genes. Cell 97:727–741
Ferrando AA, Neuberg DS, Staunton J et al (2002) Gene expression signatures define novel oncogenic pathways in T cell acute lymphoblastic leukemia. Cancer Cell 1:75–87
Furge KA, Lucas KA, Takahashi M et al (2004) Robust classification of renal cell carcinoma based on gene expression data and predicted cytogenetic profiles. Cancer Res 64:4117–4121
Furge KA, Tan MH, Dykema K et al (2007a) Identification of deregulated oncogenic pathways in renal cell carcinoma: an integrated oncogenomic approach based on gene expression profiling. Oncogene 26:1346–1350
Furge KA, Chen J, Koeman J et al (2007b) Detection of DNA copy number changes and oncogenic signaling abnormalities from gene expression data reveals MYC activation in high-grade papillary renal cell carcinoma. Cancer Res 67:3171–3176
Gettman MT, Blute ML, Spotts B et al (2001) Pathologic staging of renal cell carcinoma: significance of tumor classification with the 1997 TNM staging system. Cancer 91:354–361
Glas AM, Floore A, Delahaye LJ et al (2006) Converting a breast cancer microarray signature into a high-throughput diagnostic test. BMC Genomics 7:278
Golub TR, Slonim DK, Tamayo P et al (1999) Molecular classification of cancer: class discovery and class prediction by gene expression monitoring. Science 286:531–537

Habel LA, Shak S, Jacobs MK et al (2006) A population-based study of tumor gene expression and risk of breast cancer death among lymph node-negative patients. Breast Cancer Res 8:R25
Han KR, Bleumer I, Pantuck AJ et al (2003) Validation of an integrated staging system toward improved prognostication of patients with localized renal cell carcinoma in an international population. J Urol 170:2221–2224
Hertzberg L, Betts DR, Raimondi SC et al (2006) Prediction of chromosomal aneuploidy from gene expression data. Genes Chromosomes Cancer 46:75–86
Hieronymus H, Lamb J, Ross KN et al (2006) Gene expression signature-based chemical genomic prediction identifies a novel class of HSP90 pathway modulators. Cancer Cell 10:321–330
Hosack DA, Dennis G Jr, Sherman BT et al (2003) Identifying biological themes within lists of genes with EASE. Genome Biol 4:R70
Huang E, Ishida S, Pittman J et al (2003) Gene expression phenotypic models that predict the activity of oncogenic pathways. Nat Genet 34:226–230
Hughes TR, Roberts CJ, Dai H et al (2000a) Widespread aneuploidy revealed by DNA microarray expression profiling. Nat Genet 25:333–337
Hughes TR, Marton MJ, Jones AR et al (2000b) Functional discovery via a compendium of expression profiles. Cell 102:109–126
Hummel M, Bentink S, Berger H et al (2006) A biologic definition of Burkitt's lymphoma from transcriptional and genomic profiling. N Engl J Med 354:2419–2430
Irizarry RA, Hobbs B, Collin F et al (2003) Exploration, normalization, and summaries of high-density oligonucleotide array probe level data. Biostatistics 2:249–264
Ishida S, Huang E, Zuzan H et al (2001) Role for E2F in control of both DNA replication and mitotic functions as revealed from DNA microarray analysis. Mol Cell Biol 21:4684–4699
Joshi-Tope G, Gillespie M, Vastrik I et al (2005) Reactome: a knowledgebase of biological pathways. Nucleic Acids Res 33:D428–432
Kaposi-Novak P, Lee JS, Gomez-Quiroz L et al (2006) Met-regulated expression signature defines a subset of human hepatocellular carcinomas with poor prognosis and aggressive phenotype. J Clin Invest 116:1582–1595
Khatri P, Draghici S, Ostermeier GC et al (2002) Profiling gene expression using onto-express. Genomics 79:266–270
Kim S, Volsky DJ (2005) PAGE: Parametric Analysis of Gene Set Enrichment. BMC Bioinformatics 6:144
Koeman JM, Russell RC, Tan MH et al (2008) Somatic pairing of chromosome 19 in renal oncocytoma is associated with deregulated EGLN2-mediated oxygen sensing response. PLoS Genet 4:e1000176
Kosari F, Parker AS, Kube DM et al (2005) Clear cell renal cell carcinoma: gene expression analyses identify a potential signature for tumor aggressiveness. Clin Cancer Res 11:5128–5139
Lamb J, Crawford ED, Peck D et al (2006) The connectivity map: using gene-expression signatures to connect small molecules, genes, and disease. Science 313:1929–1935
Lander ES, Linton LM, Birren B et al (2001) Initial sequencing and analysis of the human genome. Nature 409:860–921
Li G, Barthelemy A, Feng G et al (2007) S100A1: a powerful marker to differentiate chromophobe renal cell carcinoma from renal oncocytoma. Histopathology 50:642–647
Lin F, Yang W, Betten M et al (2006) Expression of S-100 protein in renal cell neoplasms. Hum Pathol 37:462–470
Lindvall C, Furge KA, Bjorkholm M et al (2004) Combined genetic- and transcriptional profiling of acute myeloid leukemia with complex and normal karyotypes. Haematologia 89:1072–1081
Majumder PK, Febbo PG, Bikoff R et al (2004) mTOR inhibition reverses Akt-dependent prostate intraepithelial neoplasia through regulation of apoptotic and HIF-1-dependent pathways. Nat Med 10:594–601
Matar P, Rojo F, Cassia R et al (2004) Combined epidermal growth factor receptor targeting with the tyrosine kinase inhibitor gefitinib (ZD1839) and the monoclonal antibody cetuximab (IMC-C225): superiority over single-agent receptor targeting. Clin Cancer Res 10:6487–6501

Moch H, Schraml P, Bubendorf L et al (1999) High-throughput tissue microarray analysis to evaluate genes uncovered by cDNA microarray screening in renal cell carcinoma. Am J Pathol 154:981–986

Mukasa A, Ueki K, Matsumoto S et al (2002) Distinction in gene expression profiles of oligodendrogliomas with and without allelic loss of 1p. Oncogene 21:3961–3968

Okuda S, Yamada T, Hammajima M et al (2008) KEGG Atlas mapping for global analysis of metabolic pathways. Nucleic Acids Res 36:W423–426

Paik S, Shak S, Tang G et al (2004) A multigene assay to predict recurrence of tamoxifen-treated, node-negative breast cancer. N Engl J Med 351:2817–2826

Perou CM, Sorlie T, Eisen MB et al (2000) Molecular portraits of human breast tumors. Nature 406:747–752

Phillips JL, Hayward SW, Wang Y et al (2001) The consequences of chromosomal aneuploidy on gene expression profiles in a cell line model for prostate carcinogenesis. Cancer Res 61:8143–8149

Platzer P, Upender MB, Wilson K et al (2002) Silence of chromosomal amplifications in colon cancer. Cancer Res 62:1134–1138

Pollack JR, Sorlie T, Perou CM et al (2002) Microarray analysis reveals a major direct role of DNA copy number alteration in the transcriptional program of human breast tumors. Proc Natl Acad Sci USA 99:12963–12968

Quackenbush J (2006) Microarray analysis and tumor classification. N Engl J Med 354:2463–2472

Rhodes DR, Miller JC, Haab BB et al (2002) CIT: identification of differentially expressed clusters of genes from microarray data. Bioinformatics 18:205–206

Rhodes DR, Chinnaiyan AM (2005) Integrative analysis of the cancer transcriptome. Nat Genet 37:S31–37

Rocca PC, Brunelli M, Gobbo S et al (2007) Diagnostic utility of S100A1 expression in renal cell neoplasms: an immunohistochemical and quantitative RT-PCR study. Mod Pathol 20:722–708

Ross DT, Scherf U, Eisen MB et al (2000) Systematic variation in gene expression patterns in human cancer cell lines. Nat Genet 24:227–235

Samuels Y, Ericson K (2006) Oncogenic PI3K and its role in cancer. Curr Opin Oncol 18:77–82

Simon R, Radmacher MD, Dobbin K et al (2003) Pitfalls in the use of DNA microarray data for diagnostic and prognostic classification. J Natl Cancer Inst 95:14–18

Smyth GK (2004) Linear models and empirical bayes methods for assessing differential expression in microarray experiments. Stat Appl Genet Mol Biol 3:Article3

Staller P, Sulitkova J, Lisztwan J et al (2003) Chemokine receptor CXCR4 downregulated by von Hippel-Lindau tumor suppressor pVHL. Nature 425:307–311

Struski S, Doco-Fenzy M, Cornillet-Lefebvre P (2002) Compilation of published comparative genomic hybridization studies. Cancer Genet Cytogenet 135:63–90

Subramanian A, Tamayo P, Mootha VK et al (2005) Gene set enrichment analysis: a knowledge-based approach for interpreting genome-wide expression profiles. Proc Natl Acad Sci USA 102:15545–15550

Sweet-Cordero A, Mukherjee S, Subramanian A et al (2005) An oncogenic KRAS2 expression signature identified by cross-species gene-expression analysis. Nat Genet 37:48–55

Takahashi M, Rhodes DR, Furge KA et al (2001) Gene expression profiling of clear cell renal cell carcinoma: gene identification and prognostic classification. Proc Natl Acad Sci USA 98:9754–9759

Takahashi M, Sugimura J, Yang XJ et al (2003) Molecular sub-classification of kidney cancer and the discovery of new diagnostic markers. Oncogene 22:6810–6818

Tiwari G, Sakaue H, Pollack JR et al (2003) Gene expression profiling in prostate cancer cells with Akt activation reveals Fra-1 as an Akt-inducible gene. Mol Cancer Res 1:475–484

Tomlins SA, Rhodes DR, Perner S et al (2005) Recurrent fusion of TMPRSS2 and ETS transcription factor genes in prostate cancer. Science 310:644–648

Toschi L, Janne PA (2008) Single-agent and combination therapeutic strategies to inhibit hepatocyte growth factor/MET signaling in cancer. Clin Cancer Res 14:5941–5946

Troyanskaya OG, Garber ME, Brown PO et al (2002) Nonparametric methods for identifying differentially expressed genes in microarray data. Bioinformatics 18:1454–1461

Tsui KH, Shvarts O, Smith RB et al (2000) Prognostic indicators for renal cell carcinoma: a multivariate analysis of 643 patients using the revised 1997 TNM staging criteria. J Urol 163:1090–1095 (quiz 1295)

Tusher VG, Tibshirani R, Chu G (2001) Significance analysis of microarrays applied to the ionizing radiation response. Proc Natl Acad Sci USA 98:5116–5121

Van't Veer LJ, Dai H, van de Vijver MJ et al (2002) Gene expression profiling predicts clinical outcome of breast cancer. Nature 415:530–536

Venter JC, Adams MD, Myers EW et al (2001) The sequence of the human genome. Science 291:1304–1351

Virtaneva K, Wright FA, Tanner SM et al (2001) Expression profiling reveals fundamental biological differences in acute myeloid leukemia with isolated trisomy 8 and normal cytogenetics. Proc Natl Acad Sci USA 98:1124–1129

Whitfield ML, Sherlock G, Saldanha AJ et al (2002) Identification of genes periodically expressed in the human cell cycle and their expression in tumors. Mol Biol Cell 13:1977–2000

Xu XR, Huang J, Xu ZG et al (2001) Insight into hepatocellular carcinogenesis at transcriptome level by comparing gene expression profiles of hepatocellular carcinoma with those of corresponding noncancerous liver. Proc Natl Acad Sci USA 98:15089–15094

Yao M, Huang Y, Shioi K et al (2007) Expression of adipose differentiation-related protein: a predictor of cancer-specific survival in clear cell renal carcinoma. Clin Cancer Res 13:152–160

Ye Q, Qin L, Forgues M et al (2003) Predicting hepatitis B virus-positive metastatic hepatocellular carcinomas using gene expression profiling and supervised machine learning. Nat Med 9:416–423

Yu J, Zhang L, Hwang PM et al (1999) Identification and classification of p53-regulated genes. Proc Natl Acad Sci USA 96:14517–14522

Zhao H, Ljungberg B, Grankvist K et al (2006) Gene expression profiling predicts survival in conventional renal cell carcinoma. PLoS Med 3:e13

Zisman A, Pantuck AJ, Dorey F et al (2001) Improved prognostication of renal cell carcinoma using an integrated staging system. J Clin Oncol 19:1649–1657

Chapter 4
The Epigenomics of Cancer

Izuho Hatada

Abstract Epigenomics is the genome-wide study of epigenetics such as DNA methylation and histone modification. Epigenetic abnormality plays an important role in the pathogenesis of most cancers. Among these modifications, DNA methylation is the best-known and most important because of its heritable character. Here, recent advances in several technologies for genome-wide profiling of DNA methylation are reviewed.

4.1 Introduction

Genetics cannot explain monozygotic twins and cloned animals, because the same genetic information has different phenotypes. It also cannot explain the diversity of cells with the same genetic information. However, epigenetics can explain these phenomena. Epigenetics is defined as heritable information other than genetic information coded by the nucleotides adenine (A), guanine (G), cytosine (C), and thymine (T). Epigenetic information is stored using two major modifications of DNA and histone: DNA methylation, which is methylation at the C5 position of cytosine, and histone modifications, such as acetylation and methylation (Bird 2002; Jenuwein 2001). These modifications are variable and work as gene switches that regulate gene expression, giving phenotypic diversity to individuals or cells with the same genetic information. DNA methylation is associated with transcriptional repression while histone acetylation is associated with transcriptional activation. Methylation of histone H3 lysine 4 (H3K4) is associated with transcriptional activation while methylations of H3 lysine 9 (H3K9) and H3 lysine 27 (H3K27) are associated with transcriptional repression. DNA methylation is maintained by DNA (cytosine-5-)-methyltransferase 1 (Dnmt1), which preferentially methylates hemimethylated DNA after DNA replication. In mammals, most DNA methylation is located at CpG sites.

I. Hatada (✉)
Laboratory of Genome Science, Biosignal Genome Resource Center, Institute for Molecular and Cellular Regulation, Gunma University, Maebashi, 371-8511, Japan
e-mail: ihatada@showa.gunma-u.ac.jp

W.C.S. Cho (ed.), *An Omics Perspective on Cancer Research*,
DOI 10.1007/978-90-481-2675-0_4,

These sites occur at relatively low frequency in eukaryotic genomes, because most cytosines at CpG are methylated by DNA methyltransferases, and methylated cytosine can be converted to thymine by spontaneous deamination (Fryxell and Zuckerkandl 2000). However, clusters of CpG sites, called CpG islands, are frequently found in the 5′ regions of genes. CpG sites in CpG islands are usually unmethylated in normal cells, while CpG sites in other regions, which are mostly located in repetitive sequences, are usually methylated.

Epigenetics has been most studied in cancer. The global level of DNA methylation is lower in cancer than that in normal tissue counterparts (Feinberg and Tycko 2004). Loss of methylation is mostly due to the demethylation of repetitive DNA sequences, which could explain the chromosomal instability of cancers. Chromosomal instability was also shown in a mouse model representing the depletion of DNA methylation by the disruption of Dnmts (Eden et al. 2003). The first epigenetic change in genes found in tumors was hypomethylation (Feinberg and Vogelstein 1983), and presently many growth-promoting genes such as *HRAS*, *S100A4*, *PAX2*, and microRNA *let-7a-3*, are known to be activated through hypomethylation in cancers (Feinberg 2007). On the other hand, many tumor suppressor genes including *RB*, *CDKN2A*, *VHL, MLH1*, and *APC* are known to be inactivated by hypermethylation in cancers (Feinberg 2007). Hypermethylation of CpG islands in the promoter regions of tumor-suppressor genes in cancer cells is associated with a particular combination of histone markers: deacetylation, loss of H3K4 trimethylation, and gain of H3K9 methylation and H3K27 trimethylation (Ballestar et al. 2003; Jones and Baylin 2007).

Epigenetic mutations have been found to be frequent mechanisms of gene inactivation in cancers. For example, somatic mutations of *BRCA1* are extremely rare in sporadic breast cancer and ovarian cancers, while this gene is frequently inactivated by DNA methylation (Dobrovic et al. 1997). Recent studies have also suggested that epigenetic abnormalities might be the earliest events in cancer initiation (Jones and Baylin 2007). A series of genes, such as *CDKN2A*, are reported to exhibit DNA hypermethylation in the preinvasive stages of colon and other cancers, but are rarely mutated in such cancers.

4.2 Emerging Technologies for Genome-Wide Profiling of DNA Methylation

Genome-wide profiling of DNA methylation has become one of the most important and exciting approaches to cancer genomics. Usually, these techniques involved combinations of methylation detection strategies and genome-wide profiling methods (Table 4.1). We have three strategies for the detection of DNA methylation: digestion of DNA by methylation-sensitive restriction enzymes, such as *Not* I and *Hpa* II, immunoprecipitation of DNA by 5-methylcytosine antibody (MeDIP), and bisulfite conversion of DNA (Table 4.1). There are several genome-wide profiling methods, such as two-dimensional electrophoresis, subtraction, microarray, pyrosequencing, next-generation sequencing technologies, and matrix-assisted laser desorption/ionization time-of-flight (MALDI-TOF) mass spectrometry. For example, the combination

Table 4.1 Classification of genome-wide DNA methylation-profiling methods

	Methylation detection		
Profiling	Restriction enzymes	5-Methylcytosine antibody	Bisulfite treatment
2D electrophoresis	RLGS	–	–
Subtraction	MS-RDA, MCA-RDA	–	–
Microarray	MIAMI	MeDIP-chip	MSOM
MALDI-TOF mass spectrometry	–	–	MALDI-TOF
Pyrosequencing	–	–	PyroMeth
Next-generation sequencing	–	ChIP-seq	BS-seq

of a methylation-sensitive restriction enzyme and two-dimensional electrophoresis is known as restriction landmark genomic scanning (RLGS, Hatada et al. 1991, 1993). Methylation-sensitive-representational difference analysis (MS-RDA, Ushijima et al. 1997) and methylation CpG island amplification-representational difference analysis (MCA-RDA, Toyota et al. 1999) are combinations of methylation-sensitive restriction enzymes and a subtraction method. Microarray-based integrated analysis of methylation by isoschizomers (MIAMI) (Hatada et al. 2006) is a combination of methylation-sensitive restriction enzymes and microarray technology. MeDIP-chip (Weber et al. 2005) is a combination of MeDIP and microarray technology.

4.3 Methylation-Sensitive Restriction Enzyme-Based Method 1 – RLGS

RLGS (Hatada et al. 1991) was originally developed as a genome-wide profiling method to detect genetic changes, such as deletions (Hirotsune et al. 1992), in cancer. Several thousand loci in the genome can be detected simultaneously by this method, using two-dimensional gel electrophoresis combined with restriction enzyme digestions. The application of a methylation-sensitive restriction enzyme, *Not* I, makes RLGS the first technology for genome-wide methylation analysis. RLGS has been used successfully to identify new imprinted genes (Hatada et al. 1993) and several putative tumor suppressor genes, such as *SOCS-1*and *ID4* (Yoshikawa et al. 2001; Yu et al. 2005). *Not* I is a methylation-sensitive restriction enzyme whose recognition sites are rare in the genome. This enzyme cleaves the GCGGCCGC sequence at CpG sites when it is unmethylated; however, it cannot cleave the sequence when it is methylated. The first step in this technique involves cleavage with *Not* I, followed by end labeling with a radioisotope (Fig. 4.1). For the first dimension of electrophoresis with agarose gel, genomic DNA is reduced in fragment size to several kilobases with a restriction enzyme (restriction enzyme X). The second dimension of electrophoresis with acrylamide gel is performed after in-gel digestion with another restriction enzyme (restriction enzyme Y). After two-dimensional electrophoresis, unmethylated and end-labeled *Not* I sites are detected by autoradiography as spots. When two autoradiographic patterns are compared, using densitometry, the lack of a spot or decreased spot intensity in one sample indicates the presence of a differentially methylated *Not* I site. Spots can

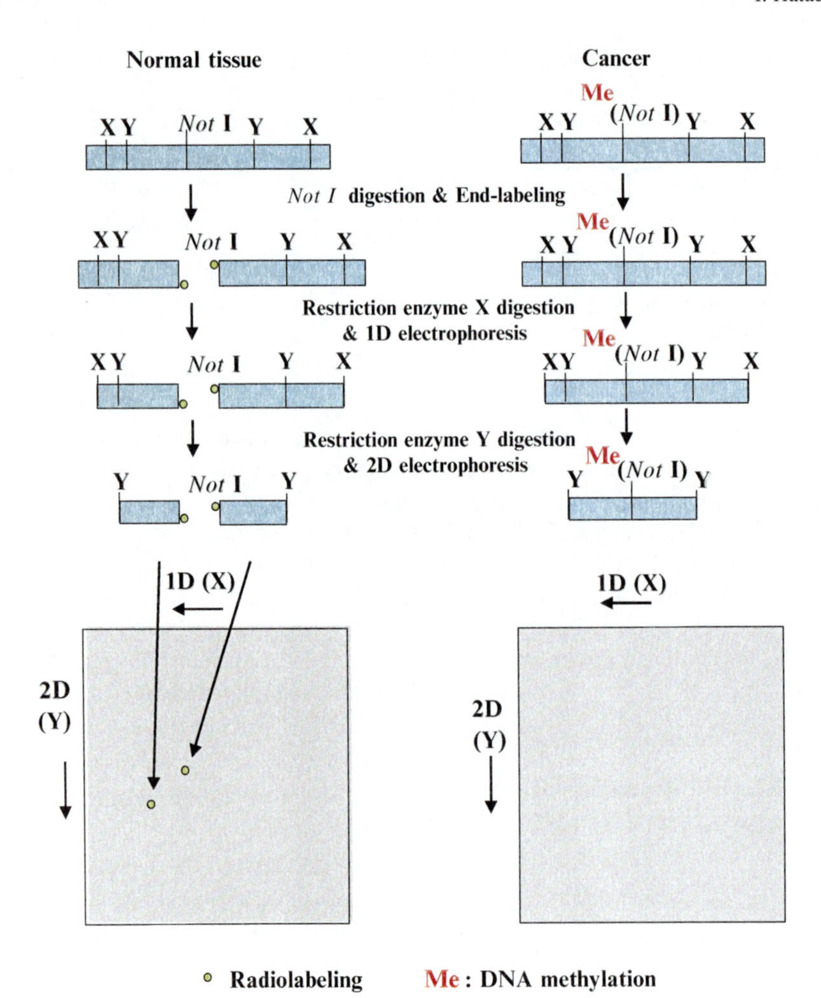

Fig. 4.1 Schematic flowchart for the RLGS. Genomic DNA is digested with a methylation-sensitive restriction enzyme, *Not* I, and labeled with a radioisotope. The labeled fragments are applied to the first dimension of electrophoresis after digestion with restriction enzyme X, to reduce to a size appropriate for fractionation. After in-gel digestion with restriction enzyme Y, the second dimension of electrophoresis is performed. Unmethylated *Not* I sites are detected as spots by autoradiography. Methylation differences can be detected by comparing the two autoradiographic patterns derived from normal tissue and cancer

be identified either by spot cloning or by virtual RLGS (vRLGS). Virtual RLGS is a virtual pattern on a two-dimensional gel, generated using the complete genome sequence and computer programs, which enable us to obtain sequence information for each spot (Matsuyama et al. 2003; Zardo et al. 2002).

There are several enduring merits of RLGS, although it was developed 18 years ago. It is quantitative enough to detect the loss of one of two copies of the *HRAS* gene in a cancer, because it directly detects end-labeled DNA and does not use PCR amplification. Another merit is that it can be applied to an organism without genome information. On the other hand, a disadvantage of the method is that its resolution is not particularly high; however, the additional application of a methylation-sensitive restriction enzyme other than *Not* I partially solves this problem.

In combination with another technique, RLGS can be applied to the simultaneous analysis of copy number and methylation status. Zardo et al. reported integrated genomic and epigenomic analysis using sequence information for spots (Zardo et al. 2002). This was performed by a combination of RLGS-based methylation analysis and high-resolution deletion maps from microarray-based comparative genomic hybridization (array CGH) in glioma. They found that certain subsets of gene-associated CpG islands were preferentially affected by convergent methylation and deletion, including genes that exhibit tumor-suppressor activity, such as *SOCS1*, a negative regulator of the JAK-STAT signaling pathway. This analysis showed that most aberrant methylation events were focal and independent of deletions, and rare convergence of these mechanisms can pinpoint biallelic gene inactivation without the use of positional cloning.

4.4 Methylation-Sensitive Restriction Enzyme-Based Method 2 – MS-RDA and MCA-RDA

MS-RDA (Ushijima et al. 1997) and MCA-RDA (Toyota et al. 1999) are subtraction-based methods using methylation-sensitive restriction enzymes for the detection of methylation. In MS-RDA, unmethylated genomic DNA fragments are amplified by adaptor-mediated PCR after cleavage with a methylation-sensitive restriction enzyme, such as *Hpa* II and *Sac* II. On the other hand, the methylated genomic DNA fragments are amplified in MCA-RDA. Amplified DNA fragments from the two samples to be compared are used for the following subtraction process. If a gene is methylated in one sample but not in the other, the fragments derived from the gene will be missing from one sample. This fragment can be isolated using a subtraction method called representational difference analysis technique (Lisitsyn et al. 1993). MS-RDA analysis of gastric cancer cell lines identified *LOX*, a negative regulator of NF-κB signaling that is silenced in gastric cancer cells (Kaneda et al. 2002a). Other genes, such as *3-OST-2*, *INSIG1*, and *p41Arp2/3* were also found to be silenced in cancers by MS-RDA analysis (Miyamoto et al. 2003; Kaneda et al. 2002b).

4.5 Methylation-Sensitive Restriction Enzyme-Based Method 3 – MIAMI

Classical microarray technology has limited genome-wide approaches to epigenomics; however, recent advances allow us to perform highly reproducible genome-wide analysis using oligonucleotide probes. Early microarrays used cloned DNA fragments or PCR-amplified DNA fragments as probes. These approaches are expensive for genome-wide analysis; however, recent progress in high-density oligonucleotide arrays makes it possible to provide a reproducible and inexpensive tool to analyze the genome. High-density oligonucleotide arrays are now commercially available. They are photolithographic masked arrays of Affymetrix, photolithographic adaptive optics arrays of NimbleGen, inkjet arrays of Agilent technology and bead arrays of Illumina. The Agilent Human promoter array covers from 8 kb upstream to 2 kb downstream of the transcriptional start sites of the genes. The Affymetrix Human promoter array covers 7.5 kb upstream to 2.4 kb downstream.

MIAMI (Hatada et al. 2006) is the first method to use such a microarray platform for genome-wide profiling of DNA methylation. Among several similar methods, MIAMI, can produce extremely accurate results. It uses methylation-sensitive restriction enzyme *Hpa* II in addition to a methylation-insensitive isoschizomer, *Msp* I, as a control to exclude effects other than methylation. *Hpa* II cleaves the unmethylated CCGG sequence; however, it cannot cleave the same sequence when internal C is methylated. On the other hand, *Msp* I cleaves both unmethylated and methylated CCGG sequences. A restriction site polymorphism at an *Hpa* II site and/or a difference in digestion at an *Hpa* II site depending on the quality of samples will give a change in *Hpa* II cleavage without a methylation difference. This can be solved by the MIAMI method, because the reliability of the methylation differences between samples calculated from the *Hpa* II cleavage difference is judged using *Msp* I cleavage difference, which is derived from the difference without methylation. For example, in the case of restriction site polymorphism at the *Hpa* II site, cleavage by methylation-insensitive *Msp* I at this site will also give differences between samples because both enzymes recognize the same recognition site. Thus, such changes can be treated as false positives.

The *Hpa* II cleavage difference, which is virtually a methylation difference, is detected as follows (Fig. 4.2). A several-fold amplification of regions located between unmethylated *Hpa* II sites is performed by *Hpa* II digestion, followed by adaptor ligation and five cycles of PCR (1st PCR). At this stage, only DNA fragments that have methylated internal *Hpa* II sites before PCR retain *Hpa* II (*Msp* I) sites. Methylated fragments become impossible to amplify in second main PCR by preceding *Msp* I digestion. The amplified and enriched unmethylated DNA fragments from two samples to be compared are labeled with Cy3 and Cy5, respectively, and co-hybridized to an oligonucleotide microarray made using Agilent inkjet technology. The ratios of fluorescence intensities are used to calculate the *Hpa* II cleavage difference. On the other hand, for the detection of *Msp* I cleavage differences, which is used to judge the reliability of the *Hpa* II cleavage difference, an identical procedure is performed, except that *Msp* I digestion is performed in the first step (Fig. 4.2). Using this method, 5.7% of gene promoters were found to be hypermethylated in a lung cancer cell line compared with those from a normal lung. This frequency is higher than in most previous reports, suggesting high sensitivity (Hatada et al. 2006). Analysis of one of the hypermethylated genes, *CIDEB* (cell death-inducing DFFA-like effector b), revealed hypermethylation in 71% of primary lung cancers. This gene activates apoptosis in mammalian cells and is located at 14q11, where LOH frequently occurs in lung

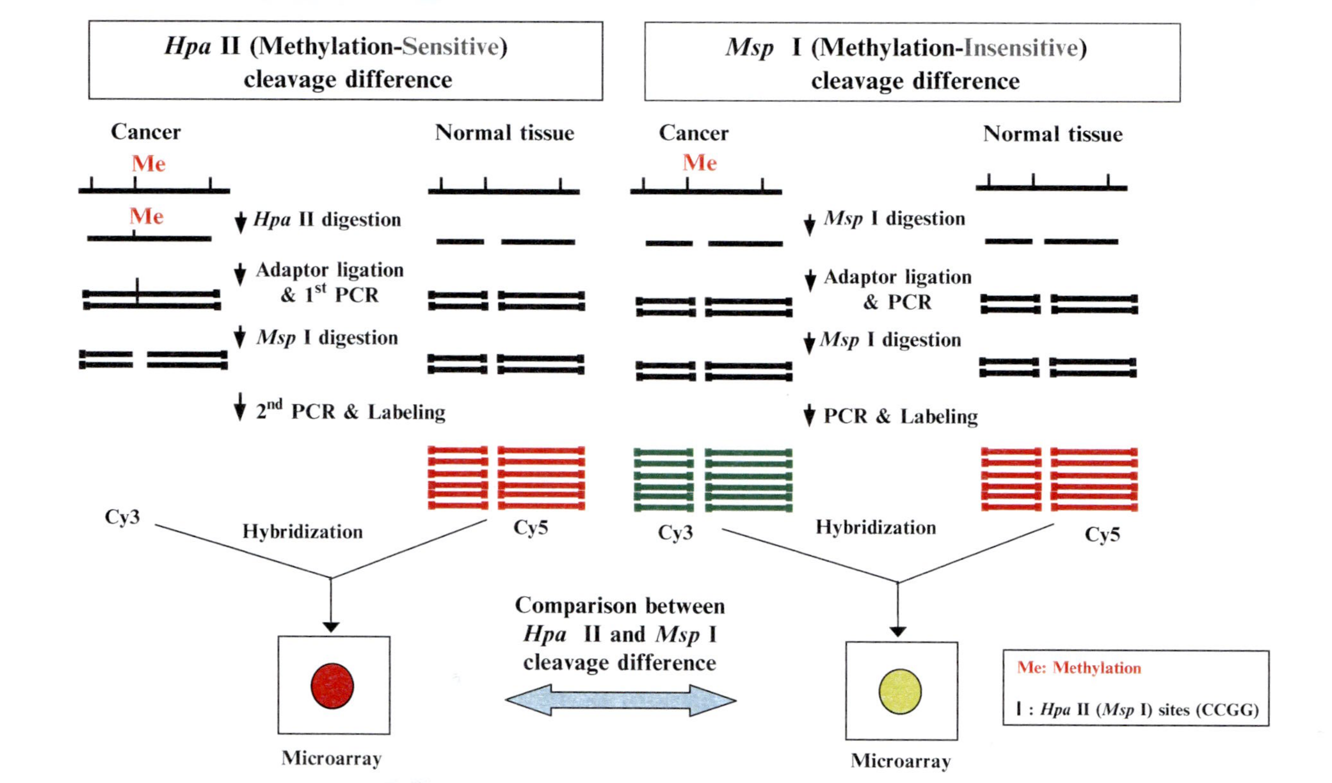

Fig. 4.2 Schematic flowchart for the MIAMI. Both the *Hpa* II (methylation-sensitive enzyme) cleavage difference and its methylation-insensitive isoschizomer *Msp* I cleavage difference are used to judge the methylation differences between cancer and normal tissue. For the *Hpa* II cleavage difference, *Hpa* II-digested genomic DNA is ligated to an adapter, followed by five cycles of PCR (First PCR), resulting in a several-fold amplification of regions located between unmethylated *Hpa* II sites. Methylated fragments become impossible to amplify in the second main PCR by preceding *Msp* I digestion. Amplified and enriched unmethylated DNA fragments from two samples are labeled with Cy3 and Cy5, respectively, and co-hybridized to an oligonucleotide microarray. For the *Msp* I cleavage difference, *Msp* I-digested genomic DNA is followed by the same procedure

cancers (Abujiang et al. 1998). In cancer, 0.6% of promoters are hypomethylated and most of their sequences show low CpG content, suggesting that CpG-poor promoters are sensitive to demethylation.

Estimation of the distribution of *Hpa* II sites in the genome is important for MIAMI, and should be the same as CpG sites in the genome. The distribution of *Hpa* II sites by the distance from transcription start sites of 1,000 genes is similar to that of CpG sites. It is also important to know the *Hpa* II site density in the genome to allow adequate tiling. There are 2,300,000 *Hpa* II sites in the human genome and the average distance between the *Hpa* II sites is 1.2 kb. Thus, the *Hpa* II site has a suitable distribution and density for genome-wide tiling analysis. One of the merits of MIAMI compared with other microarray-based methods is that it has strong signals on microarray to support its high reproducibility. On the other hand, it cannot detect the locus without *Hpa* II sites.

4.6 DNA Immunoprecipitation (MeDIP)-Based Method – MeDIP-Chip

MeDIP is immunoprecipitation with a 5-methylcytosine-specific antibody. MeDIP chip is a combination of MeDIP and microarray chip (Weber et al. 2005) (Fig. 4.3). Sonicated and denaturated genomic DNA are immunoprecipitated with a 5-methylcytosine-specific antibody. Input DNA and immunoprecipitated DNA are labeled with Cy5 and Cy3, respectively, and co-hybridized to a microarray. The methylation ratio of the genes can be compared between samples. This method was applied to cancer cell lines and 0.5–1% of gene promoters were found to be hypermethylated compared with those in normal cells.

This method was also applied to a colon cancer and revealed that large genomic segments with hypomethylation in cancer cells resided in gene-poor areas (Weber et al. 2005). MeDIP-chip was first performed with microarrays with cloned DNA; however, it has become applicable to the high-density oligonucleotide platform of NimbleGen (Weber et al. 2007). An advantage of this method compared with other microarray-based methods is that, theoretically, all methylated CpG sites can be analyzed. On the other hand, a disadvantage of this method is that the analysis of regions with low CpG density is believed to be problematic (Weber et al. 2007).

4.7 Bisulfite-Based Methods

Thirty years ago, Hayatsu found a bisulfite reaction that converts unmethylated cytosine to uracil, but not methylated cytosine (Hayatsu et al. 1970). PCR further converts uracil to thymine, therefore, unmethylated cytosine is converted to thymine while methylated cytosine is not. In combining this reaction with the quantitative pyrosequencing method, the PyroMeth (Uhlmann et al. 2002) method was developed.

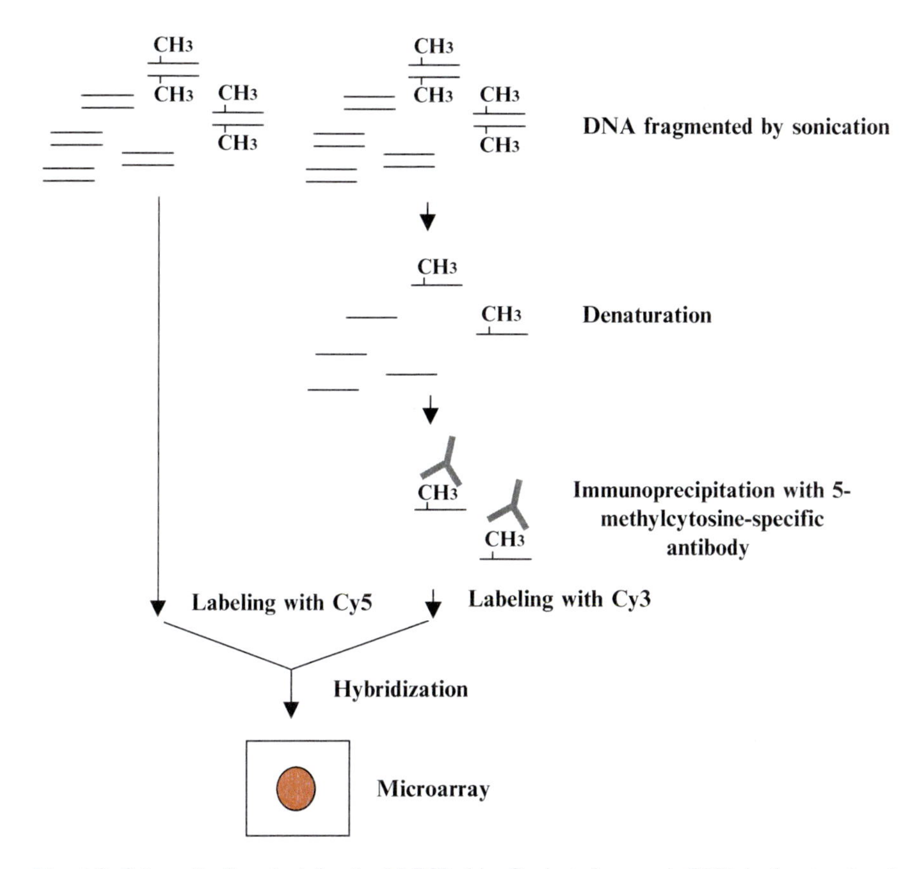

Fig. 4.3 Schematic flowchart for the MeDIP chip. Sonicated genomic DNA is denatured and immunoprecipitated with an antibody directed against 5-methyl-cytosine. Input DNA and immunoprecipitated methylated DNA are labeled with Cy5 and Cy3, respectively, and are cohybridized to a microarray

This method is direct DNA sequencing of PCR-amplified bisulfite-treated DNA by pyrosequencing, which can determine the ratio of bases in mixed PCR products, and that of C and T exactly; however, only thirty bases can be read per reaction with this method. This was improved in a method using MALDI-TOF mass spectrometry (Schatz et al. 2004). In this method (Fig. 4.4), bisulfite-treated DNA is PCR amplified using a T7 RNA polymerase promoter-tagged primer. The PCR product is transcribed into RNA and specifically cleaved at U. The cleavage products are analyzed by MALDI-TOF mass spectrometry, and a characteristic mass signal pattern can be obtained. After bisulfite treatment, a methylated template carries conserved cytosines; therefore, the reverse transcript contains guanosine residues. On the other hand, in an unmethylated template, the cytosine is converted to uracil; hence, the reverse transcript contains adenosine residues in the respective positions. The sequence changes from G to A yield a 16-Da mass shift. Spectrum analysis for the presence/absence of mass signals determines the position of methylated CpG,

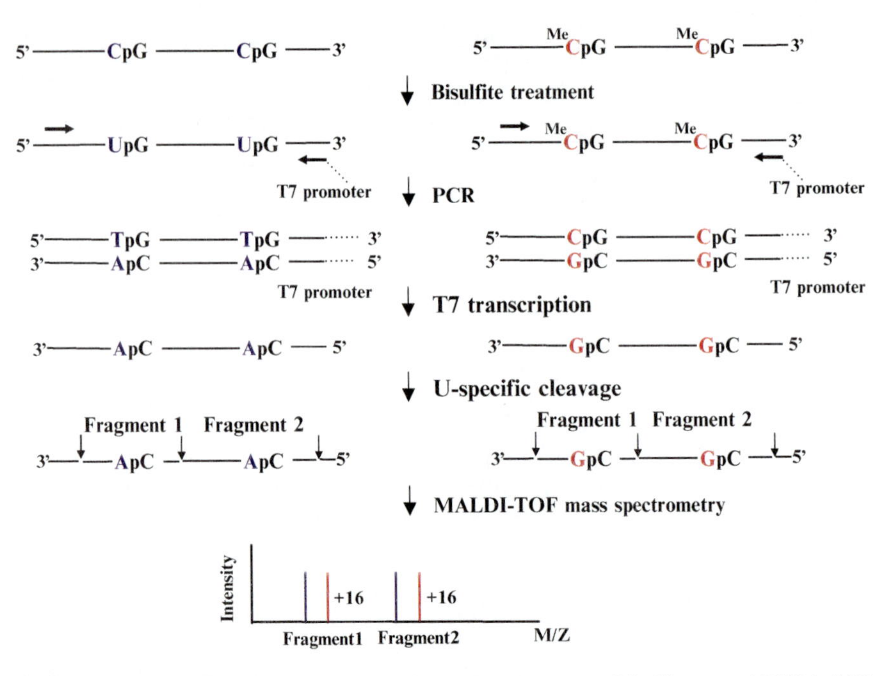

Fig. 4.4 Schematic flowchart for MALDI-TOF mass spectrometry. Bisulfite-treated DNA is PCR amplified using a T7 RNA polymerase promoter-tagged primer. The PCR product is transcribed into RNA and cleaved U specifically to give a characteristic mass signal pattern with MALDI-TOF mass spectrometry. A methylated template carries conserved cytosines after bisulfite treatment; hence, the reverse transcript contains guanosine residues. On the other hand, in an unmethylated template, the cytosine is converted to uracil; therefore, the reverse transcript contains adenosine residues in the respective positions. These sequence changes from G to A yield a 16-Da shift in mass

and the ratio of the peak areas of corresponding mass signals determines the relative methylation. Another microarray approach is the hybridization of PCR-amplified bisulfite-treated DNA with specific probes for both methylated and unmethylated DNA (Fig. 4.5) (Gitan et al. 2002; Adorjan et al. 2002). This method is called methylation-specific oligonucleotide microarray (MSOM). In this method, bisulfite-treated DNA is PCR-amplified with a pair of primers for each gene, followed by labeling with Cy5 and hybridized to a microarray with a pair of specific probes for methylated (hybridized to CG) and unmethylated DNA (hybridized to TG). The ratio of methylation can be determined by calculating at both intensities. This method was applied for tumor class prediction using machine learning techniques (Adorjan et al. 2002). Some CpG dinucleotides correlate with progression to malignancy; however, others are methylated in a tissue-specific manner, independent of malignancy. Each of the three bisulfite methods requires too many PCR amplification reactions for the genes to be analyzed. Furthermore, the reduction of sequence complexity following bisulfite conversion means that it is difficult to design enough unique probes to analyze bisulfite-converted DNA comprehensively on a genome-wide scale on microarrays.

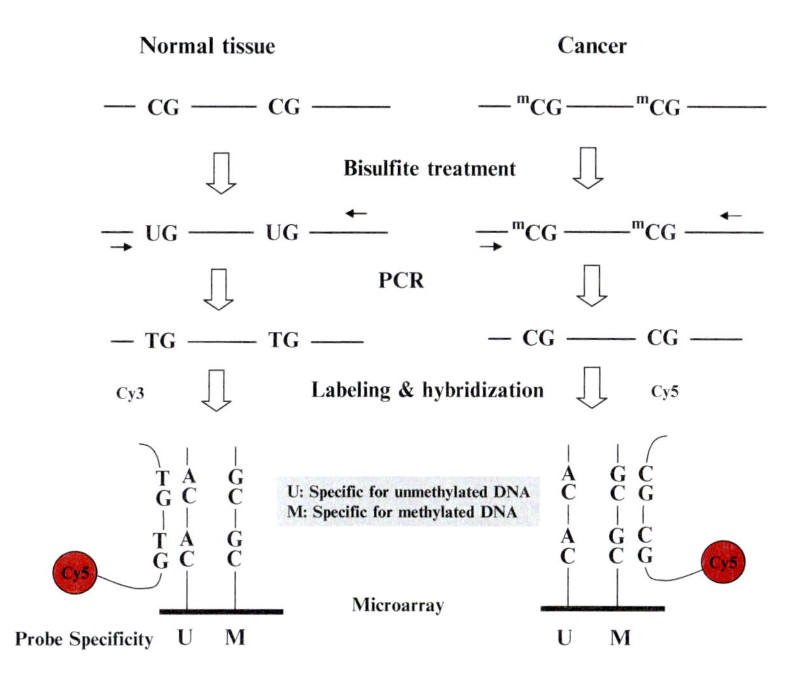

Fig. 4.5 Schematic flowchart for the MSOM. Bisulfite-treated DNA is PCR-amplified with a pair of primers for each gene. Labeling with Cy5 is followed by hybridization to a microarray with a pair of specific probes for methylated (hybridized to CG) and unmethylated DNA (hybridized to TG). By comparing both intensities, the ratio of methylation can be determined

Bisulfite sequencing of the whole genome with next-generation sequencing technology (BS-seq) is the ultimate method (Cokus et al. 2008). There are several commercially available next-generation sequencing platforms such as 454, Solexa, and SOLiD, which perform cost effective, high-throughput sequencing, thus making the sequencing of individual isolates a feasible option. For example, with the Solexa platform, a large number of DNA fragments are immobilized on a solid surface, and read simultaneously with fluorescence-labeled nucleotides. Millions of 36–50 base pair long reads can be obtained from each sample lane at a cost of less than $1,000 (USD). Deep sampling of DNA fragments allows rapid procurement of high coverage genome sequence information. This technology was successfully applied to shotgun bisulphite sequencing of the Arabidopsis genome (Cokus et al. 2008). However, the BS-seq approach is currently prohibitively expensive for the routine analysis of large genomes such as that of humans.

4.8 Comparison of the Methods

Many methods exist for DNA methylation profiling, including those not described here. It is important to choose the appropriate technique for an experiment because each method has its advantages and disadvantages (Table 4.2). The choice of method

Table 4.2 Pros and cons of DNA methylation-profiling methods

	Ease of use	Resolution	Cost/gene	Equipment	Genome info
RLGS	Difficult	Low	Low	No	Not required
MS-RDA, MCA-RDA	Easy	Low	Low	No	Not required
MIAMI	Easy	Middle	Low	Scanner	Required
MeDIP	Easy	Middle	Low	Scanner	Required
MSOM	Easy	High	Middle	Scanner	Required
PyroMeth	Easy	High	High	Pyrosequencer	Required
MALDI-TOF	Easy	High	High	Mass	Required
ChIP-seq	Easy	Middle	Low	NGS	Required
BS-seq	Easy	High	Low	NGS	Required

NGS: Next-generation sequencer.

depends on the resolution required and the budget. Usually, resolution is in inverse proportion to cost. If you regard cost as important, subtraction-based methods and RLGS are recommended; however, the resolution of these methods is not very high. For example, RLGS can analyze only several thousand loci for one gel analysis. If good resolution is required, the best can be obtained with the bisulfite-based method. However, most of these methods are time-consuming and expensive, although BS-seq with a next-generation sequencer will make this approach feasible in the near future. Therefore, if we do not have previous information about candidate genes, it is practically impossible to use these methods. Microarray-based methods other than MSOM have moderate resolution and are cost-effective, because of recent advances in a commercially available high-resolution platform. Although these methods do not have resolution at the base pair level, their resolution is sufficiently high to limit candidate genes. After finding the candidates, we can perform extensive analysis at high-resolution using bisulfite-based methods.

Microarray-based methods also have the advantage of merging data with chromatin immunoprecipitation (ChIP) on chip because both can be applied to the commercially available high-resolution platform. ChIP on chip is a combination of ChIP and microarray (Ren et al. 2000). Fragmented chromatins are immunoprecipitated by antibodies and subjected to DNA extraction followed by microarray hybridization. When specific antibodies to histone modification are used, this technique can be applied to genome-wide profiling of histone modification. These data can be merged with methylation profiling data.

4.9 Histone Modifications of Cancer

Generally, histone acetylation is associated with transcriptional activation, but the effect of histone methylation depends on the type of amino acid and its position in the histone tail (Mack 2006; Bernstein et al. 2007). For example, the methylation of histone H3 lysine 4 (H3K4) is associated with transcriptional activation, while methylations of H3 lysine 9 (H3K9) and H3 lysine 27 (H3K27) are associated with transcriptional repression. A hypermethylated CpG island in the promoter regions of tumor-suppressor genes in cancer is associated with a particular combination of histone markers: deacetylation of histones H3 and H4, loss of H3K4 methylation, and gain of H3K9 and H3K27 methylation (Ballestar et al. 2003; Jones and Baylin 2007). Aberrant regulation of histone modification is associated with carcinogenesis. In leukemias and sarcomas, chromosomal translocations involving histone-modifier genes, such as histone acetyltransferases (*CBP-MOZ*) and histone methyltransferase (*MLL1*, *NSD19*) create aberrant fusion proteins (Esteller 2006). In solid tumors, the amplification of genes for histone methyltransferase (*EZH2*, *MLL2*) and histone demethylase (*JMJD2C*) is known (Bracken et al. 2003; Esteller 2006; Cloos et al. 2006).

Recently, a hypothesis has been proposed, in which histone modification cause DNA methylation during cancer formation from stem cells (Widschwendter et al. 2007). In mouse embryonic stem cells (ESCs), many developmental genes are maintained in a state of low transcriptional activity and are available for transcriptional increase or decrease when differentiation cues are received. The promoter region of these genes is marked with a combination of active (H3K4) and repressive (H3K27) histone methylations in ES cells. This unique "bivalent" state is switched either to monovalent active (H3K4) or monovalent repressive (H3K27) after differentiation (Bernstein et al. 2006). H3K27 methylation mostly colocalized with H3K4 methylation in ESCs is maintained with Polycomb group proteins to repress the genes encoding the transcription factors required for differentiation reversibly (Ringrose and Paro 2004). Widschwendter et al. hypothesized that the acquisition of promoter DNA methylation at these repressed genes leads to permanent silencing of reversible gene repression, locking the cell into a perpetual state of self-renewal, and thereby predisposing it to subsequent malignant transformation (Widschwendter et al. 2007). By using DNA methylation and histone modification data provided by genome-wide profiling approaches such as MIAMI and ChIP on chip (Hatada et al. 2006; Lee et al. 2006), they found that hypermethylated genes in cancers are significantly enriched in polycomb targets in embryonic stem cells (Widschwendter et al. 2007), thus supporting the hypothesis.

This hypothesis suggests that epigenetic changes precede genetic mutations in cancer formation. The epigenetic change of a repair gene could cause subsequent genetic mutations. I searched the literature for repair genes among those genes with bivalent histone modifications in ESCs (Mikkelsen et al. 2007). Only four repair genes are bivalent in ESCs. Among them, *TP73* is known to be hypermethylated in several cancers (Watanabe et al. 2002; Liu et al. 2008). This gene encodes tumor suppressor *TP73*, which is a member of the *TP53* family of transcription factors involved in cellular responses to stress and development. The family members include *TP53*, *TP63*, and *TP73* and have high sequence similarity to one another, which allows *TP63* and *TP73* to transactivate *TP53*-responsive genes causing cell cycle arrest and apoptosis (Kaghad et al. 1997). *TP73* is located in 1p36, a region frequently deleted in several cancers (Kaghad et al. 1997). Interactions between

ZNF143, *TP73*, and *ZNF143* target genes are involved in DNA repair gene expression and cisplatin resistance (Wakasugi et al. 2007). Interestingly, the mutation of *TP73* is rarely observed in cancer although its epigenetic change is frequently observed. On the other hand, the genetic mutation of the *TP53* is frequently observed in cancers.

This hypothesis is also interesting in relation to "cancer stem cells", which is hypothesized to constitute the population of cells that are ultimately responsible for perpetuating the tumor. These cells have many properties common to normal stem cells, but their exact origins remain controversial (Bjerkvig et al. 2005). Currently, most researchers seem to favor the view that a range of cells in normal cell renewing systems, from the ultimate stem cells to a subsequent series of precursor and progenitor cells, have the potential to constitute the focal transformation point for individual cancers. This fact could explain the existence of many subtypes of major tumor types, such as lung and breast cancers.

4.10 Technologies for Genome-Wide Profiling of Histone Modifications

Histone modifications can be detected by ChIP. In the first step of ChIP, DNA and its associated binding proteins are cross-linked *in vivo* by chemical treatment. Following the shearing of chromatin, the DNA-protein complex is immunoprecipitated by antibodies specific for the histone modification of interest. The DNA sequences located near the modified histone binding site are therefore concentrated. These DNA sequences can be either detected by ChIP on chip (Ren et al. 2000), or by ChIP-seq (Barski et al. 2007). In ChIP on chip, the immunoprecipitated DNA is hybridized to a microarray platform. In ChIP-seq, the immunoprecipitated DNA is sequenced by next-generation DNA sequencing technology.

4.11 Epigenetic Therapy of Cancer

Drastic epigenetic change occurs in cancer and plays an important role in cancer formation. Unlike mutations, epigenetic change is reversible. This led to the idea of epigenetic therapy of cancer. Hypermethylated tumor-suppressor genes can be activated by drugs such as DNA methyltransferase inhibitors or histone deacetylase (HDAC) inhibitors. Although this strategy seems to be difficult, it can work in some tumors. Two DNA methyltransferase inhibitors, such as 5-azacytidine (Vidaza) and 5-aza-2′-deoxycytidine (decitabine), have been approved as treatments for the myelodysplastic syndrome and leukemia (Mack 2006; Müller et al. 2006; Oki et al. 2007). However, these drugs have not yet been shown to have clinical activity against solid tumors (Mack 2006). A HDAC inhibitor, suberolanilide hydroxamic acid (vorinostat), has been approved by the US Food and Drug Administration for the treatment of cutaneous T-cell lymphoma (Marks et al. 2007). Vorinostat can cause growth arrest and death of a broad variety of transformed cells, both *in vitro* and in tumor-bearing animals, at concentrations not toxic to normal cells. In clinical trials, vorinostat has shown significant anticancer activity against both hematologic and solid tumors at doses well tolerated by patients.

These drugs seem to have a nonspecific effect and impair normal cells. However, this may not be as much of a problem as it seems, because DNA methylation inhibitors only act on dividing cells, leaving nondividing normal cells unaffected. In addition, the drugs seem preferentially to activate abnormally silenced genes in cancer (Karpf et al. 1999; Liang et al. 2002).

4.12 Perspectives and Conclusion

Several emerging genome-wide analysis technologies have been introduced to epigenomics to produce new methods for genome-wide DNA methylation analysis. Recent advances in commercially available high-density oligonucledotide array platforms can tile the genome and make it possible to analyze genome-wide DNA methylation. Although bisulfite-based methods are expensive and time-consuming, breakthroughs on the horizon will realize next-generation sequencing technology applicable to BS-seq of the human genome.

DNA methylation inhibitors and HDAC inhibitors are now used for cancer therapy. A combination of both inhibitors could be more effective because of the synergistic activities of DNA methylation and HDAC inhibitors. Epigenetics seems to be more important when we think of cancer stem cells. Cancer stem cells refractory to standard chemotherapy might be stimulated to differentiate by the chronic administration of epigenetic drugs.

References

Abujiang P, Mori TJ, Takahashi T et al (1998) Loss of heterozygosity (LOH) at 17q and 14q in human lung cancers. Oncogene 17:3029–3033

Adorjan P, Distler J, Lipscher E et al (2002) Tumor class prediction and discovery by microarray-based DNA methylation analysis. Nucleic Acids Res 30:e21

Ballestar E, Paz MF, Valle L et al (2003) Methyl-CpG binding proteins identify novel sites of epigenetic inactivation in human cancer. EMBO J 22:6335–6345

Barski A, Cuddapah S, Cui K et al (2007) High-resolution profiling of histone methylations in the human genome. Cell 129:823–837

Bernstein BE, Mikkelsen TS, Xie X et al (2006) A bivalent chromatin structure marks key developmental genes in embryonic stem cells. Cell 125:315–326

Bernstein BE, Meissner A, Lander ES (2007) The mammalian epigenome. Cell 128:669–681

Bird A (2002) DNA methylation patterns and epigenetic memory. Genes Dev 16:6–21

Bjerkvig R, Tysnes BB, Aboody KS et al (2005) Opinion: the origin of the cancer stem cell: current controversies and new insights. Nat Rev Cancer 5:899–904

Bracken AP, Pasini D, Capra M et al (2003) *EZH2* is downstream of the pRB-E2F pathway, essential for proliferation and amplified in cancer. EMBO J 22:5323–5335
Cloos PA, Christensen J, Agger K et al (2006) The putative oncogene *GASC1* demethylates tri- and dimethylated lysine 9 on histone H3. Nature 442:307–311
Cokus SJ, Feng S, Zhang X et al (2008) Shotgun bisulphite sequencing of the Arabidopsis genome reveals DNA methylation patterning. Nature 452:215–219
Dobrovic A, Simpfendorfer D (1997) Methylation of the *BRCA1* gene in sporadic breast cancer. Cancer Res 57:3347–3350
Eden A, Gaudet F, Waghmare A et al (2003) Chromosomal instability and tumors promoted by DNA hypomethylation. Science 300:455
Esteller M (2006) Epigenetics provides a new generation of oncogenes and tumor-suppressor genes. Br J Cancer 94:179–183
Feinberg AP, Vogelstein B (1983) Hypomethylation distinguishes genes of some human cancers from their normal counterparts. Nature 301:89–92
Feinberg AP, Tycko B (2004) The history of cancer epigenetics. Nat Rev Cancer 4:143–153
Feinberg AP (2007) Phenotypic plasticity and the epigenetics of human disease. Nature 447:433–440
Fryxell KJ, Zuckerkandl E (2000) Cytosine deamination plays a primary role in the evolution of mammalian isochores. Mol Biol Evol 17:1371–1383
Gitan RS, Shi H, Chen CM et al (2002) Methylation-specific oligonucleotide microarray: a new potential for high-throughput methylation analysis. Genome Res 12:158–164
Hatada I, Hayashizaki Y, Hirotsune S et al (1991) A genomic scanning method for higher organisms using restriction sites as landmarks. Proc Natl Acad Sci USA 88:9523–9527
Hatada I, Sugama T, Mukai T (1993) A new imprinted gene cloned by a methylation-sensitive genome scanning method. Nucleic Acids Res 21:5577–5582
Hatada I, Fukasawa M, Kimura M et al (2006) Genome-wide profiling of promoter methylation in human. Oncogene 25:3059–3064
Hayatsu H, Wataya Y, Kazushige K (1970) The addition of sodium bisulfite to uracil and to cytosine. J Am Chem Soc 92:724–726
Hirotsune S, Hatada I, Komatsubara H et al (1992) New approach for detection of amplification in cancer DNA using restriction landmark genomic scanning. Cancer Res 52:3642–3647
Jenuwein T (2001) Re-SET-ting heterochromatin by histone methyltransferases. Trends Cell Biol 11:266–273
Jones PA, Baylin SB (2007) The epigenomics of cancer. Cell 128:683–692
Kaneda A, Kaminishi M, Yanagihara K et al (2002a) Identification of silencing of nine genes in human gastric cancers. Cancer Res 62:6645–6650
Kaneda A, Kaminishi M, Nakanishi Y et al (2002b) Reduced expression of the insulin-induced protein 1 and *p41 Arp2/3* complex genes in human gastric cancers. Int J Cancer 100:57–62
Kaghad M, Bonnet H, Yang A et al (1997) Monoallelically expressed gene related to *p53* at 1p36, a region frequently deleted in neuroblastoma and other human cancers. Cell 90:809–819
Karpf AR, Peterson PW, Rawlins JT et al (1999) Inhibition of DNA methyltransferase stimulates the expression of signal transducer and activator of transcription 1, 2, and 3 genes in colon tumor cells. Proc Natl Acad Sci USA 96:14007–14012
Lee TI, Jenner RG, Boyer LA et al (2006) Control of developmental regulators by Polycomb in human embryonic stem cells. Cell 125:301–313
Liang G, Gonzales FA, Jones PA et al (2002) Analysis of gene induction in human fibroblasts and bladder cancer cells exposed to the methylation inhibitor 5-aza-2′-deoxycytidine. Cancer Res 62:961–966
Lisitsyn N, Lisitsyn N, Wigler M (1993) Cloning the differences between two complex genomes. Science 259:946–951
Liu K, Zhan M, Zheng P (2008) Loss of *p73* expression in six non-small cell lung cancer cell lines is associated with 5′CpG island methylation. Exp Mol Pathol 84:59–63
Mack GS (2006) Epigenetic cancer therapy makes headway. J Natl Cancer Inst 98:1443–1444
Marks PA, Breslow R (2007) Dimethyl sulfoxide to vorinostat: development of this histone deacetylase inhibitor as an anticancer drug. Nat Biotechnol 25:84–90

Matsuyama T, Kimura MT, Koike K et al (2003) Global methylation screening in the Arabidopsis thaliana and Mus musculus genome: applications of virtual image restriction landmark genomic scanning (Vi-RLGS). Nucleic Acids Res 31:4490–4496

Mikkelsen TS, Ku M, Jaffe DB et al (2007) Genome-wide maps of chromatin state in pluripotent and lineage-committed cells. Nature 448:553–560

Miyamoto K, Asada K, Fukutomi T et al (2003) Methylation-associated silencing of heparan sulfate D-glucosaminyl 3-O-sulfotransferase-2 (3-OST-2) in human breast, colon, lung and pancreatic cancers. Oncogene 22:274–280

Müller CI, Rüter B, Koeffler HP et al (2006) DNA hypermethylation of myeloid cells, a novel therapeutic target in MDS and AML. Curr Pharm Biotechnol 7:315–321

Oki Y, Aoki E, Issa JP (2007) Decitabine–bedside to bench. Crit Rev Oncol Hematol 61:140–152

Ren B, Robert F, Wyrick JJ (2000) Genome-wide location and function of DNA binding proteins. Science 290:2306–2309

Ringrose L, Paro R (2004) Epigenetic regulation of cellular memory by the Polycomb and Trithorax group proteins. Annu Rev Genet 38:413–443

Schatz P, Dietrich D, Schuster M (2004) Rapid analysis of CpG methylation patterns using RNase T1 cleavage and MALDI-TOF. Nucleic Acids Res 32:e167

Toyota M, Ho C, Ahuja N et al (1999) Identification of differentially methylated sequences in colorectal cancer by methylated CpG island amplification. Cancer Res 59:2307–2312

Uhlmann K, Brinckmann A, Toliat MR et al (2002) Evaluation of a potential epigenetic biomarker by quantitative methyl-single nucleotide polymorphism analysis. Electrophoresis 23:4072–4079

Ushijima T, Morimura K, Hosoya Y et al (1997) Establishment of methylation-sensitive-representational difference analysis and isolation of hypo- and hypermethylated genomic fragments in mouse liver tumors. Proc Natl Acad Sci USA 94:2284–2289

Wakasugi T, Izumi H, Uchiumi T et al (2007) *ZNF143* interacts with *p73* and is involved in cisplatin resistance through the transcriptional regulation of DNA repair genes. Oncogene 26:5194–5203

Watanabe T, Huang H, Nakamura M et al (2002) Methylation of the *p73* gene in gliomas. Acta Neuropathol 104:357–362

Weber M, Davies JJ, Wittig D et al (2005) Chromosome-wide and promoter-specific analyses identify sites of differential DNA methylation in normal and transformed human cells. Nat Genet 37:853–862

Weber M, Hellmann I, Stadler MB et al (2007) Distribution, silencing potential and evolutionary impact of promoter DNA methylation in the human genome. Nat Genet 39:457–466

Widschwendter M, Fiegl H, Egle D et al (2007) Epigenetic stem cell signature in cancer. Nat Genet 39:157–158

Yoshikawa H, Matsubara K, Qian GS et al (2001) *SOCS-1*, a negative regulator of the JAK/STAT pathway, is silenced by methylation in human hepatocellular carcinoma and shows growth-suppression activity. Nat Genet 28:29–35

Yu L, Liu C, Vandeusen J et al (2005) Global assessment of promoter methylation in a mouse model of cancer identifies *ID4* as a putative tumor-suppressor gene in human leukemia. Nat Genet 37:265–274

Zardo G, Tiirikainen MI, Hong C et al (2002) Integrated genomic and epigenomic analyses pinpoint biallelic gene inactivation in tumors. Nat Genet 32:453–458

Chapter 5
Involvement of MicroRNAs in Human Cancer: Discovery and Expression Profiling

Massimo Negrini and George A. Calin

Abstract The initial discovery of the involvement of two microRNAs, *miR-15a* and *miR-16-1*, in human CLL opened the way to the myriad of studies that have now conclusively proved the central role of microRNAs in all human cancers. Gene expression studies revealed that hundreds of microRNAs are deregulated in cancer cells and functional studies clarified that microRNAs are involved in all the molecular and biological processes that drive tumorigenesis. These findings have greatly improved our understanding on the molecular basis of cancer and, even more importantly, laid the foundation for the exploitation of microRNAs in cancer therapy.

5.1 Introduction to MicroRNAs

MicroRNAs (miRNAs) constitute a large class of philogenetically conserved single stranded RNA molecules of ~22 nucleotides (nt) (ranging 19–25 nt), involved in post-transcriptional gene silencing. The miRNAs *lin-4* and *let-7* were the first to be discovered and shown to function in *Caenorhabditis elegans* as triggers for a cascade of gene expression that regulates developmental events by Post-Transcriptional Gene Silencing (PTGS) (Lee et al. 1993; Reinhart et al. 2000; Wightman et al. 1993). Initially believed to play a regulatory role only in worms, their importance became more apparent 7 years later, when miRNAs were identified and cloned from several organisms, including human, and their nucleotide sequences were

M. Negrini (✉)
Dipartimento di Medicina Sperimentale e Diagnostica and Centro Interdipartimentale per la Ricerca sul Cancro, Università di Ferrara, Via Luigi Borsari, 46, 44100, Ferrara, Italy
e-mail: ngm@unife.it

G.A. Calin
Department of Experimental Therapeutics, University of Texas, MD Anderson Cancer Center, Houston, TX 77030, USA
e-mail: gcalin@mdanderson.org

W.C.S. Cho (ed.), *An Omics Perspective on Cancer Research*,
DOI 10.1007/978-90-481-2675-0_5,

found to be philogenetically conserved (Lagos-Quintana et al. 2001; Lau et al. 2001; Lee and Ambros 2001; Pasquinelli et al. 2000).

Based on the most recent release of the miRBase Registry (http://microrna.sanger.ac.uk/) (Griffiths-Jones et al. 2006) (14.0, released on September 2009), there are 10,883 hairpin precursor miRNAs confirmed in primates, rodents, birds, fish, worms, flies, plants and viruses; among these, 721 are human. Computational algorithms predict that as many as 1,000 miRNA loci may exist in the human genome (Berezikov et al. 2005).

In human, miRNA genes are located in all chromosomes, with the exception of chromosome Y. Nearly 50% of known miRNAs are found in clusters and are transcribed as polycistronic transcripts. About 60% of mammalian miRNA genes are located in introns of protein-coding genes, the remaining in intergenic non-coding transcriptional units; less often in exons and anti-sense orientation with the host gene (Baskerville and Bartel 2005; Rodriguez et al. 2004) (see Kim and Nam 2006; for a review). Intergenic miRNAs and, sometimes, intronic miRNAs are transcribed by RNA polymerase II as independent units. The primary transcript (pri-miRNA) is capped and polyadenylated (Cai et al. 2004; Lee et al. 2004). Hence, their expression is regulated under the same mechanisms that control transcription of protein-coding genes.

Molecular mechanisms of miRNA maturation and action were also philogenetically conserved. MiRNA maturation begins in the nucleus, where the pri-miRNA is processed by a protein complex known as Microprocessor, which contains the nuclear RNaseIII Drosha and its cofactor DGCR8/Pasha (Han et al. 2004, 2006; Lee et al. 2003). Microprocessor action generates a precursor miRNA (pre-miRNA), a 60–70 nt long RNA with a stem-loop structure, that is rapidly exported to the cytoplasm by Exportin-5 in a Ran GTP-dependent manner. The mature miRNA(s) may reside in the 5′ arm or in the 3′ arm of the pre-miRNA stem; sometimes both arms generate mature miRNAs. Once in the cytoplasm, a second RNaseIII, Dicer, acts on the pre-miRNA to release a ~22 nt miRNA duplex in which the mature miRNA is partially paired to a miRNA* present on the pre-miRNA opposite stem strand. Usually, only the miRNA (mature miRNA) strand of the miRNA::miRNA* duplex is active and enters a protein complex, the RNA-induced silencing complex (RISC), to repress gene expression (Tang 2005). Mature miRNA guides RISC toward regions of partial complementarity in the 3′UTR of target mRNAs, and triggers either their degradation or inhibition of translation depending on the degree of complementarity between the miRNA and its target.

Recently, a deviation from the view on miRNAs acting as post-transcriptional inhibitors of gene expression has been proved by the case of *miR-369-3* and tumor necrosis factor alpha (TNF-alpha): in this case, the miRNA was found to up-regulate translation upon cell cycle arrest (Vasudevan et al. 2007). The same study proved that other miRNAs, such as *let-7* and the synthetic microRNA miRcxcr4, induce translation up-regulation of target mRNAs on cell cycle arrest, yet they repress translation in proliferating cells, suggesting that activation is a common function of microRNPs on cell cycle arrest, which leads the authors to propose that translation regulation by microRNPs oscillates between repression and activation during the cell cycle. If confirmed, this finding will lead to a change in our way of thinking on the biological effects of miRNAs.

Positive effects of miRNA action were also identified for *miR-122*, which acts to facilitate replication of the hepatitis C viral RNA by binding to the 5′ noncoding region of the virus genome (Jopling et al. 2005). However, this mechanism is presently unique. Indeed, in other known cases, cellular miRNAs interact with viral RNAs to inhibit their translation. For example, *miR-32* targets the open reading frame 2 of the primate foamy virus type 1, thereby inhibiting virus translation (Lecellier et al. 2005); *miR-28*, *miR-125b*, *miR-150*, *miR-223*, and *miR-382* target sequences in the 3′ end of HIV-1 RNA, thereby silencing almost all viral mRNAs (Huang et al. 2007).

In animal cells, post-transcriptional regulation by miRNA requires an mRNA sequence that is perfectly complementary to the "seed sequence" (positions 2–7 of the mature miRNA). Various algorithms (http://www.microrna.org/; http://www.targetscan.org/; http://pictar.bio.nyu.edu/) have been developed for predicting miRNA::targets interactions. Based on predictive algorithms, each miRNA may potentially regulate hundreds of target mRNAs (Lewis et al. 2005) and it seems so plausible that most, if not all, mRNAs are post-transcriptionally regulated by miRNAs.

Thus, by revealing a novel control level, the discovery of miRNAs has broadened our understanding on the mechanisms that regulate gene expression in multicellular organisms. The post-transcriptional control of gene expression operated by miRNAs represents a crucial part of all known regulatory pathways at the cellular and organism level. However, the miRNA/mRNA network is complex and far from being fully understood. A single miRNA can bind and regulate several different mRNAs and, at the same time, multiple miRNAs act together to regulate each single mRNA target. Furthermore, functions of miRNAs may depend on the unique set of mRNA targets transcribed in each cell type. Thus, depending on the combination of mRNAs and miRNAs, transcriptional and post-transcriptional regulation of gene expression operate together to control the fate of each cell type.

5.2 The Discovery of the Involvement of MicroRNAs in Human Cancer

Given their potential involvement in all molecular pathways, it seems obvious that miRNAs could play such an important role in human tumorigenesis. It was not the case in the early days. The first evidence of miRNA involvement in human cancer came from a study by Calin et al. (2002). The discovery occurred in Dr. Croce's laboratory by examining a recurring deletion at chromosome 13q14 in the search for a tumor suppressor gene involved in chronic lymphocytic leukemia (CLL). This discovery took advantage of two important exclusive informations. First, about 1 Mb of nucleotide sequence at chromosome 13q14 common region of deletion was generated in-house, before the full genomic sequence became publicly available through the Human Genome Project; second, a very small region of about 30–40 kb was identified in human CLL as the smallest region of deletion at 13q14, by taking advantage of two CLL cases with 13q14 translocations associated with microdeletions. Thus, by comparing the genomic sequence at chromosome 13 with gene sequences of the known miRNAs, it was simple to find that the smallest minimal common region of deletion encoded two miRNAs, *miR-15a* and *miR-16-1*, whose existence was reported just few months earlier. Subsequent analysis of their expression in CLL samples and normal CD5+ lymphocytes revealed that down-regulation of *miR-15a* and *miR-16-1*, which shares their primary transcript, was consistently associated with the deletion at chromosome 13q14. This suggested a potential role of *miR-15a* and *miR-16-1* as tumor suppressor genes.

Similarly to the initial discovery in *C. elegans*, for some time, this finding was considered unusual and its general significance was not initially appreciated by the scientific community. However, subsequent investigations confirmed the involvement and the importance of *miR-15a/miR-16-1* and other miRNAs in the pathogenesis of human cancer. The putative tumor suppressive role of *miR-15a* and *miR-16-1* was supported by the discovery in two CLL patients of a germ-line point mutation that results in reduced levels of mature *miR-15a* and *miR-16-1* (Calin et al. 2005), and the idea was further strengthened by the demonstration that *miR-15a* and *miR-16-1* negatively regulate the anti-apoptotic oncogene BCL2 at a post-transcriptional level and induce apoptosis and suppress tumorigenicity in the leukemic cell line MEG-01 (Calin et al. 2008; Cimmino et al. 2005).

5.3 Numerous MicroRNAs Are Involved in Human Cancer

Evidence now indicates that the involvement of miRNAs in cancer is much more extensive than initially expected. Initial clues came from the observation that about 50% of known miRNA genes are located at sites of recurrent deletions or amplifications in human cancers (Calin et al. 2004a, b; Zhang et al. 2006a, b). More direct evidence was provided by genome-wide expression studies.

The expression levels for hundreds of microRNAs were assessed by using high-throughput technologies from the very early studies, leaving the more traditional approaches for validation purposes only. Microarray (Liu et al. 2004), bead-based flow cytometric technique (Lu et al. 2005), qRT-PCR for miRNA precursors (Schmittgen et al. 2004), stem-loop qRT-PCR and primer-extension quantitative PCR for mature miRNA products (Chen et al. 2005; Raymond et al. 2005) were developed and applied to measure miRNA level of expression in normal and cancer tissues.

Two large multi-cancer expression profiling studies have been reported. In a study of 334 leukemias and solid cancers, Lu et al. (2005) found that miRNA-expression profiles classify human cancers based on developmental lineage and differentiation status of the tumor. This study also revealed a globally decreased miRNA expression in tumors with respect to their normal counterpart. Volinia et al. (2006) conducted a large-scale miRNome analysis on 540 samples representing six solid cancers (lung, breast, stomach, prostate, colon and pancreas) and corresponding normal tissues and established the existence of a tumor-specific miRNA signature, which comprised 43 deregulated miRNAs (26 up and 17 down-regulated).

The above studies were supported by a number of investigations on individual types of neoplasm. All the studies revealed the existence of differences in miRNA expression in neoplastic *versus* normal tissues (Bandres et al. 2006; Bottoni et al. 2007; Budhu et al. 2008; Calin et al. 2004; Ciafre et al. 2005; Cummins et al. 2006; Feber et al. 2008; Garzon et al. 2008a, b; Gottardo et al. 2007; Gramantieri et al. 2007; Guo et al. 2008; He et al. 2005a, b; Huang et al. 2008a, b; Iorio et al. 2005, 2007; Isken et al. 2008; Jiang et al. 2008; Ladeiro et al. 2008; Lawrie et al. 2007; Lin et al. 2008; Ma et al. 2007; Marcucci et al. 2008; Meng et al. 2007; Michael et al. 2003; Murakami et al. 2006; Nam et al. 2008; Ozen et al. 2008; Pallante et al. 2006; Pan et al. 2008; Porkka et al. 2007; Roldo et al. 2006; Schetter et al. 2008; Sengupta et al. 2008; Subramanian et al. 2008; Tavazoie et al. 2008; Tetzlaff et al. 2007; Visone et al. 2007; Volinia et al. 2006; Wang et al. 2007, 2008; Weber et al. 2006; Wong et al. 2008a, b; Yanaihara et al. 2006; Zhang et al. 2008).

These studies proved that each neoplasm has a distinct miRNA signature that differs from that of other neoplasms and from each normal tissue counterpart. Additionally, an important implication of miRNA studies was that, differently from mRNAs, a small number of miRNAs could uncover a large amount of diagnostic information, like assessment of tissue of origin and several bio-pathological and clinical cancer features. Besides its translational value for the development of innovative diagnostic approaches, these findings also indicated that miRNAs could act as master regulators of the overall cellular gene expression.

These studies revealed that a number of miRNAs are recurrently deregulated in human cancer (Tables 5.1 and 5.2). In most cases, deregulation was consistently in one direction, namely up- or down-regulation across different types of cancers, suggesting that the involvement of each deregulated miRNA could cause the subversion of one or more cancer-associated pathways shared by different types of cancer. Consistent with this view, the most commonly found miRNAs deregulated in cancer include *miR-21*, *miR-221/222* cluster, several members of the *miR-17-92* family clusters, *miR-210* and *miR-155*, which are up-regulated in different types of neoplasm, and *miR-143/145* cluster and several members of the *miR-199*, *miR-125* and *let-7* families, which are instead down-regulated in different types of neoplasm.

In addition to miRNAs whose deregulation is shared among different types of tumors, there are also examples of miRNAs deregulated only in specific neoplasms: for example, the liver-specific miRNA *miR-122* is consistently down-regulated in hepatocellular carcinoma only (Gramantieri et al. 2007; Kutay et al. 2006; Ladeiro et al. 2008).

There are also some apparently contradictory situations: for example, members of the *miR-181* family are up-regulated in some cancers, such as thyroid, pancreatic and prostate carcinomas (He et al. 2005a, b, Pallante et al. 2006; Volinia et al. 2006) but down-regulated in others, such as glioblastomas and pituitary adenomas (Bottoni et al. 2007; Ciafre et al. 2005). Given the role of *miR-181* in differentiation (Chen et al. 2004; Guimaraes-Sternberg et al. 2006; Naguibneva et al. 2006; Ramkissoon et al. 2006; Ryan et al. 2006). It is possible that these apparent discrepancies might reflect the original cell differentiation status and therefore, function of miRNA could depend on mRNA targets expressed in each different cell setting.

Table 5.1 miRNAs most frequently up-regulated in human cancer

Rank[a]	microRNA	Gene family	Cluster	Chrom	Start[b]	End[b]	Strand		Associated cancer
1	*miR-145*	*miR-145*	*miR-143/miR-145*	5	148.790.402	148.790.489	+	Intergenic	Breast, ovary, prostate, colon, liver, lung ca
2	*miR-199a-1-5p*	*miR-199*	–	19	10.789.102	10.789.172	–	DNM2 – intron 15	Ovary, prostate, colon, liver, lung ca
3	*miR-143*	*miR-143*	*miR-143/miR-145*	5	148.788.674	148.788.779	+	Intergenic	Breast, ovary, prostate, colon, liver, lung ca
4	*miR-125a*	*miR-125*	*miR-99b/let-7e/miR-125a*	19	56.888.319	56.888.404	+	Intergenic	Breast, ovary, prostate, liver, lung ca
5	*miR-125b-1*	*miR-125*	*miR-125b-1/let7a-2/miR-100*	11	121.475.675	121.475.762	–	Intergenic	Breast, ovary, prostate, tongue, liver ca
6	*let-7d*	*let-7*	*let-7a-1/let-7f-1/let-7d*	9	95.980.937	95.981.023	+	Intergenic	Breast, ovary, prostate, colon, liver ca
7	*miR-101*	*miR-101*	–	1	65.296.705	65.296.779	–	Intergenic	Breast, ovary, prostate, colon, liver, lung ca
8	*miR-138-2*	*miR-138*	–	16	55.449.931	55.450.014	+	Intergenic	Tongue, stomach, colon, pancreas, thyroid ca
9	*miR-199a-2-5p*	*miR-199*	*miR-214/miR-199a-2*	1	170.380.298	170.380.407	–	DNM3 – intron 14	Ovary, tongue, liver ca
10	*miR-218-2*	*miR-218*	–	5	168.127.729	168.127.838	–	SLIT3; intron 14	Prostate, stomach, colon, lung ca
11	*miR-126*	*miR-126*	–	9	138.684.875	138.684.959	+	EGFL7 – intron 7	Breast, ovary, liver, lung ca
12	*miR-195*	*miR-15*	*miR-195/miR-497*	17	6.861.658	6.861.744	–	AC027763 – intron 1	Prostate, tongue, liver ca
13	*miR-214*	*miR-214*	*miR-214/miR-199a-2*	1	170.374.561	170.374.670	–	DNM3; intron 14	Ovary, liver ca
14	*miR-26a-1*	*miR-26*	–	3	37.985.899	37.985.975	+	CTDSPL – intron 4	Ovary, prostate, lung thyroid ca

15	*let-7a-2*	*let-7*	*miR-125b-1/let7a -2/miR-100*	11	121.522.440	121.522.511	–	Intergenic	Breast, prostate, liver, lung ca
16	*miR-125b-2*	*miR-125*	*miR-99a/let-7c/ miR-125b-2*	21	16.884.428	16.884.516	+	C21orf34 – intron 6	Breast, ovary, prostate, tongue ca
17	*miR-128a*	*miR-128*	–	2	136.139.437	136.139.518	+	R3HDM1 – intron 18	Breast, prostate, colon ca, pituitary tumor
18	*miR-199b*	*miR-199*	–	9	130.046.821	130.046.930	–	DNM1; intron 15	Ovary, liver, lung ca
19	*miR-99a*	*miR-99*	*miR-99a/let-7c/ miR-125b-2*	21	16.833.280	16.833.360	+	C21orf34 – intron 6	Ovary, prostate, tongue ca
20	*miR-100*	*miR-99*	*miR-125b-1/let7a -2/miR-100*	11	121.528.147	121.528.226	–	Intergenic	Ovary, prostate, tongue ca
21	*miR-122*	*miR-122*	–	18	54.269.286	54.269.370	+	Intergenic	Liver ca
22	*let-7a-1*	*let-7*	*let-7a-1/let-7f-1/ let-7d*	9	95.978.060	95.978.139	+	Intergenic	Ovary, prostate, liver ca
23	*let-7a-3*	*let-7*	*let-7a-3/let-7b*	22	44.887.293	44.887.366	+	OTTHUMT00000 316781 – exon 5	Breast, prostate, liver ca
24	*let-7b*	*let-7*	*let-7a-3/let-7b*	22	44.888.230	44.888.312	+	OTTHUMT00000 316781 – exon 5	Ovary, prostate, liver ca
25	*let-7c*	*let-7*	*miR-99a/let-7c/ miR-125b-2*	21	16.834.019	16.834.102	+	C21orf34 – intron 6	Ovary, prostate, liver ca
26	*let-7f*	*let-7*	*let-7a-1/let -7f-1/let-7d*	9	95.978.450	95.978.536	+	Intergenic	Breast, liver, thyroid ca
27	*miR-126**	*miR-126*	–	9	138.684.875	138.684.959	+	EGFL7 – intron 7	Breast, lung ca
28	*miR-130a*	*miR-130*	–	11	57.165.247	57.165.335	+	Intergenic	Breast, liver, lung ca
29	*miR-136*	*miR-136*	*miR-136/miR-432/ miR-127/miR-433/miR-431/ miR-665*	14	100.420.792	100.420.873	+	Intergenic	Stomach, colon, liver ca
30	*miR-140*	*miR-140*	–	16	68.524.485	68.524.584	+	WWP2 – intron 16	Ovary, lung, thyroid ca
31	*miR-181c*	*miR-181*	*miR-181c/miR-181d*	19	13.846.513	13.846.622	+	Intergenic	Liver, lung ca, glioblastoma

(continued)

Table 5.1 (continued)

Rank[a]	microRNA	Gene family	Cluster	Chrom	Start[b]	End[b]	Strand		Associated cancer
32	*miR-200b*	*miR-8*	*miR-200b/miR-200a/miR-429*	1	1.092.347	1.092.441	+	Intergenic	Ovary, liver ca
33	*miR-204*	*miR-204*	–	9	72.614.711	72.614.820	–	Q9HCF6-9 – intron 6	Breast, lung ca
34	*miR-212*	*miR-132*	*miR-132/miR-212*	17	1.900.315	1.900.424	–	Intergenic	Stomach, colon, pancreas ca
35	*miR-219-1*	*miR-219*	–	6	33.283.590	33.283.699	+	Intergenic	Tongue, lung, thyroid ca
36	*miR-223*	*miR-223*	–	X	65.155.437	65.155.546	+	Intergenic	Liver ca
37	*miR-9-3*	*miR-9*	–	15	87.712.252	87.712.341	+	HsG9510-001 – intron 1	Ovary, lung, thyroid ca
38	*miR-133a*	*miR-133*	*miR-1-2/miR-133a-1*	18	17.659.657	17.659.744	–	MIB1 – intron 12	Ovary, tongue ca
39	*miR-150*	*miR-150*	–	19	54.695.854	54.695.937	–	Q6ZNZ9 – 3′UTR (exon 1) antisense	Liver ca
40	*let-7g*	*let-7*	–	3	52.277.334	52.277.417	–	WDR82 – intron 2	Prostate, liver ca
41	*miR-10b*	*miR-10*	–	2	176.723.277	176.723.386	+	Intergenic	Breast, ovary ca
42	*miR-124a-1*	*miR-124*	–	8	9.798.308	9.798.392	–	Intergenic	Ovary, lung ca
43	*miR-139*	*miR-139*	–	11	72.003.755	72.003.822	–	PDE2A – intron 2	Tongue, liver ca
44	*miR-142*	*miR-142*	–	17	53.763.592	53.763.678	–	Intergenic	Liver, thyroid ca
45	*miR-152*	*miR-148*	–	17	43.469.526	43.469.612	–	COPZ2 – intron 2	Colon, pancreas ca
46	*miR-224*	*miR-224*	*miR-224/miR-452*	X	150.877.706	150.877.786	–	GABRE – intron 6	Ovary, lung ca

[a] Rank was assessed on the basis of the number of published papers reporting the down-regulation of the miRNA. Ranking scores were determined from the following published papers (Bottoni et al. 2007; Ciafre et al. 2005; Gottardo et al. 2007; Gramantieri et al. 2007; He et al. 2005a, b; Huang et al. 2008, Iorio et al. 2005, 2007; Jiang et al. 2008; Ladeiro et al. 2008; Ma et al. 2007; Meng et al. 2007; Michael et al. 2003; Murakami et al. 2006; Nam et al. 2008; Pallante et al. 2006; Porkka et al. 2007; Roldo et al. 2006; Tavazoie et al. 2008; Volinia et al. 2006; Wang et al. 2008; Wong et al. 2008a, b; Yanaihara et al. 2006).

[b] From NCBI36.

Table 5.2 miRNAs most frequently down-regulated in human cancer

MicroRNA	MicroRNA family	Gene target symbol	Gene target name	miRNA expression in cancer	Tumor type	References
let-7a	*let-7*	*KRAS*	v-Ki-ras2 Kirsten rat sarcoma viral oncogene homolog	Down	Lung ca	Johnson et al. (2005)
let-7a	*let-7*	*NRAS*	Neuroblastoma RAS viral (v-ras) oncogene homolog	Down	Lung ca	Johnson et al. (2005)
let-7a	*let-7*	*HRAS*	v-Ha-ras Harvey rat sarcoma viral oncogene homolog	Down	Lung ca	Johnson et al. (2005)
let-7b	*let-7*	*CCNA*	cyclin A	Down	Melanoma	Schultz et al. (2008)
let-7b	*let-7*	*CCND1*	cyclin D1	Down	Melanoma	Schultz et al. (2008)
let-7b	*let-7*	*CCND3*	cyclin D3	Down	Melanoma	Schultz et al. (2008)
miR-106b	*miR-17*	*CDKN1A*	cyclin-dependent kinase inhibitor 1A (p21, Cip1)	Up	Gastric ca	Petrocca et al. (2008)
miR-106b	*miR-17*	*E2F1*	E2F transcription factor 1	Up	Gastric ca	Petrocca et al. (2008)
miR-10b	*miR-10*	*HOXD10*	Homeobox D10	Up	Breast ca	Ma et al. (2007)
miR-122	*miR-122*	*CCNG1*	cyclin G1	Down	Hepatocellular ca	Gramantieri et al. (2007)
miR-125a	*miR-125*	*ERBB2, ERBB3*	v-erb-b2 erythroblastic leukemia viral oncogene homolog 2, and homolog 3	Down	Breast ca	Scott et al. (2007)
miR-125b	*miR-125*	*ERBB2, ERBB3*	v-erb-b2 erythroblastic leukemia viral oncogene homolog 2, and homolog 3	Down	Breast ca	Scott et al. (2007)
miR-137	*miR-137*	*MITF*	Micropthalmia-associated transcription factor	Up	Melanoma	Bemis et al. (2008)
miR-141	*miR-8*	*TGFβ2*	Transforming growth factor, beta 2	Down	–	Burk et al. (2008)
miR-143	*miR-143*	*ERK5*	Mitogen-activated protein kinase 7	Down	–	Esau et al. (2004)
miR-15a	*miR-15*	*BCL2*	B-cell CLL/lymphoma 2	Down	CLL, Gastric ca	Cimmino et al. (2005); Xia et al. (2008)

(continued)

Table 5.2 (continued)

MicroRNA	MicroRNA family	Gene target symbol	Gene target name	miRNA expression in cancer	Tumor type	References
miR-16-1	*miR-15*	*BCL2*	B-cell CLL/lymphoma 2	Down	CLL, Gastric ca	Cimmino et al. (2005); Xia et al. (2008)
miR-17-92 cluster	*miR-17*	*BCL2L11*	BCL2-like 11 (apoptosis facilitator)	Up	–	Xiao et al. (2008)
miR-17-92 cluster	*miR-17*	*PTEN*	Phosphatase and tensin homolog	Up	–	Xiao et al. (2008)
miR-181	*miR-181*	*TCL1*	T-cell leukemia/lymphoma 1A	Down	T-leukemia	Pekarsky et al. (2006)
*miR-199a**	*miR-199*	*ERK2*	Mitogen-activated protein kinase 1	Down	Hepatocellular ca	Kim et al. (2008)
*miR-199a**	*miR-199*	*MET*	Met proto-oncogene (hepatocyte growth factor receptor)	Down	Hepatocellular ca	Kim et al. (2008)
miR-200a, miR-200b, miR-200c, miR-141 and miR-429	*miR-8*	*ZEB1*	Zinc finger E-box binding homeobox 1	Down	–	Burk et al. (2008); Gregory et al. (2008); Korpal et al. (2008); Park et al. (2008)
miR-200a, miR-200b, miR-200c, miR-141 and miR-429	*miR-8*	*ZEB2*	Zinc finger E-box binding homeobox 2	Down	–	Burk et al. (2008); Gregory et al. (2008); Korpal et al. (2008); Park et al. (2008)
miR-200c	*miR-8*	*TGFβ2*	Transforming growth factor, beta 2	Down	–	Burk et al. (2008)
miR-205	*miR-205*	*ZEB1*	Zinc finger E-box binding homeobox 1	Down	–	Gregory et al. (2008)
miR-205	*miR-205*	*ZEB2*	Zinc finger E-box binding homeobox 2	Down	–	Gregory et al. (2008)
miR-206	*miR-1*	*ESR*	Estrogen receptor alpha		Breast ca	Adams et al. (2007)
miR-21	*miR-21*	*PDCD4*	Programmed cell death 4	Up	Breast, Colon ca	Asangani et al. (2008); Frankel et al. (2008); Zhu et al. (2008)
miR-21	*miR-21*	*PTEN*	Phosphatase and tensin homolog	Up	Hepatocellular ca	Meng et al. (2007)
miR-21	*miR-21*	*RECK*	Suppression of tumorigenicity 5 (reversion-inducing-cysteine-rich protein with kazal motifs)	Up	Glioma	Gabriely et al. (2008)

miR-21	*miR-21*	*SERPINB5*	Maspin or serpin peptidase inhibitor	Up	Breast ca	Zhu et al. (2008)
miR-21	*miR-21*	*TIMP3*	Tissue inhibitor of metalloproteinase 3	Up	Glioma	Gabriely et al. (2008)
miR-21	*miR-21*	*TPM1*	Tropomyosin-1	Up	Breast ca	Zhu et al. (2007)
miR-214	*miR-214*	*PTEN*	Phosphatase and tensin homolog	Up	Ovarian ca	Yang et al. (2008)
miR-221	*miR-221*	*CDKN1B*	cyclin-dependent kinase inhibitor 1B (p27, Kip1)	Up	Prostate, liver, breast ca	Fornari et al. (2008); Galardi et al. (2007); Medina et al. (2008); Miller et al. (2008)
miR-221	*miR-221*	*CDKN1C*	cyclin-dependent kinase inhibitor 1C (p57, Kip2)	Up	Liver ca	Fornari et al. (2008); Medina et al. (2008)
miR-222	*miR-221*	*CDKN1B*	cyclin-dependent kinase inhibitor 1B (p27, Kip1)	Up	Breast ca	Medina et al. (2008); Miller et al. (2008)
miR-222	*miR-221*	*CDKN1C*	cyclin-dependent kinase inhibitor 1C (p57, Kip2)	Up	–	Medina et al. (2008)
miR-224	*miR-224*	*API-5*	Apoptosis inhibitor-5	Up	Hepatocellular ca	Wang et al. (2008)
miR-27a	*miR-27*	*MYT-1*	Myelin transcription factor 1	Up	Breast ca	Mertens-Talcott et al. (2007)
miR-27a	*miR-27*	*ZBTB10*	Zinc finger and BTB domain containing 10	Up	Breast ca	Mertens-Talcott et al. (2007)
miR-29	*miR-29*	*TCL1*	T-cell leukemia/lymphoma 1A	Down	T-leukemia	Pekarsky et al. (2006)
miR-335	*miR-335*	*SOX4*	SRY (sex determining region Y)-box 4	Down	Breast ca	Tavazoie et al. (2008)
miR-335	*miR-335*	*TNC*	Tenascin C	Down	Breast ca	Tavazoie et al. (2008)
miR-373	*miR-373*	*CD44*	Hyaluronate receptor	Up	Breast ca	Huang et al. (2008)
miR-378	*miR-378*	*FUS-1*	TUSC2, tumor suppressor candidate 2	Up	–	Lee et al. (2007a, b)
miR-378	*miR-378*	*SUFU*	Suppressor of fused homolog	Up	–	Lee et al. (2007a, b)
miR-520c	*miR-515*	*CD44*	Hyaluronate receptor	Up	Breast ca	Huang et al. (2008a, b)
miR-7	*miR-7*	*EGFR*	Epidermal growth factor receptor	Down	Glioblastoma	Kefas et al. (2008)

(continued)

Table 5.2 (continued)

MicroRNA	MicroRNA family	Gene target symbol	Gene target name	miRNA expression in cancer	Tumor type	References
miR-93	*miR-17*	*BCL2L11*	BCL2-like 11 (apoptosis facilitator)	Up	Gastric ca	Petrocca et al. (2008)
miR-93	*miR-17*	*CDKN1A*	cyclin-dependent kinase inhibitor 1A (p21, Cip1)	Up	Gastric ca	Petrocca et al. (2008)
miR-93	*miR-17*	*E2F1*	E2F transcription factor 1	Up	Gastric ca	Petrocca et al. (2008)

There are also some situations that are apparently contrasting with the molecular or biological function (described below) associated with a given miRNA. For example, over-expression of *miR-10b* has been involved in tumor invasion/metastasis (Ma et al. 2007), but its expression is generally down-regulated in breast cancer. A more detailed analysis revealed that *miR-10b* was indeed up-regulated in about 50% of metastatic breast cancers (Ma et al. 2007), while it was generally down-regulated in all other metastatic and non-metastatic breast cancers, indicating that the *miR-10b* is up-regulated in only a fraction of breast cancers, that the statistical analyses employed to analyze microarray data may not recognize. Another example is linked to the role of *miR-200* in epithelial-to-mesenchimal transition (EMT) (Gregory et al. 2008; Korpal et al. 2008; Park et al. 2008): while inhibition of different members of the *miR-8* family (*miR-200a*, *miR-200b*, *miR-200c* and *miR-141*), which are involved in the maintenance of epithelial traits, induces the EMT cancer progression trait and are expected to be expressed at reduced level in advanced cancers, these miRNAs were instead significantly up-regulated in ovarian, thyroid and colangiocarcinomas (Iorio et al. 2007; Meng et al. 2006; Nam et al. 2008; Pallante et al. 2006). Hence, in contrast with their potential role in cancer, their expression was frequently detected opposite to expectation in primary tumors. In this case, it may be speculated that the EMT trait may be detectable only in a small fraction of cancer cells, which have acquired a more aggressive phenotype in a heterogeneous primary tumor mass. Indeed, the EMT process appears to be only transiently acquired at the invasive tumor edge. Hence, methods capable of analyzing the overall expression of the tumor may not be able to acquire this type of information. The use of *in situ* hybridization in primary tumors should help to clarify this aspect.

In summary, microRNAome expression studies revealed a large number of miRNAs deregulated in several types of human cancers. In many cases, functional studies have now connected these miRNAs to biological and molecular cancer traits. Thus, the expression work in primary tumors appears to have identified many of the miRNAs relevant in human tumorigenesis. However, although considerable advancements have been achieved in the last 2–3 years, an understanding on the biological function and role in cancer for several of these cancer-associated miRNAs have not yet been obtained, suggesting that our knowledge needs to be further expanded by additional functional studies. Besides validating the biological significance of miRNA deregulation, functional studies may also identify miRNAs involved in human cancer and missed by expression studies.

5.4 MicroRNAs Are Central Players in Malignant Transformation Processes

As previously mentioned, several of the miRNAs whose expression is deregulated in cancer, were functionally linked to molecular pathways by the identification of gene targets involved in human cancer (Table 5.3 and Fig. 5.1). Through several studies contributed by different laboratories, miRNAs were found to play

Table 5.3 Validated targets of microRNAs deregulated in cancer

MicroRNA	MicroRNA family	Gene target symbol	Gene target name	miRNA expression in cancer	Tumor type	References
let-7a	*let-7*	*KRAS*	v-Ki-ras2 Kirsten rat sarcoma viral oncogene homolog	Down	Lung ca	Johnson et al. (2005)
let-7a	*let-7*	*NRAS*	Neuroblastoma RAS viral (v-ras) oncogene homolog	Down	Lung ca	Johnson et al. (2005)
let-7a	*let-7*	*HRAS*	v-Ha-ras Harvey rat sarcoma viral oncogene homolog	Down	Lung ca	Johnson et al. (2005)
let-7b	*let-7*	*CCNA*	cyclin A	Down	Melanoma	Schultz et al. (2008)
let-7b	*let-7*	*CCND1*	cyclin D1	Down	Melanoma	Schultz et al. (2008)
let-7b	*let-7*	*CCND3*	cyclin D3	Down	Melanoma	Schultz et al. (2008)
miR-106b	*miR-17*	*CDKN1A*	cyclin-dependent kinase inhibitor 1A (p21, Cip1)	Up	Gastric ca	Petrocca et al. (2008)
miR-106b	*miR-17*	*E2F1*	E2F transcription factor 1	Up	Gastric ca	Petrocca et al. (2008)
miR-10b	*miR-10*	*HOXD10*	Homeobox D10	Up	Breast ca	Ma et al. (2007)
miR-122	*miR-122*	*CCNG1*	cyclin G1	Down	Hepatocellular ca	Gramantieri et al. (2007)
miR-125a	*miR-125*	*ERBB2, ERBB3*	v-erb-b2 erythroblastic leukemia viral oncogene homolog 2, and homolog 3	Down	Breast ca	Scott et al. (2007)
miR-125b	*miR-125*	*ERBB2, ERBB3*	v-erb-b2 erythroblastic leukemia viral oncogene homolog 2, and homolog 3	Down	Breast ca	Scott et al. (2007)
miR-137	*miR-137*	*MITF*	Micropthalmia-associated transcription factor	Up	Melanoma	Bemis et al. (2008)
miR-141	*miR-8*	*TGFβ2*	Transforming growth factor, beta 2	Down	–	Burk et al. (2008)
miR-143	*miR-143*	*ERK5*	Mitogen-activated protein kinase 7	Down	–	Esau et al. (2004)
miR-15a	*miR-15*	*BCL2*	B-cell CLL/lymphoma 2	Down	CLL, Gastric ca	Cimmino et al. (2005); Xia et al. (2008)
miR-16-1	*miR-15*	*BCL2*	B-cell CLL/lymphoma 2	Down	CLL, Gastric ca	Cimmino et al. (2005); Xia et al. (2008)

miR-17-92 cluster	*miR-17*	*BCL2L11*	BCL2-like 11 (apoptosis facilitator)	Up	–	Xiao et al. (2008)
miR-17-92 cluster	*miR-17*	*PTEN*	Phosphatase and tensin homolog	Up	–	Xiao et al. (2008)
miR-181	*miR-181*	*TCL1*	T-cell leukemia/lymphoma 1A	Down	T-leukemia	Pekarsky et al. (2006)
*miR-199a**	*miR-199*	*ERK2*	Mitogen-activated protein kinase 1	Down	Hepatocellular ca	Kim et al. (2008)
*miR-199a**	*miR-199*	*MET*	Met proto-oncogene (hepatocyte growth factor receptor)	Down	Hepatocellular ca	Kim et al. (2008)
miR-200a, miR-200b, miR-200c, miR-141 and miR-429	*miR-8*	*ZEB1*	Zinc finger E-box binding homeobox 1	Down	–	Burk et al. (2008); Gregory et al. (2008); Korpal et al. (2008); Park et al. (2008)
miR-200a, miR-200b, miR-200c, miR-141 and miR-429	*miR-8*	*ZEB2*	Zinc finger E-box binding homeobox 2	Down	–	Burk et al. (2008); Gregory et al. (2008); Korpal et al. (2008); Park et al. (2008)
miR-200c	*miR-8*	*TGFβ2*	Transforming growth factor, beta 2	Down	–	Burk et al. (2008)
miR-205	*miR-205*	*ZEB1*	Zinc finger E-box binding homeobox 1	Down	–	Gregory et al. (2008)
miR-205	*miR-205*	*ZEB2*	Zinc finger E-box binding homeobox 2	Down	–	Gregory et al. (2008)
miR-206	*miR-1*	*ESR*	Estrogen receptor alpha	–	Breast ca	Adams et al. (2007)
miR-21	*miR-21*	*PDCD4*	Programmed cell death 4	Up	Breast, Colon ca	Asangani et al. (2008); Frankel et al. (2008); Zhu et al. (2008)
miR-21	*miR-21*	*PTEN*	Phosphatase and tensin homolog	Up	Hepatocellular ca	Meng et al. (2007)
miR-21	*miR-21*	*RECK*	Suppression of tumorigenicity 5 (reversion-inducing-cysteine-rich protein with kazal motifs)	Up	Glioma	Gabriely et al. (2008)
miR-21	*miR-21*	*SERPINB5*	Maspin or serpin peptidase inhibitor	Up	Breast ca	Zhu et al. (2008)
miR-21	*miR-21*	*TIMP3*	Tissue inhibitor of metalloproteinase 3	Up	glioma	Gabriely et al. (2008)
miR-21	*miR-21*	*TPM1*	Tropomyosin-1	Up	Breast ca	Zhu et al. (2007)
miR-214	*miR-214*	*PTEN*	Phosphatase and tensin homolog	Up	Ovarian ca	Yang et al. (2008)

(continued)

Table 5.3 (continued)

MicroRNA	MicroRNA family	Gene target symbol	Gene target name	miRNA expression in cancer	Tumor type	References
miR-221	*miR-221*	*CDKN1B*	cyclin-dependent kinase inhibitor 1B (p27, Kip1)	Up	Prostate, liver, breast ca	Fornari et al. (2008); Galardi et al. (2007); Medina et al. (2008); Miller et al. (2008)
miR-221	*miR-221*	*CDKN1C*	cyclin-dependent kinase inhibitor 1C (p57, Kip2)	Up	Liver ca	Fornari et al. (2008); Medina et al. (2008)
miR-222	*miR-221*	*CDKN1B*	cyclin-dependent kinase inhibitor 1B (p27, Kip1)	Up	Breast ca	Medina et al. (2008); Miller et al. (2008)
miR-222	*miR-221*	*CDKN1C*	cyclin-dependent kinase inhibitor 1C (p57, Kip2)	Up	–	Medina et al. (2008)
miR-224	*miR-224*	*API-5*	Apoptosis inhibitor-5	Up	Hepatocellular ca	Wang et al. (2008)
miR-27a	*miR-27*	*MYT-1*	Myelin transcription factor 1	Up	Breast ca	Mertens-Talcott et al. (2007)
miR-27a	*miR-27*	*ZBTB10*	Zinc finger and BTB domain containing 10	Up	Breast ca	Mertens-Talcott et al. (2007)
miR-29	*miR-29*	*TCL1*	T-cell leukemia/lymphoma 1A	Down	T-leukemia	Pekarsky et al. (2006)
miR-335	*miR-335*	*SOX4*	SRY (sex determining region Y)-box 4	Down	Breast ca	Tavazoie et al. (2008)
miR-335	*miR-335*	*TNC*	tenascin C	Down	Breast ca	Tavazoie et al. (2008)
miR-373	*miR-373*	*CD44*	Hyaluronate receptor	Up	Breast ca	Huang et al. (2008a, b)
miR-378	*miR-378*	*FUS-1*	TUSC2, tumor suppressor candidate 2	Up	–	Lee et al. (2007a, b)
miR-378	*miR-378*	*SUFU*	Suppressor of fused homolog	Up	–	Lee et al. (2007a, b)
miR-520c	*miR-515*	*CD44*	Hyaluronate receptor	Up	Breast ca	Huang et al. (2008a, b)
miR-7	*miR-7*	*EGFR*	Epidermal growth factor receptor	Down	Glioblastoma	Kefas et al. (2008)
miR-93	*miR-17*	*BCL2L11*	BCL2-like 11 (apoptosis facilitator)	Up	Gastric ca	Petrocca et al. (2008)
miR-93	*miR-17*	*CDKN1A*	cyclin-dependent kinase inhibitor 1A (p21, Cip1)	Up	Gastric ca	Petrocca et al. (2008)
miR-93	*miR-17*	*E2F1*	E2F transcription factor 1	Up	Gastric ca	Petrocca et al. (2008)

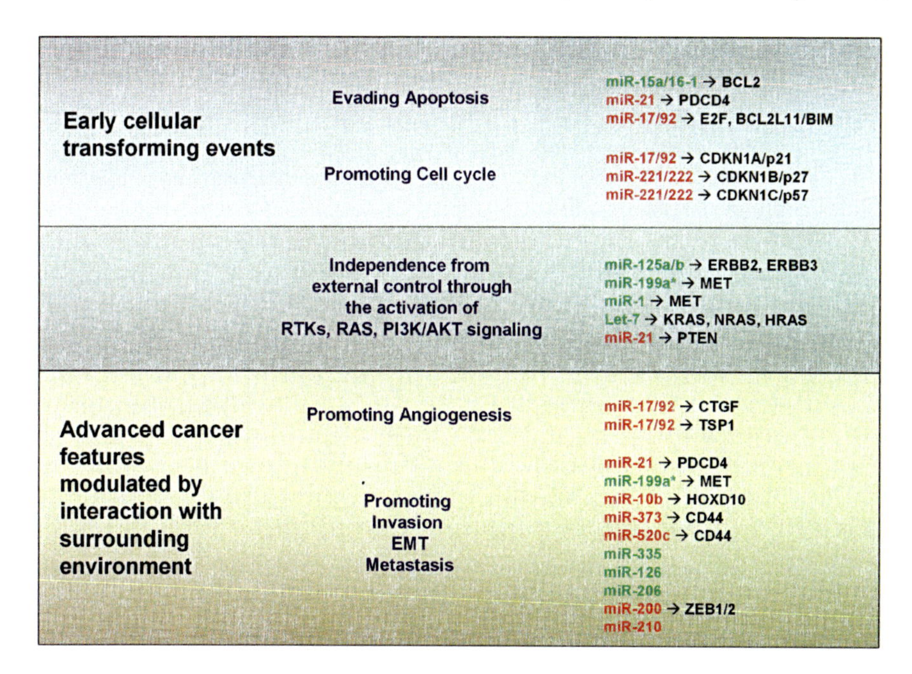

Fig. 5.1 miRNAs are involved in all human cancer traits. Early and late events are here schematically shown. MiRNAs in green are generally down-regulated in cancer, thereby leading to target up-regulation; conversely, the miRNAs in red are generally up-regulated in cancer and their targets repressed

important roles in cell growth, death, differentiation, angiogenesis, invasion of surrounding tissues and metastasis to distant sites (Hanahan and Weinberg 2000). Moreover, since miRNAs could regulate multiple mRNA targets at the same time, the deregulation of a single miRNA was shown to affect, in some instances, various cancer traits.

Independence of growth factors provides cancer cell the ability to grow without tissue or organ control. Growth factor receptors represent the interface that allows cells to respond to external proliferation stimuli. *miR-125a* and *miR-125b* are down-regulated in breast cancer (Iorio et al. 2005). It was shown that these two miRNAs regulate the expression of the receptor tyrosine kinases ERBB2 and ERBB3 (Scott et al. 2007). Consistently with the suppression of ErbB signaling, ectopic over-expression of *miR-125a* or *miR-125b* in SK-BR-3 cells impaired *in vitro* anchorage-dependent growth and reduced migration and invasion capacities, suggesting that down-regulation of *miR-125* in breast cancer may promote tumor cell motility and invasiveness.

Another tyrosine kinase receptor that is involved in conferring motility and invasive potential to cancer cells is MET, which becomes physiologically activated by the hepatocyte growth factor (HGF). HGF is the most potent growth factor for hepatocytes and, by binding to its tyrosine kinase receptor, MET, promotes proliferation, regeneration, migration, survival and angiogenesis and it is involved in the control of invasive growth both during tumorigenesis and in embryonic development (Benvenuti and Comoglio 2007; Birchmeier et al. 2003). Recent reports indicated that MET is post-transcriptionally regulated by *miR-199a/a** and *miR-1* (Datta et al. 2008; Grady et al. 2008). Genes for both miRNAs are methylated in hepatocellular carcinomas (HCCs). Since members of the *miR-199* family emerged as frequently down-regulated in several human cancers (Gramantieri et al. 2007; Iorio et al. 2007; Jiang et al. 2008; Meng et al. 2007; Murakami et al. 2006; Nam et al. 2008; Pallante et al. 2006; Porkka et al. 2007), this mechanism could favour the activation of the MET oncogenic potential.

The activation of tyrosine kinase receptors (RTKs) initiates a downstream cascade of events that lead cells to proliferate. Crucial elements of this signaling transduction pathway are members of the RAS family of oncogenes. A molecular link between miRNA deregulation and RAS expression has been established. The 3′UTRs of the KRAS, NRAS and HRAS mRNAs contain multiple complementary sites for binding of *let-7* members, and forced expression of *let-7* in human cancer cells reduces RAS protein levels (Johnson et al. 2005). Since *let-7* is frequently down-regulated in several human cancers (Gramantieri et al. 2007; Iorio et al. 2005, 2007; Porkka et al. 2007; Yanaihara et al. 2006), this mechanism could lead to the activation of the RAS pathway.

It has also been shown that *let-7* can also repress the HMGA2 oncogene (Calin et al. 2007; Mayr et al. 2007), which encodes for a high-mobility group protein oncogenic in a variety of tumors, including benign mesenchymal tumors and lung cancers. The effect of *let-7* on HMGA2 was dependent on multiple target sites in the 3′ untranslated region (UTR). The disrupted repression promotes anchorage-independent growth and the growth-suppressive effect of *let-7* on lung cancer cells was rescued by over-expression of the HMGA2 ORF without a 3′UTR (Calin et al. 2007).

The importance of *let-7* down-regulation in cancer was also supported by studies by Takamizawa et al. and Akao et al. (Akao et al. 2006; Takamizawa et al. 2004), who showed that *let-7* can suppress the growth of A549 lung cancer cells and DLD-1 colon cancer cells *in vitro*.

Signal pathways from activated RTKs include also the phosphatidyl inositol 3-phosphated kinase (PI3K)/AKT signalling, which leads to the activation of AKT kinases, which phosphorylate several protein targets that in turn promote cell survival. This pathway is controlled by the tumor suppressor lipid-phosphatase PTEN. It was shown that PTEN is a direct target of *miR-21* (Meng et al. 2007), a miRNA that is over-expressed in most of human cancers (Ciafre et al. 2005; He et al. 2005a, b; Iorio et al. 2005; Nam et al. 2008; Volinia et al. 2006; Wong et al. 2008a, b; Yanaihara et al. 2006). Thus, PTEN could be repressed by the over-expression of *miR-21*, which would lead to cell survival through PI3K-AKT pathway activation.

miR-21 can also down-regulate the tumor suppressor Programmed Cell Death 4 (Pdcd4) (Asangani et al. 2008; Frankel et al. 2008). Pdcd4 is believed to have a role in TGF-beta induced apoptosis. However, it may also have other functions. It is up-regulated in senescent fibroblasts and it may inhibit proliferation, possibly through the indirect suppression of CDK1/cdc2 kinase. Moreover, it acts as a negative regulator of intravasation, initial step for cancer cell metastasis. Anti-*miR-21*-transfected RKO cells showed an increase of Pdcd4-protein and reduced invasion, while over-expression of *miR-21* in Colo206f significantly reduced Pdcd4-protein amounts and increased invasion. Analyses of primary colorectal cancers revealed that an inverse correlation between *miR-21* and Pdcd4-protein exists, suggesting that *miR-21*/Pdcd4 interaction may be relevant for invasion/intravasation/metastasis of cancer cells.

The anti-apoptotic role of *miR-21* was supported by experiments based on transfection of cultured glioblastoma and breast cancer cells with anti-miRNA oligonucleotides (AMOs): inhibition of *miR-21* was accompanied by suppression of cell growth *in vitro*, associated with increased apoptosis (Chan et al. 2005). *miR-21* is over-expressed in colangiocarcinoma and its inhibition by AMOs increases sensitivity to the chemotherapeutic agent gemcitabine (Meng et al. 2006). *miR-21* also down-regulates the tropomyosin 1 (TPM1) (Ahituv et al. 2007), which suppresses anchorage-independent growth of MCF-7 breast cancer cells, further supporting the oncogenic function of *miR-21*.

Defective apoptosis is a trait of cancer cells. The *miR-21* example indicates that miRNAs can affect this essential element of tumorigenesis. Moreover, as mentioned above, *miR-15a* and *miR-16-1* act as regulators of the anti-apoptotic BCL2 oncoprotein. These miRNAs are encoded by genes located in a chromosomal region deleted in more than 50% of CLLs. In the leukemic cell line MEG-01, expression of *miR-15a* and/or *miR-16-1* leads to BCL2 down-regulation and increased apoptosis (Cimmino et al. 2005). Hence, the loss of expression of these miRNAs, by removing a control over BCL2 expression, may be relevant in the pathogenesis of human CLL. Indeed, BCL2 is highly expressed in CLL; however, unlike in follicular lymphoma, its activation is not associated with translocation to the IgH-encoding locus. BCL2 activation in CLL thus appears at least in part linked to the reduced expression of *miR-15a* and *miR-16-1*.

Another important link between miRNAs and apoptosis and cell proliferation pathways is given by the *miR-17-92* cluster, which acts together with *MYC* to accelerate tumor development in a mouse B-cell lymphoma model (He et al. 2005a, b). These lymphomas, differently from those arising in the *MYC*-only system, are characterized by the absence of apoptosis, which suggests that various *miR-17-92* family members regulate a pro-apoptotic gene. Interestingly, two miRNAs encoded by the cluster, *miR-17-5p* and *miR-20a*, negatively regulate the expression of E2F1 (O'Donnell et al. 2005), a transcription factor that promotes cell cycle progression, but is also a strong inducer of apoptosis when abnormally expressed. The absence of apoptosis might thus be linked to the tight control on E2F1 by the *miR-17* family. A more recent report, reveals that E2F1, E2F2, and E2F3 directly bind the promoter of the *miR-17-92* cluster, activating its transcription, and *miR-20a*, a member of the *miR-17–92* cluster, modulates the translation of the E2F2 and E2F3 mRNAs (Sylvestre et al. 2007). These results suggest the existence of a feed-back regulatory loop involving *miR-20a* and E2F that protects cells from apoptosis induced by excessive E2F expression.

Additionally, the pro-apoptotic *BCL2L11/BIM* gene was also shown to be a direct target of multiple members of the various *miR-17-92* clusters (Koralov et al. 2008; Petrocca et al. 2008), suggesting that the observed anti-apoptotic mechanism could be possibly more related to the repression of Bim rather than of E2F1. Indeed, haploinsufficiency for Bim can accelerate lymphomagenesis in Eu-myc transgenic mice (Egle et al. 2004), similarly to the overexpression of the *miR-17-92* cluster.

In human, the *miR-17-92* family includes 15 homologous miRNAs, which are encoded by 3 gene clusters on chromosomes 7, 13 and X (Mendell 2008; Tanzer and Stadler 2004). It is significant that members of all three clusters were up-regulated in different types of cancer, symptomatic of the importance of these miRNAs in cancer development. The role of these miRNAs as oncogenes has been proved by several evidences: expression studies revealed that these miRNAs are over-expressed in different types of hematopoietic and solid malignancies (Volinia et al. 2006). The cluster on chromosome 13 is amplified in human B-cell lymphomas (Ota et al. 2004), which leads to increased expression of various miRNA members. In human solid tumors, expression of the *miR-17-92* cluster at chromosome 13 is up-regulated in small-cell lung cancer, and ectopic over-expression of this cluster enhances lung cancer cell growth (Dews et al. 2006; Hayashita et al. 2005). Over-expression of the paralog on chromosome 7 renders gastric cancer insensitive to TGF-β-mediated cell cycle arrest, by interfering with the expression of CDKN1A/p21 (Petrocca et al. 2008). Importantly, *miR-17-92* up-regulation can increase tumor angiogenesis, through the down-regulation of the anti-angiogenic factors thrombospondin-1 (Tsp1) and connective tissue growth factor (CTGF), both predicted targets of the *miR-17-92* microRNAs (Dews et al. 2006). These findings indicate that members of the *miR-17* family are involved in multiple biological functions, most prominently apoptosis and cell cycle regulation, that affect different cancer traits.

A direct role of miRNAs in controlling cell growth by directly acting on elements of the cell cycle machinery was provided by studies on *miR-221/222* cluster. *miR-221*, which was recently shown to be induced by MYCN (Schulte et al. 2008) and repressed by p53 (Tarasov et al. 2007), emerged as a significantly up-regulated miRNA in glioblastoma, pancreatic, hepatocellular, kidney, bladder, prostate and thyroid cancer (Calin et al. 2007; Ciafre et al. 2005; Galardi et al. 2007; Gottardo et al. 2007; He et al. 2005a, b; Lee et al. 2007a, b; Pallante et al. 2006), indicative of an oncogenic function affecting a trait commonly altered in several human neoplasms. Its oncogenic function was substantiated by the discovery of its ability to modulate the expression of the cyclin-dependent kinase inhibitors CDKN1B/p27 (Corsten et al. 2007; Galardi et al. 2007) and CDKN1C/p57 (Fornari et al. 2008; Medina et al. 2008), two key controllers of cell cycle progression, implicating an important role of *miR-221/222* in promoting cell cycle progression in cancer cells.

P53 is a tumor suppressor protein at the cross-road of a variety of signalling pathways that may induce cell cycle arrest or apoptosis, whose main function is to safeguard cellular integrity. This tumor suppressor gene encodes a transcription factor that is post-transcriptionally activated by DNA damage, oxidative stress, or activation of oncogenes (Vousden and Lane 2007). P53 activation leads the transcription of genes, whose products in turn induce cell cycle arrest, which can be transient or permanent (senescence), or promote apoptosis in cases where stress cannot be overcome. Biological outcome is largely dependent on cellular background.

*miR-34*s form an evolutionarily conserved miRNA family, with three members in vertebrate genomes (*miR-34a*, *miR-34b*, and *miR-34c*), organized in two separate loci: *miR-34a*, located at chromosome 1p16, is encoded by its own transcript, whereas *miR-34b* and *miR-34c,* at 11q23, share a common primary transcript. The two loci show little phylogenetic conservation, except in the miRNA-encoding sequences and in short promoter proximal regions, each containing a consensus p53-binding site. A series of reports from several laboratories proved that members of the *miR-34* family are induced by p53 (Adams et al. 2007; Bommer et al. 2007; Corney et al. 2007; Perkins et al. 2007; Raver-Shapira et al. 2007; Tarasov et al. 2007; Tazawa et al. 2007). Through various induction stimuli, such as DNA damage and oncogenic activation, *miR-34a* and *miR-34b*/*miR-34c* loci were shown to be directly regulated by interaction of p53, thus adding miRNAs among the numerous p53-regulated genes. Importantly, biological responses to ectopic expression of *miR-34* included senescence, cell cycle arrest or apoptosis depending on cellular model, indicating that, as cell cycle arrest and apoptosis are common endpoints of p53 activation, *miR-34* genes represent important effectors of tumor suppression by p53. Various gene targets compatible with the observed biological effects were also identified. For examples, cyclin E2, cyclin-dependent kinase 4 and hepatocyte growth factor receptor MET, were down-regulated by *miR-34* expression. It is interesting to note that p53 was also reported to induce the down-regulation of cyclin-dependent kinases (CDK4) and cyclins (Cyclin E2) (Spurgers et al. 2006), suggesting that down-regulation of these proteins could be mediated by induction of *miR-34*.

A direct link between *miR-34a* and tumorigenesis emerged from earlier studies of Welch et al. (Welch et al. 2007), who showed that the *miR-34a* gene is often lost in human neuroblastoma. This study proved that primary neuroblastomas and cell lines often exhibit low *miR-34a* expression. Importantly, enforced expression of *miR-34a* in these cells inhibited proliferation and activated cell death pathways. Because of the chromosomal location of the *miR-34* genes, at 1p36 and 11q23, which are frequently deleted in a variety of human carcinomas, it is possible that these genomic abnormalities, by affecting *miR-34*, could indirectly alter the p53 function in these cells.

*miR-34*s were not the only miRNAs modulated by p53. Notably, quantitative analyses indicated that 34 miRNAs were significantly induced by p53, whereas 16 miRNAs were repressed (Tarasov et al. 2007). Some of these differentially regulated miRNAs were connected to cancer: among the induced miRNAs were *miR-15*/*16*, which target the oncogene product Bcl2, and *let-7*, which downregulates RAS and HMGA2 (Calin and Croce 2006; Calin et al. 2007; Johnson et al. 2005; Mayr et al. 2007). Among the miRNAs repressed by p53 was *miR-221*, which promotes cell cycle progression by down-regulating the CDK inhibitors p27 and p57 (Corsten et al. 2007; Fornari et al. 2008; Galardi et al. 2007; Medina et al. 2008). These miR-mediated regulations are likely to contribute to tumor suppression activity of p53.

Another link between miRNAs and p53 was shown by a study on *miR-372* and *miR-373*, which have been shown to cooperate with oncogenic RAS to transform primary human cells (Voorhoeve et al. 2006). This study proved that miRNAs can confer protection to oncogene-activated p53 pathway. It was shown that primary human cells undergo growth arrest and senescence in response to mitogenic signals from oncogenes such as RAS, by the activation of the p53 pathway, a response that is reversed by the presence of non-functional p53. Voorhoeve e coll. demonstrated that ectopic expression of miR372/373 was sufficient to allow transformation in the presence of wt p53. Thus, the study demonstrated that miR372/373 confers protection to oncogene-activated p53 pathway, but not to DNA damage p53-dependent cellular response. Interestingly, this is a characteristic found in testicular germ cell tumors (TGCTs), where, in contrast with other types of tumors, the *miR-372/373* cluster is indeed highly expressed in most TGCTs, suggesting a role in the development of these tumors.

Altered apoptosis and cell cycle are early events in cancer, which progresses to more advanced stages by acquiring additional traits, like angiogenesis, invasion and metastasis. The role of *miR-17-92* up-regulation in tumor angiogenesis has been previously mentioned (Dews et al. 2006). The role of miRNAs in invasion and metastasis was demonstrated by different studies (Adams et al. 2007; Huang et al. 2008a, b; Tavazoie et al. 2008). In a study from Robert Weinberg's laboratory, it was shown that *miR-10b* initiates tumor invasion and metastasis (Adams et al. 2007). Although *miR-10b* is down-regulated in most breast cancers in comparison with normal breast tissue (Iorio et al. 2005), this miRNA is instead over-expressed in about 50% of metastatic breast cancers. The authors proved that *miR-10b* was responsible for initiating tumor invasion and metastasis. They found that Twist, a metastasis-promoting transcription factor, can induce *miR-10b* expression by binding to an upstream element, and they proved that HOXD10, a homeobox transcription factor that promotes or maintains a differentiated phenotype in epithelial cells, is a target of *miR-10b*. The importance of HOXD10 as an effector of *miR-10b* was proven by showing that the ectopic expression of an unresponsive-to-*miR-10b*-HOXD10 could abrogate *miR-10b*-induced cell motility and invasiveness. Finally, the authors showed that RhoC, a G-protein involved in metastasis that is repressed by HOXD10, becomes strongly expressed in response to *miR-10b* expression. Importantly, reduction of RhoC expression by siRNA caused repression of *miR-10b*-induced cell migration and invasion, implying RhoC as a downstream effector of *miR-10b*.

Two additional miRNAs, *miR-373* and *miR-520c*, were also found to function as metastasis-promoting miRNAs in breast cancer (Huang et al. 2008a, b). *miR-373* was previously associated with testicular cancer (Voorhoeve et al. 2006), but not to metastasis. Similarly to *miR-10b*, *miR-373* and *miR-520c* did not affect cell proliferation, but promoted an *in vitro* migratory and invasive phenotype of MCF7 cells. Furthermore, MCF7 cells over-expressing *miR-373* or *miR-520c* developed metastatic nodules, which were absent in control cells. Interestingly, *miR-373* and *miR-520c* could regulate an overlapping set of gene targets, among which, CD44, which encodes a cell surface receptor for hyaluronan that could reduce the migratory properties of MCF7 cells. In primary breast carcinomas, *miR-373* was highly expressed in tumors presenting lymph-node metastasis and an inverse correlation with CD44 expression was revealed.

In addition to metastasis-promoting miRNAs, *miR-335*, *miR-126* and *miR-206* were found to function as metastasis-suppressor microRNAs (Tavazoie et al. 2008). These miRNAs were identified by comparing miRNA expression of metastatic nodules *versus* unselected breast cancer parental cells. These microRNAs were consistently down-regulated in metastatic foci.

Invasion and metastasis of cancer cells present similarities with the embryonic process of EMT (Berx et al. 2007). The EMT program allows detachment of cells from each other and increases cell mobility, both of which are necessary for tumor cell dissemination. During this process, cells down-regulate E-cadherin and up-regulate N-cadherin and Vimentin expression. In cancer, EMT seems to be transiently activated at the invasive tumor edge by microenvironment factors. Expression of *miR-200* was found to be a marker for cells that express E-cadherin but lack expression of Vimentin. *miR-200* was also found to directly target the mRNA of the E-cadherin transcriptional repressors ZEB1 and ZEB2 (Burk et al. 2008; Gregory et al. 2008; Korpal et al. 2008; Park et al. 2008). Expression of ZEB1 promotes metastasis of tumor cells in a mouse xenograft model (Spaderna et al. 2008). Ectopic expression of *miR-200* caused up-regulation of E-cadherin in cancer cell lines and reduced their motility; conversely, inhibition of *miR-200* reduced E-cadherin expression, increased expression of Vimentin, and induced EMT. In spite of the fact that these miRNAs were found significantly up-regulated in ovarian, thyroid and colangiocarcinomas (Iorio et al. 2007; Meng et al. 2006; Nam et al. 2008; Pallante et al. 2006), these findings maintain their relevance in tumor biology and clinical significance. In fact, down-regulation of *miR-200* may occur in a small fraction of cells present at the edge of tumor mass and may be missed by approaches that analyze the overall tumor expression.

As their role has been linked to any cancer trait, it is not surprising that miRNAs may also be important in normal and cancer stem cell maintenance. In this regard, *let-7* was shown to play an important role in human and mouse mammary cell stemness. A paper by Ibarra et al. (2007) proved a role of *let-7* in the maintenance of a mammary mouse progenitor cells, and a paper by Yu et al. (2007) proved that the fraction of cells CD44+/CD24^{low} of the breast cancer cell line SK-BR-3, which exhibits an increased potential of *in vitro* self-renewal and differentiation and capacity of *in vivo* tumor formation and metastasis, is also characterized by low level of *let-7*. Yu et al demonstrated that the low level of *let-7* is required for the maintenance of self-renewal and differentiation properties and verified that the reduced level of *let-7* is also present in CSCs directly enriched from primary tumors. Among targets of *let-7*, HMGA2 is responsible for maintaining cells in an undifferentiated status, a situation that mirrors what is observed in normal embryonic stem cells. This area of investigation is at a very early stage and it is therefore likely that new important discoveries on the role of miRNAs in the maintenance of normal and cancer stem cells will increase our understanding in this important area of investigation in the near future.

The above studies clearly proved the importance of miRNAs in most, if not all, pathways associated with cancer traits (Fig. 5.1). In addition to advance our understanding on the molecular basis of cancer, it is now expected that this new information could be translated into useful diagnostic and therapeutic tools.

5.5 miRNA Expression Signatures for Cancer Classification, Prognostic Stratification and Therapy Response

The findings discussed above clearly indicate that miRNAs aberrant expression can influence cancer phenotype. If so, specific miRNA expression signatures could potentially be used to classify tumors and reveal distinct subtypes in term of prognosis or therapy response. Indeed, in several instances, the expression of individual or groups of miRNAs was associated with distinct bio-pathological and clinical features.

In CLL, a specific expression signature consisting of 13 miRNAs could distinguish cases of CLL with poor prognosis (high ZAP-70/unmutated IgVH) *versus* those with good prognosis (low ZAP-70/mutated IgVH). The same microRNA signature was also associated with disease progression (Calin et al. 2005).

In AML, miRNA expression was closely associated with selected cytogenetic and molecular abnormalities, such as t(11)(q23), isolated trisomy 8, and FLT3-ITD mutations; patients with high expression of *miR-191* and *miR-199a* were shown to bear a significantly worse overall and event-free survival (Garzon et al. 2008).

Higher levels of *miR-155* were found in DLBCLs with an activated B cell phenotype than with germinal center phenotype. Because patients with an activated B cell-type DLBCL have a poorer clinical prognosis, quantification of this miRNA may be clinically useful (Eis et al. 2005).

In hepatocellular carcinoma (HCC), Ladeiro and colleagues (Ladeiro et al. 2008) revealed that distinct miRNA expression signatures were associated with histological features (tumor/nontumor, $p < 0.001$; benign/malignant tumors, $p < 0.01$; inflammatory adenoma and focal nodular hyperplasia, $p < 0.01$), clinical characteristics [hepatitis B virus (HBV) infection, $p < 0.001$; alcohol consumption, $p < 0.05$], and oncogene/tumor suppressor gene mutations [beta-catenin, $p < 0.01$; hepatocyte nuclear factor 1alpha (HNF1alpha), $p < 0.01$]. Budhu and colleagues identified a specific expression signature consisting of 20 miRNAs that was associated with prognosis in HCC. In their entire cohort of samples, the 20-miRNA tumor signature was a significant independent predictor of survival ($p = 0.009$) and, in early stage HCCs, was significantly associated with both survival and relapse ($p = 0.022$ and 0.002, respectively), thus providing a simple profiling method to assist in identifying patients with HCC who are likely to develop metastases/recurrence (Budhu et al. 2008).

In lung cancer, Yanaihara and colleagues found that high expression of *miR-155* ($p = 0.006$) and low expression of *let-7a-2* ($p = 0.033$) were associated with poor overall survival (Yanaihara et al. 2006); while Yu et al. identified a 5-set miRNA signature (*miR-221*, *let-7a*, *miR-137*, *miR-372* and *miR-182**) able to predict overall survival ($p = 0.007$) and disease-free survival ($p = 0.037$) (Yu et al. 2008). In spite of the different technologies employed to assess miRNA expression, it appears significant that both studies revealed that the down-regulation of *let-7a* represents a predictor of a poor prognosis. Furthermore, this conclusion is also in agreement with the report by Takamizawa e colleagues, who found that the down-regulation of *let-7* in non-small cell lung cancer was associated with poor prognosis and reduced post-operative survival (Takamizawa et al. 2004).

In oesophageal squamous cell carcinoma, low expression of *miR-103/107* correlated with a high overall survival rate and high expression correlated with a low overall survival rate ($p = 0.041$) (Guo et al. 2008).

In colon cancer, high *miR-21* expression was associated with poor survival and poor therapeutic outcome, independently of clinical covariates, including TNM staging, and was associated with a poor therapeutic outcome (Schetter et al. 2008).

In breast cancer, various miRNAs could predict disease-free and overall survival. Low expression of *miR-335* or *miR-126* was significantly associated with poor metastasis-free survival (Tavazoie et al. 2008) and the level of *miR-210* expression exhibited an inverse correlation with disease-free and overall survival (Camps et al. 2008). The prognostic value of *miR-210* was also confirmed in an independent study (Foekens et al. 2008), which also revealed that up-regulation of *miR-210* is associated with tumor aggressiveness in estrogen receptor positive/lymph node negative breast cancers and early relapse in estrogen receptor negative/lymph node negative breast cancers. These results suggest the potential use of these miRNAs in prognostic stratification of breast cancer patients. In addition, although their usefulness as prognostic markers in a clinical setting needs to be assessed by more direct studies, over-expression of *miR-10b* (Adams et al. 2007), *miR-373* and *miR-520c* (Huang et al. 2008a, b) were associated with breast cancer metastasis, as previously described.

miRNAs are also emerging as predictors of therapy response. In colangiocarcinoma, the expression of *miR-21* could modulate gemcitabine-induced apoptosis (Meng et al. 2006), while *miR-214* could induce cisplatin resistance in ovarian cancer (Yang et al. 2008). In breast cancer, over-expression of *miR-221/222* was found to confer resistance to tamoxifen (Miller et al. 2008). Interestingly, *miR-128b* was shown to regulate EGFR and its deletion in non-small-cell lung cancer correlated with clinical response and survival following treatment with the EGFR inhibitor gefitinib (Weiss et al. 2008), suggesting that expression of *miR-128b* could improve selection of patients eligible for gefitinib treatment.

5.6 miRNAs in Anti-cancer Therapy

miRNAs not only can be used as diagnostic markers, but they may become therapeutic targets or therapeutic molecules themselves, the anti-miRNA agents and the miRNA-mimics agents have been recently added to the expanding list of anticancer ammunition. The conceptual basis for their use has been established by *in vitro* and animal models.

In 1998 Mello and Fire discovered the RNA interference (RNAi) in vertebrates (Fire et al. 1998). Similar to miRNAs, RNAi is a form of posttranscriptional gene silencing in which double-stranded RNA (dsRNA), named short interfering RNAs (siRNAs), catalyzes the degradation of complementary mRNA targets. A siRNA is a dsRNA homologous to an mRNA of a target gene. The processing of the siRNAs is similar of that of miRNAs (Kim and Rossi 2007; McManus and Sharp 2002). A siRNA can be generated by short hairpin RNA (shRNA), which represents a siRNA precursor hairpin molecule expressed from a vector (Bernards et al. 2006). DNA cassettes encoding RNA polymerase III promoter-driven shRNAs allow long-term expression of therapeutic RNAs in targeted cells. This process immediately attracted investigators all over the world, for the possibility to specifically inhibit the gene of interest, including oncogenes, with a relatively simple approach.

Song and colleagues used anti-Fas siRNA to protect mice from induced acute liver injury and, similarly, Zender and colleagues used anti-caspase-8 siRNA to protect mice against Fas ligand-induced liver injury (Kim et al. 2003; Wirth et al. 2003). Both studies demonstrated a survival benefit using a siRNA approach without significant side effects, and a high hepatic uptake following siRNA systemic administration.

Oncogenes expressed at abnormally high levels represent the obvious targets for anti-cancer siRNA-directed therapy. Several examples have been reported in the literature. It was shown that the targeting of the *BCR-ABL* chimeric oncogene, derived from the chromosomal translocation associated with the Philadelphia (Ph') chromosome in chronic myeloid leukemia, resulted in BCR/ABL knockdown and was accompanied by strong induction of *in vitro* apoptotic cell death (Wilda et al. 2002). Likewise, siRNA targeting of an activated *K-RAS* or viral oncogenes have also been proven to have an anti-cancer effect (Brummelkamp et al. 2002; Butz et al. 2003; Sabbioni et al. 2007; Yoshinouchi et al. 2003). These studies proved the efficacy of the approach in specifically targeting activated oncogenes and proved that this approach can be highly selective, a critical element in anti-neoplastic therapeutic intervention. The results from preclinical studies were also promising. For instance, in a mouse model of ovarian cancers, the administration of liposomal delivered siRNA targeting *Eph2* combined with paclitaxel, determined a reduction of the tumor size greater than 50% on either intravenous or intraperitoneal routes of delivery (Landen et al. 2005). In another example, adenovirus-mediated siRNA against a *K-ras* mutated messenger (K-ras codon 12 GGT to GTT) markedly decreased *K-ras* gene expression and inhibited cellular proliferation of lung cancer cells that express the relevant mutation, but produced only minimal growth inhibition on cells that lack the specific abnormality (Zhang et al. 2006a, b). These reports support the idea that siRNAs, alone or combined with chemotherapy, could stimulate a powerful anti-cancer activity.

However, the siRNA cancer therapy is still shadowed by few, but significant issues. The first involves the low bio-availability. Another is represented by "off-target-effects", meaning that, in addition to the complementary target, a specific siRNA can induce the silencing of several other imperfect complementary mRNAs. Finally, a major concern is the stimulation by siRNA duplexes of the innate immune system and the production of high amounts of interferon (for a review see Bumcrot et al. 2006).

miRNAs could have multiple advantages over siRNAs. First, since deregulation of a miRNA may affect multiple pathways, its re-establishment may have a restoring effect on the same multiple pathways, a result that cannot be achieved by the siRNA approach. Second, since miRNAs are normal cellular molecules, the interactions that are produced are likely to be physiological, thus possibly avoiding potential deleterious off-target effects.

Several studies have established the usefulness of miRNA-based therapy in cancer. The induction of apoptosis by *miR-15a* and *miR-16-1* in CLL (Cimmino et al. 2005), the inhibition of growth of cancer cells by *let-7* (Akao et al. 2006; Takamizawa et al. 2004), the reduced migration and invasion capacities induced by *miR-125* in breast cancer cells (Scott et al. 2007) are results that indicate the potential value of miRNA molecules in cancer therapy. However, because of limitations in methods for *in vivo* delivery, up to now there has not been any report of using agents that mimic miRNAs in animal or clinical models.

More successful has been the use of AMOs. The use of anti-*miR-21* AMOs to elicit a pro-apoptotic response in glioblastoma and breast cancer cells are examples (Chan et al. 2005). Significantly, the use of the anti-*miR-21* AMOs increases susceptibility of colangiocarcinoma cells to gemcitabine (Meng et al. 2006), suggesting that miRNA-based therapy may be effectively combined with chemotherapy.

Even the efficiency of *in vivo* delivery of AMOs seems to have found effective solutions through the use of cholesterol-conjugation of "antagomirs" (Krutzfeldt et al. 2005, 2007) or the use of "locked" nucleotides for the production of LNA-modifed oligonucleotides (Chan et al. 2005; Lecellier et al. 2005).

The "antagomir" represents a RNA therapeutic molecule originally designed to inhibit miRNAs (Krutzfeldt et al. 2005, 2007). These are chemically modified and cholesterol-conjugated single-stranded 23-nt RNA molecules complementary to the targeted miRNA. The modifications were introduced to increase the stability of the RNA and protect it from degradation. When intravenously administered to mice, antagomirs against *miR-122* (antago-*miR-122*), a miRNA highly expressed in liver induced a marked, specific and persistent reduction of endogenous miRNA gene. One important aspect is that the silencing of miRNAs by these new agents is followed by measurable effects, for example the decrease in plasma cholesterol levels after antago-*miR-122* administration. The only tissue where antagomirs did not act when injected systemically was the brain, probably due to the difficulty of crossing the blood-brain barrier, but they efficiently target miRNAs when injected locally into the mouse cortex. One clear advantage with respect to siRNA technology is that antagomirs did not induce an immune response.

Mixed DNA/LNA AMOs have been used to inhibit *miR-21*, which results in an increased apoptotic death in glioblastomas cells (Chan et al. 2005). LNA-based oligonucleotides have been shown to be non-toxic at dosages of less than 5 mg/kg/day in mice and these produce anti-tumor effects *in vivo* (Fluiter et al. 2003, 2005).

Recently, studies performed in mouse and African green monkeys models assessed safety and efficacy of the approach (Elmen et al. 2008a, b). Efficient silencing of *miR-122* was achieved by three doses of 10 mg/kg LNA-antimiR, leading to a long-lasting and reversible decrease in total plasma cholesterol without any evidence for associated toxicities or histopathological changes in the liver of the animals. Thus, by proving feasibility, safety and efficacy for the use of AMOs in a pre-clinical setting, these studies established the basis for their use as therapeutic molecules in clinical trials.

The potentiality for therapeutic implementation of small RNAs or AMOs in clinical practice is enormous. The development of animal models will certainly help to establish the role of miRNAs in tumorigenesis and represent useful models for *in vivo* testing of anti-cancer AMOs and miRNAs.

5.7 Concluding Remarks

The discovery of the aberrant expression of miRNAs in human cancer has opened the way to an understanding of a central event in tumorigenesis. Gene expression studies have revealed a large number of deregulated miRNAs, whose biological functions have been partly deciphered to reveal that cancer traits are controlled by miRNAs. Further functional studies and animal models will certainly expand our understanding on their role in normal and disease processes. Moreover, the future will see whether miRNAs will find their place also in the clinical setting. miRNAs signatures useful for cancer prognostic stratification and response to therapy have already been discovered and the use of AMOs as anti-cancer agents represents an exciting new field full of great expectations.

References

Adams BD, Furneaux H, White BA (2007) The micro-ribonucleic acid (miRNA) *miR-206* targets the human estrogen receptor-alpha (ERalpha) and represses ERalpha messenger RNA and protein expression in breast cancer cell lines. Mol Endocrinol 21:1132–1147

Ahituv N, Zhu Y, Visel A et al (2007) Deletion of ultraconserved elements yields viable mice. PLoS Biol 5:e234

Akao Y, Nakagawa Y, Naoe T (2006) *let-7* microRNA functions as a potential growth suppressor in human colon cancer cells. Biol Pharm Bull 29:903–906

Asangani IA, Rasheed SA, Nikolova DA et al (2008) MicroRNA-21 (*miR-21*) post-transcriptionally downregulates tumor suppressor Pdcd4 and stimulates invasion, intravasation and metastasis in colorectal cancer. Oncogene 27:2128–2136

Bandres E, Cubedo E, Agirre X et al (2006) Identification by real-time PCR of 13 mature microRNAs differentially expressed in colorectal cancer and non-tumoral tissues. Mol Cancer 5:29

Baskerville S, Bartel DP (2005) Microarray profiling of microRNAs reveals frequent coexpression with neighboring miRNAs and host genes. RNA 11:241–247

Bemis LT, Chen R, Amato CM et al (2008) MicroRNA-137 targets microphthalmia-associated transcription factor in melanoma cell lines. Cancer Res 68:1362–1368

Benvenuti S, Comoglio PM (2007) The MET receptor tyrosine kinase in invasion and metastasis. J Cell Physiol 213:316–325
Berezikov E, Guryev V, van de Belt J et al (2005) Phylogenetic shadowing and computational identification of human microRNA genes. Cell 120:21–24
Bernards R, Brummelkamp TR, Beijersbergen RL (2006) shRNA libraries and their use in cancer genetics. Nat Methods 3:701–706
Berx G, Raspe E, Christofori G et al (2007) Pre-EMTing metastasis? Recapitulation of morphogenetic processes in cancer. Clin Exp Metastasis 24:587–597
Birchmeier C, Birchmeier W, Gherardi E et al (2003) Met, metastasis, motility and more. Nat Rev Mol Cell Biol 4:915–925
Bommer GT, Gerin I, Feng Y et al (2007) p53-mediated activation of miRNA34 candidate tumor-suppressor genes. Curr Biol 17:1298–1307
Bottoni A, Zatelli MC, Ferracin M et al (2007) Identification of differentially expressed microRNAs by microarray: a possible role for microRNA genes in pituitary adenomas. J Cell Physiol 210:370–377
Brummelkamp TR, Bernards R, Agami R (2002) Stable suppression of tumorigenicity by virus-mediated RNA interference. Cancer Cell 2:243–247
Budhu A, Jia HL, Forgues M et al (2008) Identification of metastasis-related microRNAs in hepatocellular carcinoma. Hepatology 47:897–907
Bumcrot D, Manoharan M, Koteliansky V et al (2006) RNAi therapeutics: a potential new class of pharmaceutical drugs. Nat Chem Biol 2:711–719
Burk U, Schubert J, Wellner U et al (2008) A reciprocal repression between ZEB1 and members of the *miR-200* family promotes EMT and invasion in cancer cells. EMBO Rep 9:582–589
Butz K, Ristriani T, Hengstermann A et al (2003) siRNA targeting of the viral E6 oncogene efficiently kills human papillomavirus-positive cancer cells. Oncogene 22:5938–5945
Cai X, Hagedorn CH, Cullen BR (2004) Human microRNAs are processed from capped, polyadenylated transcripts that can also function as mRNAs. RNA 10:1957–1966
Calin GA, Dumitru CD, Shimizu M et al (2002) Frequent deletions and down-regulation of micro-NA genes *miR15* and *miR16* at 13q14 in chronic lymphocytic leukemia. Proc Natl Acad Sci USA 99:15524–15529
Calin GA, Liu CG, Sevignani C et al (2004a) MicroRNA profiling reveals distinct signatures in B cell chronic lymphocytic leukemias. Proc Natl Acad Sci USA 101:11755–11760
Calin GA, Sevignani C, Dumitru CD et al (2004b) Human microRNA genes are frequently located at fragile sites and genomic regions involved in cancers. Proc Natl Acad Sci USA 101:2999–3004
Calin GA, Ferracin M, Cimmino A et al (2005) A MicroRNA signature associated with prognosis and progression in chronic lymphocytic leukemia. N Engl J Med 353:1793–1801
Calin GA, Croce CM (2006) Genomics of chronic lymphocytic leukemia microRNAs as new players with clinical significance. Semin Oncol 33:167–173
Calin GA, Liu CG, Ferracin M et al (2007) Ultraconserved regions encoding ncRNAs are altered in human leukemias and carcinomas. Cancer Cell 12:215–229
Calin GA, Cimmino A, Fabbri M et al (2008) *miR-15a* and *miR-16-1* cluster functions in human leukemia. Proc Natl Acad Sci USA 105:5166–5171
Camps C, Buffa FM, Colella S et al (2008) *hsa-miR-210* is induced by hypoxia and is an independent prognostic factor in breast cancer. Clin Cancer Res 14:1340–1348
Chan JA, Krichevsky AM, Kosik KS (2005) *MicroRNA-21* is an antiapoptotic factor in human glioblastoma cells. Cancer Res 65:6029–6033
Chen CZ, Li L, Lodish HF et al (2004) MicroRNAs modulate hematopoietic lineage differentiation. Science 303:83–86
Chen C, Ridzon DA, Broomer AJ et al (2005) Real-time quantification of microRNAs by stem-loop RT-PCR. Nucleic Acids Res 33, e179
Ciafre SA, Galardi S, Mangiola A et al (2005) Extensive modulation of a set of microRNAs in primary glioblastoma. Biochem Biophys Res Commun 334:1351–1358
Cimmino A, Calin GA, Fabbri M et al (2005) *miR-15* and *miR-16* induce apoptosis by targeting BCL2. Proc Natl Acad Sci USA 102:13944–13949

Corney DC, Flesken-Nikitin A, Godwin AK et al (2007) *MicroRNA-34b* and *microRNA-34c* are targets of p53 and cooperate in control of cell proliferation and adhesion-independent growth. Cancer Res 67:8433–8438

Corsten MF, Miranda R, Kasmieh R et al (2007) MicroRNA-21 knockdown disrupts glioma growth *in vivo* and displays synergistic cytotoxicity with neural precursor cell delivered S-TRAIL in human gliomas. Cancer Res 67:8994–9000

Cummins JM, He Y, Leary RJ et al (2006) The colorectal microRNAome. Proc Natl Acad Sci USA 103:3687–3692

Datta J, Kutay H, Nasser MW et al (2008) Methylation mediated silencing of MicroRNA-1 gene and its role in hepatocellular carcinogenesis. Cancer Res 68:5049–5058

Dews M, Homayouni A, Yu D et al (2006) Augmentation of tumor angiogenesis by a Myc-activated microRNA cluster. Nat Genet 38:1060–1065

Egle A, Harris AW, Bouillet P et al (2004) Bim is a suppressor of Myc-induced mouse B cell leukemia. Proc Natl Acad Sci USA 101:6164–6169

Eis PS, Tam W, Sun L et al (2005) Accumulation of *miR-155* and BIC RNA in human B cell lymphomas. Proc Natl Acad Sci USA 102:3627–3632

Elmen J, Lindow M, Schutz S et al (2008a) LNA-mediated microRNA silencing in non-human primates. Nature 452:896–899

Elmen J, Lindow M, Silahtaroglu A et al (2008b) Antagonism of microRNA-122 in mice by systemically administered LNA-antimiR leads to up-regulation of a large set of predicted target mRNAs in the liver. Nucleic Acids Res 36:1153–1162

Esau C, Kang X, Peralta E et al (2004) MicroRNA-143 regulates adipocyte differentiation. J Biol Chem 279:52361–52365

Feber A, Xi L, Luketich JD et al (2008) MicroRNA expression profiles of esophageal cancer. J Thorac Cardiovasc Surg 135:255–260

Fire A, Xu S, Montgomery MK et al (1998) Potent and specific genetic interference by double-stranded RNA in Caenorhabditis elegans. Nature 391:806–811

Fluiter K, ten Asbroek AL, de Wissel MB et al (2003) *In vivo* tumor growth inhibition and biodistribution studies of locked nucleic acid (LNA) antisense oligonucleotides. Nucleic Acids Res 31:953–962

Fluiter K, Frieden M, Vreijling J et al (2005) On the *in vitro* and *in vivo* properties of four locked nucleic acid nucleotides incorporated into an anti-H-Ras antisense oligonucleotide. Chembiochem 6:1104–1109

Foekens JA, Sieuwerts AM, Smid M et al (2008) Four miRNAs associated with aggressiveness of lymph node-negative, estrogen receptor-positive human breast cancer. Proc Natl Acad Sci USA 105:13021–13026

Fornari F, Gramantieri L, Ferracin M et al (2008) *miR-221* controls CDKN1C/p57 and CDKN1B/p27 expression in human hepatocellular carcinoma. Oncogene 27:5651–5661

Frankel LB, Christoffersen NR, Jacobsen A et al (2008) Programmed cell death 4 (PDCD4) is an important functional target of the microRNA *miR-21* in breast cancer cells. J Biol Chem 283:1026–1033

Gabriely G, Wurdinger T, Kesari S et al (2008) MicroRNA 21 promotes glioma invasion by targeting matrix metalloproteinase regulators. Mol Cell Biol 28:5369–5380

Galardi S, Mercatelli N, Giorda E et al (2007) *miR-221* and *miR-222* expression affects the proliferation potential of human prostate carcinoma cell lines by targeting p27kip1. J Biol Chem 282:23716–23724

Garzon R, Garofalo M, Martelli MP et al (2008a) Distinctive microRNA signature of acute myeloid leukemia bearing cytoplasmic mutated nucleophosmin. Proc Natl Acad Sci USA 105:3945–3950

Garzon R, Volinia S, Liu CG et al (2008b) MicroRNA signatures associated with cytogenetics and prognosis in acute myeloid leukemia. Blood 111:3183–3189

Gottardo F, Liu CG, Ferracin M et al (2007) Micro-RNA profiling in kidney and bladder cancers. Urol Oncol 25:387–392

Grady WM, Parkin RK, Mitchell PS et al (2008) Epigenetic silencing of the intronic microRNA *hsa-miR-342* and its host gene EVL in colorectal cancer. Oncogene 27:3880–3888

Gramantieri L, Ferracin M, Fornari F et al (2007) Cyclin G1 is a target of *miR-122a*, a microRNA frequently down-regulated in human hepatocellular carcinoma. Cancer Res 67:6092–6099

Gregory PA, Bert AG, Paterson EL et al (2008) The *miR-200* family and *miR-205* regulate epithelial to mesenchymal transition by targeting ZEB1 and SIP1. Nat Cell Biol 10:593–601

Griffiths-Jones S, Grocock RJ, van Dongen S et al (2006) miRBase: microRNA sequences, targets and gene nomenclature. Nucleic Acids Res 34:D140–144

Guimaraes-Sternberg C, Meerson A, Shaked I et al (2006) MicroRNA modulation of megakaryoblast fate involves cholinergic signaling. Leuk Res 30:583–595

Guo Y, Chen Z, Zhang L et al (2008) Distinctive microRNA profiles relating to patient survival in esophageal squamous cell carcinoma. Cancer Res 68:26–33

Han J, Lee Y, Yeom KH et al (2004) The Drosha-DGCR8 complex in primary microRNA processing. Genes Dev 18:3016–3027

Han J, Lee Y, Yeom KH et al (2006) Molecular basis for the recognition of primary microRNAs by the Drosha-DGCR8 complex. Cell 125:887–901

Hanahan D, Weinberg RA (2000) The hallmarks of cancer. Cell 100:57–70

Hayashita Y, Osada H, Tatematsu Y et al (2005) A polycistronic microRNA cluster, *miR-17–92*, is overexpressed in human lung cancers and enhances cell proliferation. Cancer Res 65: 9628–9632

He H, Jazdzewski K, Li W et al (2005a) The role of microRNA genes in papillary thyroid carcinoma. Proc Natl Acad Sci USA 102:19075–19080

He L, Thomson JM, Hemann MT et al (2005b) A microRNA polycistron as a potential human oncogene. Nature 435:828–833

Huang J, Wang F, Argyris E et al (2007) Cellular microRNAs contribute to HIV-1 latency in resting primary CD4+ T lymphocytes. Nat Med 13:1241–1247

Huang Q, Gumireddy K, Schrier M et al (2008a) The microRNAs *miR-373* and *miR-520c* promote tumor invasion and metastasis. Nat Cell Biol 10:202–210

Huang YS, Dai Y, Yu XF et al (2008b) Microarray analysis of microRNA expression in hepatocellular carcinoma and non-tumorous tissues without viral hepatitis. J Gastroenterol Hepatol 23:87–94

Ibarra I, Erlich Y, Muthuswamy SK et al (2007) A role for microRNAs in maintenance of mouse mammary epithelial progenitor cells. Genes Dev 21:3238–3243

Iorio MV, Ferracin M, Liu CG et al (2005) MicroRNA gene expression deregulation in human breast cancer. Cancer Res 65:7065–7070

Iorio MV, Visone R, Di Leva G et al (2007) MicroRNA signatures in human ovarian cancer. Cancer Res 67:8699–8707

Isken F, Steffen B, Merk S et al (2008) Identification of acute myeloid leukemia associated microRNA expression patterns. Br J Haematol 140:153–161

Jiang J, Gusev Y, Aderca I et al (2008) Association of MicroRNA expression in hepatocellular carcinomas with hepatitis infection, cirrhosis, and patient survival. Clin Cancer Res 14:419–427

Johnson SM, Grosshans H, Shingara J et al (2005) *RAS* is regulated by the *let-7* microRNA family. Cell 120:635–647

Jopling CL, Yi M, Lancaster AM et al (2005) Modulation of hepatitis C virus RNA abundance by a liver-specific MicroRNA. Science 309:1577–1581

Kefas B, Godlewski J, Comeau L et al (2008) *microRNA-7* inhibits the epidermal growth factor receptor and the Akt pathway and is down-regulated in glioblastoma. Cancer Res 68:3566–3572

Kim DH, Rossi JJ (2007) Strategies for silencing human disease using RNA interference. Nat Rev Genet 8:173–184

Kim S, Lee UJ, Kim MN et al (2008) MicroRNA *miR-199A** regulates the Met proto-oncogene and the downstream extracellular signal-regulated kinase 2 (ERK2). J Biol Chem 283:18158–18166

Kim TY, Jong HS, Song SH et al (2003) Transcriptional silencing of the DLC-1 tumor suppressor gene by epigenetic mechanism in gastric cancer cells. Oncogene 22:3943–3951

Kim VN, Nam JW (2006) Genomics of microRNA. Trends Genet 22:165–173

Koralov SB, Muljo SA, Galler GR et al (2008) Dicer ablation affects antibody diversity and cell survival in the B lymphocyte lineage. Cell 132:860–874

Korpal M, Lee ES, Hu G et al (2008) The *miR-200* family inhibits epithelial-mesenchymal transition and cancer cell migration by direct targeting of E-cadherin transcriptional repressors ZEB1 and ZEB2. J Biol Chem 283:14910–14914

Krutzfeldt J, Rajewsky N, Braich R et al (2005) Silencing of microRNAs *in vivo* with 'antagomirs'. Nature 438:685–689

Krutzfeldt J, Kuwajima S, Braich R et al (2007) Specificity, duplex degradation and subcellular localization of antagomirs. Nucleic Acids Res 35:2885–2892

Kutay H, Bai S, Datta J et al (2006) Downregulation of *miR-122* in the rodent and human hepatocellular carcinomas. J Cell Biochem 99:671–678

Ladeiro Y, Couchy G, Balabaud C et al (2008) MicroRNA profiling in hepatocellular tumors is associated with clinical features and oncogene/tumor suppressor gene mutations. Hepatology 47:1955–1963

Lagos-Quintana M, Rauhut R, Lendeckel W et al (2001) Identification of novel genes coding for small expressed RNAs. Science 294:853–858

Landen CN Jr, Chavez-Reyes A, Bucana C et al (2005) Therapeutic EphA2 gene targeting *in vivo* using neutral liposomal small interfering RNA delivery. Cancer Res 65:6910–6918

Lau NC, Lim LP, Weinstein EG et al (2001) An abundant class of tiny RNAs with probable regulatory roles in Caenorhabditis elegans. Science 294:858–862

Lawrie CH, Soneji S, Marafioti T et al (2007) MicroRNA expression distinguishes between germinal center B cell-like and activated B cell-like subtypes of diffuse large B cell lymphoma. Int J Cancer 121:1156–1161

Lecellier CH, Dunoyer P, Arar K et al (2005) A cellular microRNA mediates antiviral defense in human cells. Science 308:557–560

Lee DY, Deng Z, Wang CH et al (2007a) MicroRNA-378 promotes cell survival, tumor growth, and angiogenesis by targeting SuFu and Fus-1 expression. Proc Natl Acad Sci USA 104:20350–20355

Lee EJ, Gusev Y, Jiang J et al (2007b) Expression profiling identifies microRNA signature in pancreatic cancer. Int J Cancer 120:1046–1054

Lee RC, Feinbaum RL, Ambros V (1993) The *C. elegans* heterochronic gene *lin-4* encodes small RNAs with antisense complementarity to lin-14. Cell 75:843–854

Lee RC, Ambros V (2001) An extensive class of small RNAs in Caenorhabditis elegans. Science 294:862–864

Lee Y, Ahn C, Han J et al (2003) The nuclear RNase III Drosha initiates microRNA processing. Nature 425:415–419

Lee Y, Kim M, Han J et al (2004) MicroRNA genes are transcribed by RNA polymerase II. EMBO J 23:4051–4060

Lewis BP, Burge CB, Bartel DP (2005) Conserved seed pairing, often flanked by adenosines, indicates that thousands of human genes are microRNA targets. Cell 120:15–20

Lin SL, Chiang A, Chang D et al (2008) Loss of *miR-146a* function in hormone-refractory prostate cancer. RNA 14:417–424

Liu CG, Calin GA, Meloon B et al (2004) Anoligonucleotide microchip for genome-wide microRNA profiling in human and mouse tissues. Proc Natl Acad Sci U S A 101:9740–9744

Lu J, Getz G, Miska EA et al (2005) MicroRNA expression profiles classify human cancers. Nature 435:834–838

Ma L, Teruya-Feldstein J, Weinberg RA (2007) Tumor invasion and metastasis initiated by microRNA-10b in breast cancer. Nature 449:682–688

Marcucci G, Radmacher MD, Maharry K et al (2008) MicroRNA expression in cytogenetically normal acute myeloid leukemia. N Engl J Med 358:1919–1928

Mayr C, Hemann MT, Bartel DP (2007) Disrupting the pairing between *let-7* and Hmga2 enhances oncogenic transformation. Science 315:1576–1579

McManus MT, Sharp PA (2002) Gene silencing in mammals by small interfering RNAs. Nat Rev Genet 3:737–747

Medina R, Zaidi SK, Liu CG et al (2008) MicroRNAs 221 and 222 bypass quiescence and compromise cell survival. Cancer Res 68:2773–2780

Mendell JT (2008) miRiad roles for the *miR-17–92* cluster in development and disease. Cell 133:217–222

Meng F, Henson R, Lang M et al (2006) Involvement of human micro-RNA in growth and response to chemotherapy in human cholangiocarcinoma cell lines. Gastroenterology 130:2113–2129

Meng F, Henson R, Wehbe-Janek H et al (2007) *MicroRNA-21* regulates expression of the PTEN tumor suppressor gene in human hepatocellular cancer. Gastroenterology 133:647–658

Mertens-Talcott SU, Chintharlapalli S, Li X et al (2007) The oncogenic *microRNA-27a* targets genes that regulate specificity protein transcription factors and the G2-M checkpoint in MDA-MB-231 breast cancer cells. Cancer Res 67:11001–11011

Michael MZ, O'Connor SM, van Holst Pellekaan NG et al (2003) Reduced accumulation of specific microRNAs in colorectal neoplasia. Mol Cancer Res 1:882–891

Miller TE, Ghoshal K, Ramaswamy B et al (2008) *MicroRNA-221/222* confers tamoxifen resistance in breast cancer by targeting p27(Kip1). J Biol Chem 283:29897–29903

Murakami Y, Yasuda T, Saigo K et al (2006) Comprehensive analysis of microRNA expression patterns in hepatocellular carcinoma and non-tumorous tissues. Oncogene 25:2537–2545

Naguibneva I, Ameyar-Zazoua M, Polesskaya A et al (2006) The microRNA *miR-181* targets the homeobox protein Hox-A11 during mammalian myoblast differentiation. Nat Cell Biol 8:278–284

Nam EJ, Yoon H, Kim SW et al (2008) MicroRNA expression profiles in serous ovarian carcinoma. Clin Cancer Res 14:2690–2695

O'Donnell KA, Wentzel EA, Zeller KI et al (2005) c-Myc-regulated microRNAs modulate E2F1 expression. Nature 435:839–843

Ota A, Tagawa H, Karnan S et al (2004) Identification and characterization of a novel gene, C13orf25, as a target for 13q31–q32 amplification in malignant lymphoma. Cancer Res 64:3087–3095

Ozen M, Creighton CJ, Ozdemir M et al (2008) Widespread deregulation of microRNA expression in human prostate cancer. Oncogene 27:1788–1793

Pallante P, Visone R, Ferracin M et al (2006) MicroRNA deregulation in human thyroid papillary carcinomas. Endocr Relat Cancer 13:497–508

Pan Q, Luo X, Chegini N (2008) Differential expression of microRNAs in myometrium and leiomyomas and regulation by ovarian steroids. J Cell Mol Med 12:227–240

Park SM, Gaur AB, Lengyel E et al (2008) The *miR-200* family determines the epithelial phenotype of cancer cells by targeting the E-cadherin repressors ZEB1 and ZEB2. Genes Dev 22:894–907

Pasquinelli AE, Reinhart BJ, Slack F et al (2000) Conservation of the sequence and temporal expression of *let-7* heterochronic regulatory RNA. Nature 408:86–89

Pekarsky Y, Santanam U, Cimmino A et al (2006) Tcl1 expression in chronic lymphocytic leukemia is regulated by *miR-29* and *miR-181*. Cancer Res 66:11590–11593

Perkins DO, Jeffries CD, Jarskog LF et al (2007) microRNA expression in the prefrontal cortex of individuals with schizophrenia and schizoaffective disorder. Genome Biol 8:R27

Petrocca F, Visone R, Onelli MR et al (2008) E2F1-regulated microRNAs impair TGFbeta-dependent cell-cycle arrest and apoptosis in gastric cancer. Cancer Cell 13:272–286

Porkka KP, Pfeiffer MJ, Waltering KK et al (2007) MicroRNA expression profiling in prostate cancer. Cancer Res 67:6130–6135

Ramkissoon SH, Mainwaring LA, Ogasawara Y et al (2006) Hematopoietic-specific microRNA expression in human cells. Leuk Res 30:643–647

Raymond CK, Roberts BS, Garrett-Engele P et al (2005) Simple, quantitative primer-extension PCR assay for direct monitoring of microRNAs and shortinterfering RNAs. RNA 11:1737–1744
Raver-Shapira N, Marciano E, Meiri E et al (2007) Transcriptional activation of *miR-34a* contributes to p53-mediated apoptosis. Mol Cell 26:731–743
Reinhart BJ, Slack FJ, Basson M et al (2000) The 21-nucleotide *let-7* RNA regulates developmental timing in *Caenorhabditis elegans*. Nature 403:901–906
Rodriguez A, Griffiths-Jones S, Ashurst JL et al (2004) Identification of mammalian microRNA host genes and transcription units. Genome Res 14:1902–1910
Roldo C, Missiaglia E, Hagan JP et al (2006) MicroRNA expression abnormalities in pancreatic endocrine and acinar tumors are associated with distinctive pathologic features and clinical behavior. J Clin Oncol 24:4677–4684
Ryan DG, Oliveira-Fernandes M, Lavker RM (2006) MicroRNAs of the mammalian eye display distinct and overlapping tissue specificity. Mol Vis 12:1175–1184
Sabbioni S, Callegari E, Spizzo R et al (2007) Anticancer activity of an adenoviral vector expressing short hairpin RNA against BK virus T-ag. Cancer Gene Ther 14:297–305
Schetter AJ, Leung SY, Sohn JJ et al (2008) MicroRNA expression profiles associated with prognosis and therapeutic outcome in colon adenocarcinoma. JAMA 299:425–436
Schmittgen TD, Jiang J, Liu Q et al (2004) A high-throughput method to monitor the expression of microRNA precursors. Nucleic Acids Res 32:e43
Schulte JH, Horn S, Otto T et al (2008) MYCN regulates oncogenic MicroRNAs in neuroblastoma. Int J Cancer 122:699–704
Schultz J, Lorenz P, Gross G et al (2008) MicroRNA *let-7b* targets important cell cycle molecules in malignant melanoma cells and interferes with anchorage-independent growth. Cell Res 18:549–557
Scott GK, Goga A, Bhaumik D et al (2007) Coordinate suppression of ERBB2 and ERBB3 by enforced expression of micro-RNA *miR-125a* or *miR-125b*. J Biol Chem 282: 1479–1486
Sengupta S, den Boon JA, Chen IH et al (2008) MicroRNA 29c is down-regulated in nasopharyngeal carcinomas, up-regulating mRNAs encoding extracellular matrix proteins. Proc Natl Acad Sci USA 105:5874–5878
Spaderna S, Schmalhofer O, Wahlbuhl M et al (2008) The transcriptional repressor ZEB1 promotes metastasis and loss of cell polarity in cancer. Cancer Res 68:537–544
Spurgers KB, Gold DL, Coombes KR et al (2006) Identification of cell cycle regulatory genes as principal targets of p53-mediated transcriptional repression. J Biol Chem 281: 25134–25142
Subramanian S, Lui WO, Lee CH et al (2008) MicroRNA expression signature of human sarcomas. Oncogene 27:2015–2026
Sylvestre Y, De Guire V, Querido E et al (2007) An E2F/*miR-20a* autoregulatory feedback loop. J Biol Chem 282:2135–2143
Takamizawa J, Konishi H, Yanagisawa K et al (2004) Reduced expression of the *let-7* microRNAs in human lung cancers in association with shortened postoperative survival. Cancer Res 64:3753–3756
Tang G (2005) siRNA and miRNA: an insight into RISCs. Trends Biochem Sci 30:106–114
Tanzer A, Stadler PF (2004) Molecular evolution of a microRNA cluster. J Mol Biol 339:327–335
Tarasov V, Jung P, Verdoodt B et al (2007) Differential regulation of microRNAs by p53 revealed by massively parallel sequencing: *miR-34a* is a p53 target that induces apoptosis and G1-arrest. Cell Cycle 6:1586–1593
Tavazoie SF, Alarcon C, Oskarsson T et al (2008) Endogenous human microRNAs that suppress breast cancer metastasis. Nature 451:147–152
Tazawa H, Tsuchiya N, Izumiya M et al (2007) Tumor-suppressive *miR-34a* induces senescence-like growth arrest through modulation of the E2F pathway in human colon cancer cells. Proc Natl Acad Sci USA 104:15472–15477

Tetzlaff MT, Liu A, Xu X et al (2007) Differential expression of miRNAs in papillary thyroid carcinoma compared to multinodular goiter using formalin fixed paraffin embedded tissues. Endocr Pathol 18:163–173

Vasudevan S, Tong Y, Steitz JA (2007) Switching from repression to activation: microRNAs can up-regulate translation. Science 318:1931–1934

Visone R, Pallante P, Vecchione A et al (2007) Specific microRNAs are downregulated in human thyroid anaplastic carcinomas. Oncogene 26:7590–7595

Volinia S, Calin GA, Liu CG et al (2006) A microRNA expression signature of human solid tumors defines cancer gene targets. Proc Natl Acad Sci USA 103:2257–2261

Voorhoeve PM, le Sage C, Schrier M et al (2006) A genetic screen implicates miRNA-372 and miRNA-373 as oncogenes in testicular germ cell tumors. Cell 124:1169–1181

Vousden KH, Lane DP (2007) p53 in health and disease. Nat Rev Mol Cell Biol 8:275–283

Wang T, Zhang X, Obijuru L et al (2007) A micro-RNA signature associated with race, tumor size, and target gene activity in human uterine leiomyomas. Genes Chromosomes Cancer 46:336–347

Wang Y, Lee AT, Ma JZ et al (2008) Profiling microRNA expression in hepatocellular carcinoma reveals microRNA-224 up-regulation and apoptosis inhibitor-5 as a microRNA-224-specific target. J Biol Chem 283:13205–13215

Weber F, Teresi RE, Broelsch CE et al (2006) A limited set of human microRNA is deregulated in follicular thyroid carcinoma. J Clin Endocrinol Metab 91:3584–3591

Weiss GJ, Bemis LT, Nakajima E et al (2008) EGFR regulation by microRNA in lung cancer: correlation with clinical response and survival to gefitinib and EGFR expression in cell lines. Ann Oncol 19:1053–1059

Welch C, Chen Y, Stallings RL (2007) MicroRNA-34a functions as a potential tumor suppressor by inducing apoptosis in neuroblastoma cells. Oncogene 26:5017–5022

Wightman B, Ha I, Ruvkun G (1993) Posttranscriptional regulation of the heterochronic gene lin-14 by *lin-4* mediates temporal pattern formation in *C. elegans*. Cell 75:855–862

Wilda M, Fuchs U, Wossmann W et al (2002) Killing of leukemic cells with a BCR/ABL fusion gene by RNA interference (RNAi). Oncogene 21:5716–5724

Wirth T, Zender L, Schulte B et al (2003) A telomerase-dependent conditionally replicating adenovirus for selective treatment of cancer. Cancer Res 63:3181–3188

Wong QW, Lung RW, Law PT et al (2008a) MicroRNA-223 is commonly repressed in hepatocellular carcinoma and potentiates expression of Stathmin1. Gastroenterology 135:257–269

Wong TS, Liu XB, Wong BY et al (2008b) Mature *miR-184* as potential oncogenic microRNA of squamous cell carcinoma of tongue. Clin Cancer Res 14:2588–2592

Xia L, Zhang D, Du R et al (2008) *miR-15b* and *miR-16* modulate multidrug resistance by targeting BCL2 in human gastric cancer cells. Int J Cancer 123:372–379

Xiao C, Srinivasan L, Calado DP et al (2008) Lymphoproliferative disease and autoimmunity in mice with increased *miR-17–92* expression in lymphocytes. Nat Immunol 9:405–414

Yanaihara N, Caplen N, Bowman E et al (2006) Unique microRNA molecular profiles in lung cancer diagnosis and prognosis. Cancer Cell 9:189–198

Yang H, Kong W, He L et al (2008) MicroRNA expression profiling in human ovarian cancer: *miR-214* induces cell survival and cisplatin resistance by targeting PTEN. Cancer Res 68:425–433

Yoshinouchi M, Yamada T, Kizaki M et al (2003) *In vitro* and *in vivo* growth suppression of human papillomavirus 16-positive cervical cancer cells by E6 siRNA. Mol Ther 8:762–768

Yu F, Yao H, Zhu P et al (2007) *let-7* regulates self renewal and tumorigenicity of breast cancer cells. Cell 131:1109–1123

Yu SL, Chen HY, Chang GC et al (2008) MicroRNA signature predicts survival and relapse in lung cancer. Cancer Cell 13:48–57

Zhang L, Huang J, Yang N et al (2006a) microRNAs exhibit high frequency genomic alterations in human cancer. Proc Natl Acad Sci USA 103:9136–9141

Zhang L, Volinia S, Bonome T et al (2008) Genomic and epigenetic alterations deregulate microRNA expression in human epithelial ovarian cancer. Proc Natl Acad Sci USA 105:7004–7009

Zhang Z, Jiang G, Yang F et al (2006b) Knockdown of mutant *K-ras* expression by adenovirus-mediated siRNA inhibits the *in vitro* and *in vivo* growth of lung cancer cells. Cancer Biol Ther 5:1481–1486

Zhu S, Si ML, Wu H et al (2007) MicroRNA-21 targets the tumor suppressor gene tropomyosin 1 (TPM1). J Biol Chem 282:14328–14336

Zhu S, Wu H, Wu F et al (2008) MicroRNA-21 targets tumor suppressor genes in invasion and metastasis. Cell Res 18:350–359

Chapter 6
Functional Proteomics in Oncology: A Focus on Antibody Array-Based Technologies

Marta Sanchez-Carbayo

Abstract Protein–protein interactions, post-translational modifications, and interaction between protein and DNA or RNA can all shift the activity of a protein from what would have been predicted by its level of transcription. Functional proteomics studies the interaction of proteins within their cellular environment to determine how a given protein accomplishes its specific cellular task. Accordingly, the promise of functional proteomics is that by chronicling the function of aberrant or over-expressed proteins, it will be possible to characterize the mechanism of the disease-sustaining proteins. The further understanding of the disease networks will lead to targeted cancer therapy and specific biomarkers for diagnosis, prognosis or therapeutic response prediction based on disease specific proteins. In the context of other proteomic technologies, targeted antibody arrays are strongly contributing for functional proteomics analyses. This chapter describes how such strategies reported to date that may assist in the diagnosis, surveillance, prognosis, and potentially for predictive and therapeutic purposes for patients affected with solid and haematological neoplasias.

6.1 Functional Proteomics in Oncology: Concepts

Cancer can be described as a genetic disease, driven by the multistep accumulation of genetic and epigenetic factors. These molecular alterations result in uncontrolled cellular proliferation, cell cycle deregulation, decrease in cell death or apoptosis, blockage of differentiation, invasion, and metastatic spread. The particular genetic and protein expression alterations that occur as part of the crosstalk between these pathways, will in great part determine the biological behavior of the tumor including

M. Sanchez-Carbayo (✉)
Tumor Markers Group, Spanish National Cancer Research Center, 310A, Melchor Fernandez Almagro 3, E-28029, Madrid, Spain
e-mail: mscarbayo@cnio.es

W.C.S. Cho (ed.), *An Omics Perspective on Cancer Research*,
DOI 10.1007/978-90-481-2675-0_6, © Springer Science+Business Media B.V. 2010

its ability to grow, recur, progress and metastasize. The advent of high-throughput methods of molecular analysis can comprehensively survey the genetic and protein profiles characteristic of distinct tumor types and identify targets and pathways that may underlie a particular clinical behavior. The driving force behind oncoproteomics is the belief that certain protein signatures or patterns are associated with a particular malignancy and clinical behavior. If so, the correlation of clinical parameters with defined protein expression patterns that reflect the mutated genetic program that caused or was involved in cancer progression, would allow tumor stratification, predict disease progression and even define improved tailored therapeutic modalities. The technological challenges to achieve these goals are significant since the human proteome is not defined. One potential solution to finding cancer-associated protein signatures is functional proteomic antibody array-based techniques.

While the amino acid sequence of a protein is uniquely determined by a nucleotide sequence, the genetic code of a protein is not a complete predictor of the function of a protein. Many *in vivo* factors can alter the activity level or function of a protein as cells are influenced by a complex system of communication with other cells and factors in their microenvironment. Protein–protein interactions, post-translational modifications, and interaction between protein and DNA or RNA can all shift the activity of a protein from what would have been predicted by its level of transcription. Functional proteomics studies the interaction of proteins within their cellular environment to determine how a given protein accomplishes its specific cellular task. Accordingly the promise of functional proteomics is that by chronicling the function of aberrant or over-expressed proteins, it will be possible to characterize the mechanism of the disease-sustaining proteins. The further understanding of the disease networks will lead to targeted cancer therapy and specific biomarkers for diagnosis, prognosis or therapeutic response prediction based on disease specific proteins. In addition, the response of proteins to molecular targeted therapy could be monitored to determine the efficacy of the targeted therapy and potential viable future therapies involving the same protein pathway (Azad et al. 2006).

Several high-throughput techniques are available today for functional proteomics. These techniques can be applied to *in vitro*, *in vivo* and clinical samples to further characterize protein functions in a multiplexed manner. Immunocapture through immunoblotting, precipitation, and histochemistry and protein and tissue microarrays are tools usually applied to clinical samples (tissue and body fluids). Immunoprecipitation can identify interactions between proteins and can be applied if the clinical sample is of adequate size and stability. Unknown partner proteins in a multiprotein complex can be identified using SDS-PAGE followed by mass spectrometry (MS) analysis and peptide mass fingerprinting as is done routinely for non-clinical samples. MS cannot only provide sequence from which to identify the protein, it is precise enough to detect co- and post-translational changes such as phosphorylation, glycosylation, acetylation, and alternate cleavage sites. In this chapter, antibody array-based technologies will be described for functional proteomic analyses.

6.2 Antibody Array-Based Techniques in the Context of Other Functional Proteomic Approaches

It is important to correctly classify antibody array-based targeted proteomic approaches in the context of other strategies that may be undertaken to investigate the cancer proteome for functional analyses of the proteins under study (Haab et al. 2001; Kingsmore 2006; Chan et al. 2004; Angenendt et al. 2002; Kopf and Zharhary 2007; Sanchez-Carbayo 2006; Borrebaeck and Wingren 2007). The terminology of untargeted and targeted proteomics refers to whether the proteins to be measured are known and considered in the experimental design (targeted) and the number of proteins that can be detected and characterized (decided at front in targeted approaches). Untargeted platforms such as two-dimensional electrophoresis (2D) and mass spectrometry are best suited for first pass comparisons of proteomes unknown at front in the experimental design to identify relatively few, novel and known proteins that may exhibit the greatest differences in abundance. These techniques in their low- and high-resolution versions were initially considered the mainstay or standard of proteomic technologies (Kopf and Zharhary 2007; Sanchez-Carbayo 2006; Borrebaeck and Wingren 2007). Targeted platforms measure and quantify known proteins of interest identified previously, and are suited for analyses of quantitative differences in abundance among known protein families and pathways. Tissue arrays and multiplexed western blots are considered targeted proteomic approaches (Borrebaeck and Wingren 2007). However, antibody and protein microarrays are considered the main targeted techniques used for large-scale analysis of many samples and known proteins. These two latter represent the most versatile among the proteomics techniques available to date, since antigens, peptide, complex protein solutions or antibodies can be immobilized to capture and quantify the presence of specific either proteins or antibodies, respectively (Kopf and Zharhary 2007; Sanchez-Carbayo 2006; Borrebaeck and Wingren 2007). Immobilization of proteins either as purified or phage-displayed protein versions or in format of complex protein solutions have led to tumor-associated antigen (TAAs) or reverse-phase arrays (Wang et al. 2005; Anderson and Labaer 2005; Nishizuka et al. 2003; Petricoin et al. 2005). TAAs arrays utilized on serum specimens enhance the detection of autoantibodies against TAAs, which are being utilized for cancer diagnosis and patient outcome stratification and the characterization of protein–antibody interactions. The rationale of TAAs arrays in clinical practice is related to the presence in the cancer sera of antibodies which react with a unique group of autologous cellular antigens or TAAs (Wang et al. 2005; Anderson and Labaer 2005). Complex protein extracts can also be spotted onto membranes and probed with antibodies targeting specific proteins and pathways on the so-called reverse-phase arrays (Nishizuka et al. 2003; Petricoin et al. 2005). Overall, the versatility of targeted platforms allows controlling and estimating the reproducibility, scalability and precise antibody and protein quantification, leading to high sensitivity and coverage. One of the major advantages of the antibody arrays approach is that it allows experimental designs to address specific hypothesis, and biological

Table 6.1 Main characteristics of array-based proteomic techniques

Technique	Printed molecule	Pitfalls	Most frequent application
Antibody (forward-phase) arrays	Highly specific antibodies	Availability of antibodies Cross-reactivity	Protein profiling Biomarker discovery Signaling Post-translational modifications
Bead-based multiplexed arrays	Antibodies coating differentially identifiable beads	Degree of multiplexing limited by number of differentially identifiable beads	Protein profiling Cytokine Signaling Biomarker discovery
Reverse-phase arrays	Lysate protein extracts	Limited number of analytes analyzed even with multisectored slides Crossreactivity	Protein profiling Biomarker discovery Signaling Post-translational modifications
Antigen arrays	Purified proteins and peptides	Significance of Autoabs in progression is controversial	Antibody profiling Immune response evaluation Biomarker discovery

interpretation of the results obtained, making them critical for functional proteomic analyses in oncology. However, the number of proteins amenable for these analyses depends on the availability of antibodies with high affinity and specificity to bind a target protein (Kopf and Zharhary 2007; Sanchez-Carbayo 2006; Borrebaeck and Wingren 2007). Because of the little overlap between studies conducted with targeted and untargeted approaches using the same specimens, confirmation of the advantages and pitfalls of these types of high-throughput technologies for the functional proteomics remains an elusive goal. Overall, any of these proteomic strategies are impacting on functional proteomic studies and the discovery of cancer specific candidates (Table 6.1). In this chapter, these proteomic technologies have only been summarized to set up the main differences among them.

6.3 Antibody Array Formats

6.3.1 Current Formats

Depending on whether the antibodies are immobilized on a planar or spherical surface, antibody arrays have been classified into planar and suspension/bead formats, respectively (Fig. 6.1). Innovation in the immobilization surfaces and detection strategies are leading to an increasing number of planar arrays and bead-based antibody array technologies. Planar antibody arrays represent the most versatile

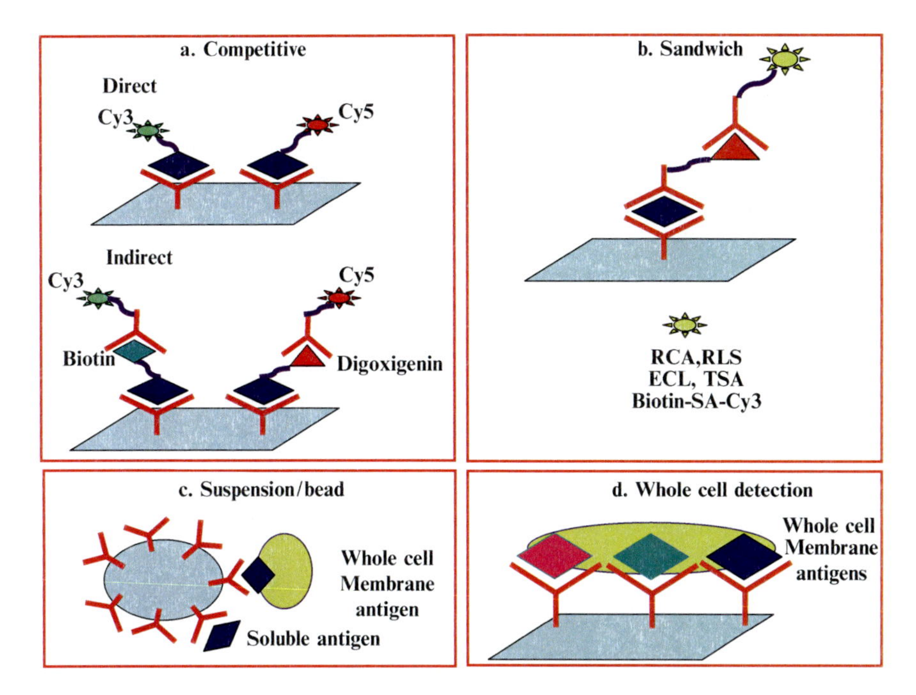

Fig. 6.1 Main formats of planar and suspension antibody arrays. RCA: rolling circle amplification; RLS: resonance light scattering; ECL: enhanced chemiluminescence; TSA: tyramide signal amplification; SA: streptavidin

type, as shown along the clinical applications presented for the discovery of targets, functional networks and biomarker candidates below. The main planar label-based formats comprise one- antibody and sandwich assays. One-antibody and sandwich assays present advantages and pitfalls over each other. In both formats, the target protein is always captured by one (or more) immobilized "capture" antibody in the array. In one-antibody label-based assays, the targeted proteins are detected through labelling with a tag. In sandwich assays, a second not-immobilized "detection" antibody interacts with a different epitope for a given monomeric protein enabling detection by forming a "sandwich" (Fig. 6.1). In the direct labelling, proteins are labelled with a fluorophore (including cyanines such as Cy3 or Cy5). In the indirect labelling, proteins are labelled with a tag that is later detected by a labelled antibody (Sanchez-Carbayo 2006). By multiplexing with different fluorescent labels for each sample, one-antibody label-based assays may allow the incubation of more than one sample simultaneously. These assays can be designed to be competitive if the analytes belonging to the co-incubated test and reference solutions compete for binding at the antibodies. The competition in one antibody (two-colour) assays is ratiometric and does not imply that the analytes are saturating the antibodies. This competition has been suggested to lead to improvements in linearity of response and dynamic range as compared to non-competitive assays (Angenendt et al. 2002; Kopf and Zharhary 2007; Sanchez-Carbayo 2006; Borrebaeck and Wingren 2007).

The main disadvantage is related to the potential disruption of the analyte–antigen interaction by the label, which may also limit the detection, as well as the sensitivity and specificity.

In the sandwich label-based format, immobilized "capture" antibodies capture unlabeled proteins, which are detected by another "detection" antibody using several methods to generate the signal for detection (Fig. 6.1b). The use of these two "capture" and "detection" antibodies against different epitopes of a given analyte increases the specificity for the target protein to be measured as compared to label-based assays. The reduced background of these assays increases also the sensitivity. The sandwich format allows only non-competitive assays, since only one sample can be incubated on each array (Haab et al. 2002; Kingsmore 2006; Chan et al. 2004; Angenendt et al. 2002; Kopf and Zharhary 2007; Sanchez-Carbayo 2006; Borrebaeck and Wingren 2007). This format requires standard curves of known concentrations of analytes to achieve accurate calibration of concentrations. As compared to label-based assays, sandwich arrays are more difficult to develop in a multiplexed manner since matched pairs of antibodies and purified antigens may not be available for each target, and the potential cross-reactivity among detection antibodies increasing with additional analytes. The practical size of multiplexed sandwich assays limits to 30–50 different targets (Kopf and Zharhary 2007; Sanchez-Carbayo 2006; Borrebaeck and Wingren 2007). This contrasts with one-antibody assays where only availability of antibodies and space on the substrate limit the number of targets analyzed.

Proteins in suspension can then be detected using bead/suspension arrays (Fig. 6.1c) (Lash et al. 2006; De Jager and Rijkers 2006; Waterboer et al. 2006). These arrays use different fluorescent beads, each coated with a different antibody and spectrally resolvable from each other (Lash et al. 2006; De Jager and Rijkers 2006; Waterboer et al. 2006). The beads are incubated with a sample to allow protein binding to the capture antibodies, and the mixture is incubated with a cocktail of detection antibodies, each corresponding to one of the capture antibodies. The detection antibodies are tagged to allow fluorescent detection. The beads are passed through a flow cytometer system, and each bead is probed by two lasers, one to read to the colour, or identity of the beam and another to read the amount of detection antibody on the bead (Lash et al. 2006; De Jager and Rijkers 2006; Waterboer et al. 2006). Multiplexed bead-based flow-cytometry assays represent an active area of development. Differentially identifiable beads coated with either proteins, autoantigens or antibodies can identify a variety of bound antibodies or proteins using a flow cytometer system (Lash et al. 2006; De Jager and Rijkers 2006; Waterboer et al. 2006). Other antibody array approaches have been developed as modifications of the one-antibody and sandwich label-based arrays. These alternate strategies allow detection of proteins on whole cells without protein isolation (Fig. 6.1c, d) (Kingsmore 2006; Chan et al. 2004). Advances in instrumentation and bead chemistries are making this approach very valuable for the detection of circulating cancer cells. As another version of this concept, suspensions of cells can be incubated on antibody arrays, and the amount of cells that bound each antibody can be quantified by dark field microscopy. These arrays have the potential of characterizing multiple membrane proteins in specific cell populations or changes in cell surfaces induced by drug therapies (Belov et al. 2001).

6.3.2 Emerging Formats for Functional Proteomics

Several examples can be provided to delineate recent remarkable innovations achieved to monitor specific post-translational modifications as well as to increase the limits of detection or enable the technology to profile protein extracts obtained from very few individual cells. In a first example, antibody arrays are adapted to detect differences in the content of glycans (sugars or carbohydrates) of proteins. These carbohydrate post-translational modifications on proteins are known to be important determinants of protein function in both normal and disease biology. Antibody array designs have been developed to allow efficient, multiplexed study of glycans on individual proteins from complex mixtures (Dotan et al. 2006; Chen et al. 2007). Once multiple proteins are captured using antibody microarrays, these post-translational modifications can be detected using lectins or glycan-binding antibodies (Chen et al. 2007). In pancreatic cancer, profiling of both protein and glycan variation in multiple serum samples using parallel sandwich and glycan-detection assays, has identified the cancer-associated glycan alteration on proteins in the serum of pancreatic cancer patients (Chen et al. 2007). These antibody arrays for glycan detection are opening a novel field of glycobiology research in the context of neoplastic diseases for functional proteomics and the discovery of potential targets and cancer biomarker candidates.

High sensitivity, in the femtomolar range, allowing protein quantification from limited sample quantities (only six cells) can be achieved by the so-called antibody "ultramicroarrays" (Nettikadan et al. 2006). These arrays were initially tested for the detection of interleukin-6 (IL-6) and prostate specific antigen (PSA), finding detection levels using purified proteins in the attomolar range (Nettikadan et al. 2006). Remarkably, this strategy should enable proteomic analysis of clinical specimens available in very limited quantities such as those collected by laser capture microdissection.

Another critical technical development that is being applied to antibody arrays increasing the limits of detection is quantum dot technology. By offering remarkable photostability and brightness and low photobleaching, quantum dots allow detection of proteins in biological specimens (serum, plasma, body fluids) at pg/ml concentration, as has been shown to detect several cytokines (Zajac et al. 2007). Models of quantum dot probes include conjugation of nanocrystals to antibody specific to selected markers and the use of streptavidin (SA) coated quantum dots and biotinylated detector antibody (Zajac et al. 2007). By allowing monitoring of changes in protein concentration in physiological range in body fluids, the methodology can potentially be applied to other types of planar and suspension arrays.

Another technical innovation allowing detection of proteins at picomolar concentrations utilizes surface plasmon resonance imaging (SPRI) measurements of

RNA aptamer microarrays. The adsorption of proteins onto the RNA microarray is detected by the formation of a surface aptamer-protein-antibody complex. The SPRI response signal is then amplified using a localized precipitation reaction catalyzed by the enzyme horseradish peroxidase that is conjugated to the antibody. This enzymatically amplified SPRI methodology has initially been characterized for the detection of human thrombin at the fM concentration range. The appropriate thrombin aptamer for the sandwich assay can be identified from a microarray using several potential thrombin aptamer candidates. The SPRI method has also been optimized to detect the protein vascular endothelial growth factor (VEGF) at a biologically relevant pM concentration. This incipient technology shows a potential for increasing this sensitivity for detecting proteins in body fluids (Li et al. 2007). The sensitivity achieved for VEGF allows its measurement in the serum for selecting or monitoring antiangiogenic therapies for breast, lung or colorectal cancer. In the same line of research, an independent study using a 17-multiplexed photoaptamer-based array has exhibited limits of detection below 10 fM for several analytes including the VEGF and endostatin, among others in serum samples. Since photoaptamers covalently bind to their target analytes before fluorescent signal detection, the arrays can be vigorously washed to remove background proteins, providing the potential for superior signal-to-noise ratios and lower limits of quantification in biological matrices. Interestingly, the affinity of the capture reagent can be directly correlated to the limit of detection for the analyte on the array (Bock et al. 2004).

6.4 Strategies and Applications of Antibody Arrays for Functional Proteomics

The increasing number of strategies of antibody arrays is improving, emerging and challenging in their applications in functional proteomic research. Significant contributions of proteomics research using antibody arrays reported to date have derived from a wide spectrum of experimental designs using different specimens varying from cells, tissues and bodily fluids. Representative examples of these strategies comprising from single experiments to comparison of relatively low or medium size datasets obtained under different conditions (e.g. normal, preneoplastic, inflammation, cancer) are described in this section.

6.4.1 Cell Culture

Protein profiling studies of cultured cells using antibody arrays are allowing in-depth analyses of cancer biology. Since many of these cancer cells derived from human tumors, they resemble human disease and may also lead to functional proteomic analyses depending on the experimental design. The use of antibody arrays for high-throughput profiling of cultured cells is also useful to evaluate signaling pathways including tyrosine kinases networks (Gembitsky et al. 2004). Antibody arrays can profile enzyme activities using both protein extracts and cell culture supernatants. As compared to initial gel-based strategies utilized to assess the functional state of enzymes, they represent a convenient platform to evaluate activity-based protein profiling with high sensitivity and specificity and reducing sample consumption. While gel based strategies basically enable protein discovery, antibody microarrays may define new patterns of expression of known proteins. The presence of phosphorylated and unphosphorylated forms of proteins can be assessed in cell cultured systems using antibody arrays if adequate antibodies are available to address the specific posttranslational modifications of the target proteins under study (Gembitsky et al. 2004; Ivanov et al. 2004).

Cytokine profiles of cell lysates have also been analyzed by means of cytokine arrays and compared to those obtained on body fluids and tissue extracts (Lin et al. 2003a, b). Commercially available cytokine arrays have been applied to conditioned media of cancer cells to dissect functional cytokine secreted signatures associated to the over-expression of critical breast cancer target genes in breast cancer cells. This strategy revealed that the enhanced synthesis and secretion of members of the IL-8 chemokine family may represent a new pathway involved in the metastatic progression and endocrine resistance of HER2-over-expressing breast carcinomas (Vazquez-Martin et al. 2007). Not only this *in vitro* strategy served to identify a potentially relevant signalling pathway but also identified a cancer protein specific signature with clinical applications.

An independent study has screened the native cytokine expression patterns in human breast cancer cell lines associated to the expression of the estrogen receptor (ER) using cytokine arrays. ER positive cells expressed low levels of IL-8 whereas ER negative cells expressed high levels of IL-8. Such profiling served to monitor functional analyses blocking IL-8-mediated tumor cell invasion and angiogenesis using a neutralizing antibody against IL-8 as well as the exogenous over-expression of this gene, which substantially inhibited IL-8 expression. The combination of several *in vitro* strategies monitored by cytokine profiling using antibody arrays served to link the functional role of IL-8 in the development and progression of human breast cancer in association with the ER status (Lin et al. 2003a, b, 2004).

Cytokine profiles of cell supernatants of other tumor types such as the Jurkat (T-cell leukemia), and the A549 (non-small cell lung cancer) cells have also been monitored by means of cytokine arrays. The cytokine/chemokine response was evoked after cell stimulation with tumor necrosis factor alpha (TNF-alpha), phorbol-12-myristate-13-acetate, and phytohaemagglutinin. Stimulated cells showed an increase in the expression level of many of the 41 test analytes, including IL-8 and TNF-alpha in the treated cells (Garcia et al. 2007). This strategy shows the ability of antibody array analysis of cell-culture supernatants for the functional profiling of the release of cellular inflammatory mediators.

Antibody arrays can also be utilized for functional proteomics to monitor signal transduction mediated by complex networks of interacting proteins in mammalian cells and screen cancer drugs. Microarrays created in 96-well microtiter plate formats multiplexed the measurement of amounts and modification states of signal transduction proteins in crude cell lysates. These arrays have been applied to monitor the activation, uptake, and signaling of ErbB receptor tyrosine kinases in cancer cells. This strategy was used to characterize thee action of the epidermal growth factor receptor inhibitor PD153035 on cells (Nielsen et al. 2003). Thus, the integration of microplate and microarray methods for crude cell lysates may identify small molecules with specific inhibitory profiles against specific signaling networks. The technology is yielding comprehensive information about the mechanism of action and the efficacy of existing and novel cancer compounds in preclinical studies for the treatment of human cancer.

An interesting strategy reported on leukemias and lymphomas cells has allowed biological immunophenotyping by means of a "DotScan" antibody array, where these cells are incubated and captured based on their membrane protein expression patterns. The antibody array was initially with a set of 88 immobilized antibodies and later on using a higher number of 147 antibodies (Belov et al. 2001, 2006). Interestingly, a high number of leukemias and lymphoma cells, as well as clinical samples were analyzed and classified by their surface protein profiles (Belov et al. 2006). The relevance of these strategies relies on the possibility of antibody arrays to capture cells based on the protein expression pattern of surface proteins. Moreover this approach might potentially lead to a molecular classification of human blood malignancies. Thus, cell binding assays on antibody arrays might permit the rapid immunophenotyping of human living cells. The throughput of the analysis, however, is still limited due to the ability to perform parallel and quantitative detection of cells captured on the array. This limitation can be addressed using imaging techniques based on surface plasmon resonance (SPR). In addition to monitoring capture of proteins on antibody microarrays, SPR is being optimized for cell capture allowing more sophisticated functional proteomic analyses (Kato et al. 2007).

Protein extracts from cancer cells have been utilized to optimize and develop technological innovations of antibody arrays. The availability of cultured material allows reproducibility analyses and testing the analytical properties of a given novel innovation in antibody array technologies. Remarkably, dilution analyses varying from 100 to 4 prostatic LNCaP cells served to optimized ultramicroarrays that allow reproducible protein detection from the lysate of an average of just six cells of two known serum proteins such as secreted IL6 or PSA (Nettikadan et al. 2006). By resembling human disease, this sensitivity improvement using cultured cells suggests the potential utility of the protein quantification of these molecules or others in human body fluids or protein extracts of laser microdissected neoplastic prostatic populations (Nettikadan et al. 2006). Another example of technology optimization establishing limiting factors of labelling methods using protein extracts of breast cancer cells dealt with competitive binding assays under different conditions with one-colour or two-colour fluorescence detection methods. These analyses revealed that antibody cross-reactivity, target protein truncation and abundance, as well as the cellular compartment of origin are major factors that affect protein profiling on antibody arrays (Yeretssian et al. 2005).

Protein extracts from murine and human lung cell lines have served to identify protein phosphatase 1 (PP1) interacting proteins (PIP) that are important in cell proliferation and cell survival by means of antibody arrays. This comparative approach identified 31 potential novel PIPs and confirmed 11 of 17 well-known PIPs included as controls. Interestingly, validation analysed by co-immunoprecipitation confirmed that nine of these proteins associated with PP1. By exposing these cells to nicotine, the association of PP1 with these proteins could be modulated. Thus, novel interactions with PP1 were identified and were consisting with the PP1 role at facilitating cell cycle arrest and/or apoptosis (Flores-Delgado et al. 2007). The important observation is that protein profiling using carefully selected antibodies served to design functional analyses to characterize the relevance of the post-translational modifications of these proteins along cell cycle.

6.4.2 Tissue Specimens

It is also feasible to characterize functional proteomic profiles of protein extracts of tissue specimens using antibody arrays. By comparing malignant and normal counterparts it is possible to identify differentially expressed proteins associated with disease progression. This strategy has been performed in lung cancer comparing tumor samples from patients with squamous cell lung carcinoma and normal lung tissue controls with a high number of antibodies printed on antibody arrays (Bartling et al. 2005). Among the differentially expressed proteins, up-regulated proteins were shown to correlate with a high mRNA expression obtained from paired gene microarray data. Thus, using a tumor profiling strategy, antibody microarrays served to identify functional networks and biomarker candidates in lung cancer (Bartling et al. 2005). In line with this strategy, it is possible to characterize protein profiles of neoplastic subpopulations obtained from frozen resected tumor specimens using laser capture microdissection (Hudelist et al. 2004). Microdissection is especially critical for data interpretation in heterogeneous tumors such as breast or prostatic cancer. For example, profiling of protein extracts of breast tumor *versus* the adjacent normal breast tissue identified a number of proteins with increased expression levels in malignant specimens such as casein kinase Ie, p53 or annexin XI. Decreased expressed proteins in the malignant tissue included the multifunctional regulator 14-3-3. Immunohistochemistry in paraffin-embedded normal and malignant sections deriving from the same patient using antibodies against these proteins served to validate the data obtained using the antibody microarrays (Hudelist et al. 2004). In this exercise, protein profiling of a single neoplastic patient using a commercially available microarray served to identify molecular functional determinants of cancer progression in breast cancer. It seems reasonable to insist on that the clinical validation with high number of specimens on independent sets of clinical material is critical to verify the clinical significance of cancer-specific discovery analyses.

The results of protein profiling of tumor protein extracts using antibody arrays can be validated in several manners in order to confirm that potential identified functional networks and biomarker candidates are cancer specific. On one hand, gene profiling of matched tumors can prove that the increased protein expression is associated with increased transcript profiles (Bartling et al. 2005). At the protein level, it can also be tested that the differential expression of proteins can be detected using an independent method such as immunoblotting (Bartling et al. 2005). Clinical validation of differential protein expression patterns can be confirmed by immunohistochemistry using the same antibodies that were printed on the antibody arrays on paraffin-embedded normal and malignant tissues providing high reliability on the results found by protein profiling. If tissue arrays with well-characterized independent set of tumors are available, it is possible to evaluate clinico-pathological correlations of novel cancer-specific proteins with tumor stratification, disease progression and clinical outcome (Sanchez-Carbayo 2006).

The use of comprehensive gene profiling analyses using tissue material can identify tumor targets relevant of specific neoplasias for antibody arrays design. Such approach can be applied in antibody-based proteomics to generate protein-specific affinity antibodies to functionally explore the human proteome. Specific protein epitope signature tags (PrEST) can be identified and used to raise mono-specific, polyclonal antibodies, and be subsequently analyzed on paraffin-embedded sections of malignant and normal tissue. Genome-based, affinity proteomics, using PrEST-induced antibodies, is an efficient way to rapidly identify a number of disease-associated protein candidates of previously both known and unknown identity (Ek et al. 2006). A descriptive and comprehensive protein atlas for tissue distribution and subcellular localization of human proteins in both normal and cancer tissues is being created (Uhlén et al. 2005). The subsequent antibodies generated can be used for analysis of corresponding proteins in a wide range of assay platforms, including: i) immunohistochemistry for detailed tissue profiling; ii) specific affinity reagents for various functional protein assays; and iii) capture ("pull-down") reagents for purification of specific proteins and their associated complexes for structural and biochemical analyses (Uhlén et al. 2005).

A critical part in functional proteomics research deals with optimization of sample preparation for comprehensive protein measurements. Proteases inhibitors can be added in order to overcome accelerated protein degradation due to the presence of secreted proteases. Novel tissue sample handling approaches to enrich (> 95% purity) of epithelial cells from fresh human tissue samples include the use of an epithelial cell surface antibody. This purification method showed several advantages for proteomic analyses on tissue specimens since a large quantity of cells available for downstream analysis were available and it showed an increased reproducibility (Kellner et al. 2004). Flow cytometry, sorting analyses, pulldowns of protein extracts, or spectrometry techniques represent alternative approaches to enrich cell populations of interest before protein profiling using antibody arrays.

Thus, quality control is a critical consideration as proteins and modifications such as phosphorylation may be unstable in improperly handled clinical samples.

Optimal outcome can be found when clinical samples are flash frozen immediately upon removal from the patient. Small samples are recommended to be directly embedded in an optimal cutting temperature (OCT-like) medium and frozen *in situ* so that thawing is retarded when samples are removed from freezer storage. Furthermore inclusion of protease inhibitor cocktails that may include phosphatase and other inhibitors in any fixative or lysis background may further protect frozen samples.

As commented above, tissue microarrays using core specimens of tissue paraffin archived blocks which are recast to create whole-cell microarrays of tumor specimens are considered targeted proteomic approaches that may complement antibody array functional proteomic analyses. Once tissue specimens are placed on the tissue array, they can be analyzed concurrently with immunohistochemistry, fluorescence *in situ* hybridization, and RNA-RNA *in situ* hybridization. It is necessary to mention that for RNA analyses, prior formalin fixation and paraffin embedding may limit these later techniques. Newer protocols are under development to improve protein and RNA resolution from these fixed archival samples.

6.4.3 *Body Fluids*

The initial report applying antibody microarrays in serum cancer for the discovery of biomarker candidates was performed using direct labelling methods for prostate cancer, comparing several substrates for antibody printing (Miller et al. 2003). As part of optimization analyses, data from "reverse-labelled" experiment sets accurately predicted the agreement between antibody microarrays and enzyme-linked immunosorbent assay measurements (Miller et al. 2003). Comparison of protein profiles of patients with prostate cancer and control serum samples identified five proteins (von Willebrand Factor, immunoglobulin M, α1-antichymotrypsin, Villin and immunoglobulin G) that had significantly different levels between the prostate cancer samples and the controls. This initial study using direct labelling protocols is one of the critical analyses that led to multiple developments enabling the immediate use of high-density antibody and protein microarrays (Miller et al. 2003). The use of amplification protocols, such as the two-colour rolling circle amplification (RCA) method served to improve the detection of low-abundant proteins. This method has also been shown to provide adequate reproducibility and accuracy for protein profiling on serum specimens and clinical applications (Schweitzer et al. 2002; Zhou et al. 2004; Shao et al. 2003). Sandwich assays can also measure protein abundances in body fluids using amplification detection methods such as resonance light scattering (RLS) (Saviranta et al. 2004), enhanced chemiluminescence (ECL) (Huang et al. 2004), or the tyramide signal amplification (TSA) method (Varnum et al. 2004) (Fig. 6.1, reviewed in Sanchez-Carbayo 2006). A recent report designed antibody arrays for bladder cancer by selecting antibodies against targets differentially expressed in bladder tumors *versus* their respective normal urothelium identified by gene profiling (Sanchez-Carbayo et al. 2006). Serum protein profiles obtained by two independent sets of antibody arrays served to segregate bladder cancer patients from controls. Protein profiles provided predictive information by stratifying patients with bladder tumors based on their overall survival. In addition, serum proteins, such as c-met, that were top ranked at identifying bladder cancer patients were associated with pathological stage, tumor grade, and survival when validated by immunohistochemistry of tissue microarrays containing bladder tumors (Sanchez-Carbayo et al. 2006). Such strategy provides experimental evidence for the use of several integrated technologies strengthening the discovery process of cancer-specific biomarker candidates and functional proteomic analyses of disease progression.

Cytokine profiling on serum and plasma specimens represents one of the most described applications of antibody arrays technology, especially for autoimmune diseases. In neoplastic diseases, they have been evaluated to a lower extent, although the implementation of cytokine antibody arrays is increasing in many aspects of cancer research, such as the discovery of biomarker candidates, molecular mechanisms of cancer development, preclinical studies and the effects of cancer compounds (Celis et al. 2005). All of these are considered critical aspects for functional proteomics. Studies in clinical material and *in vitro* systems have revealed the potential of cytokine profiling using antibody arrays for characterizing haematological neoplasias (Borrebaeck and Wingren 2007; Wang et al. 2005; Anderson and Labaer 2005), or in serum of patients with breast cancer (Vazquez-Martin et al. 2007). Cytokine profiles can support differentiation between cancer patients from control subjects and also stratify patients with leukemia based on clinical outcome. Several reports have also compared the reproducibility and differences among the several technologies available for multiplexing cytokine measurements, including not only planar antibody arrays but also bead-based technologies (Lash et al. 2006; De Jager and Rijkers 2006; Waterboer et al. 2006).

The tumor interstitial fluid (TIF) which perfuses the tumor environment has also been utilized for protein functional proteomic profiling using antibody arrays. Analysis of the TIF could identify factors present in the tumor microenvironment that may be associated with tumor growth and progression. TIFs collected from small pieces of freshly dissected invasive breast carcinomas have been analyzed by cytokine-specific antibody arrays. The approach provided a snapshot of more than 1,000 proteins – either secreted, shed by membrane vesicles, or externalized due to cell death – produced by the complex network of cell types that make up the tumor microenvironment. Considering that the protein composition of the TIF reflects the physiological and pathological state of the tissue, it should provide a new and potentially rich resource for the discovery of diagnostic biomarker candidates and for identifying more selective targets for therapeutic intervention (Celis et al. 2004, 2005). Interestingly, labelling and hybridization methods have been optimized for multiple protein detection on cerebrospinal fluid specimens, characterized by low protein concentrations (Romeo et al. 2005). Non-invasive body fluids such the saliva, sputum or urine specimens represent potential samples for clinical application of antibody arrays.

It is required to optimize labelling and hybridization protocols to the sensitivities required for such specimens.

6.5 Conclusions

The parallel analysis of multiple proteins in small sample volumes is being applied to measure multiple protein abundances for functional proteomic analyses using antibody arrays. Application on biological specimens is serving to address disease progression, clinical subtypes and outcomes in exploratory analyses. Modifications to antibody arrays are leading to functional protein profiling strategies that may also result into novel cancer targets and biomarker candidates such as: (a) detecting specific protein post-translational modifications; (b) measurement of enzyme activities; (c) quantification of protein cell-surface expression; d) characterizing signaling pathways; (e) the development and characterization of antibodies including identification of binding partners to proteins derived from functional studies for drug discovery or novel epitope mapping for determining regions of proteins than bind specific antibodies.

The use of antibody array methods not only results in added benefit for cancer diagnostics and patient stratification but also provides complementary information for the characterization of the biology underlining tumorigenesis and tumor progression. Protein profiling using antibody arrays is contributing to reveal the importance of monitoring multiple cell signaling endpoints and thus, mapping specific cellular networks not only in protein extracts from cell lines but also form tissue or body fluid specimens. Changes in glycan contents, phosphorylation status or cleaved states of key signaling proteins can easily be evaluated using antibody arrays as well. It is possible to test whether one pathway might become blocked with chemotherapeutic agents. Analyses of these pathways might reveal relevant information for designing individual targeted therapies and/or combinatorial strategies directed at multiple nodes in a cell signaling cascade. This strategy might be tested to predict response to novel drug therapies using the protein extracts of the tumors or in body fluids specimens.

Antibody-based microarrays represents a rapidly emerging technology for functional proteomic analyses that is advancing from the first proof-of-concept studies to increasing protein profiling applications in cancer biomarker development. The increasing number, scope and effectiveness of the formats, methods and applications of antibody arrays are likely to markedly accelerate the characterization of cancer-specific pathways, networks and post-translational modifications. Identifying cancer-associated protein changes may lead to the discovery of cancer-associated targets and biomarker candidates that may assist in disease predisposition, diagnosis, prognosis, patient monitoring and possibly for therapeutic purposes on various sample types, such as serum, plasma, and other bodily fluids; cell culture supernatants; tissue culture lysates; and resected tumor specimens. As standards do not yet exist that bridge all of these applications, the current recommended best practice for clinical validation of results is to approach study design in an iterative process using independent sets of human clinical material and to integrate data from several measurement technologies. The main problems described in poorly delineated experimental designs include lack of uniform patient inclusion and exclusion criteria, low patient numbers, poorly supporting clinical data, absence of standardized sample preparation, and limited analytical verification providing estimations of the intra and inter-assay reproducibility.

Several challenges and limitations remain to be improved in the design and application of antibody arrays for functional proteomics: (a) The mechanisms by which proteins or antibodies are immobilized in substrates such as nitrocellulose are poorly understood for certain technological innovations; (b) the limited dynamic ranges of two or three orders of magnitude for certain labelling protocols can be increased; (c) achieving accuracy and reproducibility similar to clinical immunoassays at the very low pico/femtomolar detection level; (d) the immunoreactivity might be affected by the molecular protein complexity and potential protein denaturation; (e) lack of standards and calibrators for all the antibody and reagents utilized; (f) development of high-affinity and highly-specific antibodies are not available for all the potential target antigens under study.

The highly increasing technical modalities of antibody arrays are requiring standardized processes for storing and retrieving data obtained from different technologies by different research groups. In this regard, it is necessary to acknowledge the multi-institutional effort of the Human Proteome Organization (HUPO) towards the standardization of protocols for critical parameters in serum or plasma proteomic analyses, including protein profiling using antibody arrays. Initial studies provided guidance on pre-analytical variables that can alter the analysis of blood-derived samples, including choice of sample type, stability during storage, use of protease inhibitors, and clinical standardization. It is also critical to standardize statistical strategies for high confidence protein identification and data analysis. These efforts and strategies towards integrating proteomic datasets would lead towards accurate and comprehensive representation of human proteomes.

Thus, the most significant contribution of functional proteomics research using antibody arrays for the discovery of molecular networks, targets and cancer biomarker candidates is expected to derive not from single experiments, but from the synthesis and comparison of large datasets obtained under different conditions (e.g. normal, inflammation, cancer) and in different *in vitro* and clinical material from various tissues and organs. The technology will continue providing unique opportunities in cancer diagnostics, patient stratification, predicting clinical outcome and therapeutic response. Continued progress in the technology will surely lead to extensions of these applications and the development of new ways of using the methods. Further innovations in the technology and in the experimental strategies will further broaden the scope of the applications and the type of information that can be gathered. In the near future, the detailed characterization of the specific protein expression profiles or protein atlases of each tumor will also serve to better detect, monitor and stratify the clinical outcome risk of each specific cancer patient so that they may be benefit of tailored interventions based on the aggressiveness of their diseases.

References

Anderson KS, Labaer J (2005) The sentinel within: exploiting the immune system for cancer biomarkers. J Proteome Res 4:1123–1133

Angenendt P, Glokler J, Murphy D et al (2002) Toward optimized antibody microarrays: a comparison of current microarray support materials. Anal Biochem 309:253–260

Azad NS, Rasool N, Annunziata CM et al (2006) Proteomics in clinical trials and practice: present uses and future promise. Mol Cell Proteomics 5:1819–1829

Bartling B, Hofmann HS, Boettger T et al (2005) Comparative application of antibody and gene array for expression profiling in human squamous cell lung carcinoma. Lung Cancer 49:145–154

Belov L, de la Vega O, dos Remedios CG et al (2001) Immunophenotyping of leukemia using a cluster of differentiation antibody microarray. Cancer Res 61:4483

Belov L, Mulligan SP, Barber N et al (2006) Classification of human leukemias and lymphomas using extensive immunophenotypes obtained by cell capture on an antibody microarray. Brit J Haem 135:184–197

Bock C, Coleman M, Collins B et al (2004) Photoaptamer arrays applied to multiplexed proteomic analysis. Proteomics 4:609–618

Borrebaeck CA, Wingren C (2007) High-throughput proteomics using antibody microarrays: an update. Expert Rev Mol Diagn 7:673–686

Celis JE, Gromov P, Cabezón T et al (2004) Proteomic characterization of the interstitial fluid perfusing the breast tumor microenvironment: a novel resource for biomarker and therapeutic target discovery. Mol Cell Proteomics 3:327–344

Celis JE, Moreira JM, Cabezón T et al (2005) Identification of extracellular and intracellular signaling components of the mammary adipose tissue and its interstitial fluid in high risk breast cancer patients: toward dissecting the molecular circuitry of epithelial-adipocyte stromal c. Mol Cell Proteomics 4:492–522

Chan SM, Ermann J, Su L et al (2004) Protein microarrays for multiplex analysis of signal transduction pathways. Nat Med 10:1390–1396

Chen S, LaRoche T, Hamelinck D et al (2007) Multiplexed analysis of glycan variation on native proteins captured by antibody microarrays. Nat Methods 4:437–444

De Jager W, Rijkers GT (2006) Solid-phase and bead-based cytokine immunoassay: a comparison. Methods 38:294–303

Dotan N, Altstock RT, Schwarz M et al (2006) Anti-glycan antibodies as biomarkers for diagnosis and prognosis. Lupus 15:442–450

Ek S, Andréasson U, Hober S et al (2006) From gene expression analysis to tissue microarrays – a rational approach to identify therapeutic and diagnostic targets in lymphoid malignancies. Mol Cell Proteomics 5:1072–1081

Flores-Delgado G, Liu CW, Sposto R et al (2007) A limited screen for protein interactions reveals new roles for protein phosphatase 1 in cell cycle control and apoptosis. J Proteome Res 6:1165–1175

Garcia BH II, Hargrave A, Morgan A et al (2007) Antibody microarray analysis of inflammatory mediator release by human leukemia T-cells and human non small cell lung cancer cells. J Biomol Tech 18:245–251

Gembitsky DS, Lawlor K, Jacovina A et al (2004) A prototype antibody microarray platform to monitor changes in protein tyrosine phosphorylation. Mol Cell Proteomics 3:1102–1118

Haab BB, Dunham MJ, Brown PO (2001) Protein microarrays for highly parallel detection and quantitation of specific proteins and antibodies in complex solutions. Genome Biol 2:RESEARCH0004.

Huang R, Lin Y, Shi Q et al (2004) Enhanced protein profiling arrays with ELISA-based amplification for high-throughput molecular changes of tumor patients' plasma. Clin Cancer Res 10:598–609

Hudelist G, Pacher-Zavisin M, Singer CF et al (2004) Use of high-throughput protein array for profiling of differentially expressed proteins in normal and malignant breast tissue. Breast Cancer Res Treat 86:281–291

Ivanov SS, Chung AS, Yuan ZL et al (2004) Antibodies immobilized as arrays to profile protein post-translational modifications in mammalian cells. Mol Cell Proteomics 3:788–795

Kato K, Ishimuro T, Arima Y et al (2007) High-throughput immunophenotyping by surface plasmon resonance imaging. Anal Chem 79:8616–8623

Kellner U, Steinert R, Seibert V et al (2004) Epithelial cell preparation for proteomic and transcriptomic analysis in human pancreatic tissue. Pathol Res Pract 200:155–163

Kingsmore SF (2006) Multiplexed protein measurement: technologies and applications of protein and antibody arrays. Nat Rev Drug Discov 5:310–321

Kopf E, Zharhary D (2007) Antibody arrays – an emerging tool in cancer proteomics. Int J Biochem Cell Biol 39:1305–1317

Lash GE, Scaife PJ, Innes BA et al (2006) Comparison of three multiplex cytokine analysis systems: Luminex, SearchLight and FAST Quant. J Immunol Meth 309:205–208

Li Y, Lee HJ, Corn RM (2007) Detection of protein biomarkers using RNA aptamer microarrays and enzymatically amplified surface plasmon resonance imaging. Anal Chem 79:1082–1088

Lin Y, Huang R, Cao X et al (2003a) Detection of multiple cytokines by protein arrays from cell lysate and tissue lysate. Clin Chem Lab Med 41:139–145

Lin Y, Huang R, Chen L et al (2004) Identification of interleukin-8 as estrogen receptor-regulated factor involved in breast cancer invasion and angiogenesis by protein arrays. Int J Cancer 109:507–515

Lin Y, Huang R, Chen LP et al (2003b) Profiling of cytokine expression by biotin-labelled-based protein arrays. Proteomics 3:1750–1757

Miller JC, Zhou H, Kwekel J et al (2003) Antibody microarray profiling of human prostate cancer sera: antibody screening and identification of potential biomarkers. Proteomics 3:56–63

Nettikadan S, Radke K, Johnson J et al (2006) Detection and quantification of protein biomarkers from fewer than 10 cells. Mol Cell Proteomics 5:895–901

Nielsen UB, Cardone MH, Sinskey AJ et al (2003) Profiling receptor tyrosine kinase activation by using Ab microarrays. Proc Natl Acad Sci USA 100:9330–9335

Nishizuka S, Charboneau L, Young L et al (2003) Proteomic profiling of the NCI-60 cancer cell lines using new high-density reverse-phase lysate microarrays. Proc Natl Acad Sci USA 100:14229–14234

Petricoin EF 3rd, Bichsel VE, Calvert VS et al (2005) Mapping molecular networks using proteomics: a vision for patient-tailored combination therapy. J Clin Oncol 23:3614–3621

Romeo MJ, Espina V, Lowenthal M et al (2005) CSF proteome: a protein repository for potential biomarker identification. Expert Rev Proteomics 2:57–70

Sanchez-Carbayo M (2006) Antibody arrays: technical considerations and clinical applications in cancer. Clin Chem 52:1651–1659

Sanchez-Carbayo M, Socci ND, Lozano JJ et al (2006) Profiling bladder cancer using targeted antibody arrays. Am J Pathol 168:93–103

Saviranta P, Okon R, Brinker A et al (2004) Evaluating sandwich immunoassays in microarray format in terms of the ambient analyte regime. Clin Chem 50:1907–1920

Schweitzer B, Roberts S, Grimwade B et al (2002) Multiplexed protein profiling on microarrays by rolling-circle amplification. Nat Biotechnol 20:359–365

Shao W, Zhou Z, Laroche I et al (2003) Optimization of rolling- circle amplified protein microarrays for multiplexed protein profiling. J Biomed Biotechnol 5:299–307

Uhlén M, Björling E, Agaton C et al (2005) A human protein atlas for normal and cancer tissues based on antibody proteomics. Mol Cell Proteomics 4:1920–1932

Varnum SM, Woodbury RL, Zangar RC (2004) A protein microarray ELISA for screening biological fluids. Methods Mol Biol 264:161–172

Vazquez-Martin A, Colomer R, Menendez JA (2007) Protein array technology to detect HER2 (erbB-2)-induced 'cytokine signature' in breast cancer. Eur J Cancer 43:1117–1124

Wang X, Yu J, Sreekumar A et al (2005) Autoantibody signatures in prostate cancer. N Engl J Med 353:1224–1235

Waterboer T, Sehr P, Pawlita M (2006) Suppression of non-specific binding in serological Luminex assays. J Immunol Methods 309:200–204

Yeretssian G, Lecocq M, Lebon G et al (2005) Competition on nitrocellulose-immobilized antibody arrays: from bacterial protein binding assay to protein profiling in breast cancer cells. Mol Cell Proteomics 4:605–617

Zajac A, Song D, Qian W et al (2007) Protein microarrays and quantum dot probes for early cancer detection. Colloids Surf B Biointerfaces 58:309–314

Zhou H, Bouwman K, Schotanus M et al (2004) Two-colour, rolling-circle amplification on antibody microarrays for sensitive, multiplexed serum-protein measurements. Genome Biol 5:R28

Chapter 7
Protein Graphs in Cancer Prediction

Humberto González-Díaz, Giulio Ferino, Francisco J. Prado-Prado, Santiago Vilar, Eugenio Uriarte, Alejandro Pazos, and Cristian R. Munteanu

Abstract The consequences of breast, colon and prostate cancer create the necessity of new, simpler and faster theoretical models that may allow earlier cancer detection. The present work has built several Quantitative Protein (or Proteome) – Disease Relationships (QPDRs). QPDRs, similar to Quantitative Structure Activity Relationship (QSAR) models, are based on topological indices (TIs) and/or connectivity indices (CIs) of graphs. In particular, we used Star graphs and Lattice networks of protein sequence or MS outcomes of blood proteome in order to predict the proteins related to breast and colon cancer and to improve the diagnostic potential of the PSA biomarker for prostate cancer. The advantages of this method are the simplicity, fast calculations and few resources needed (free software programmes, such as MARCH-INSIDE and S2SNet). Thus, this ideal theoretical scheme can be easily extended to other types of diseases or even other fields, such as Genomics or Systems Biology.

7.1 Introduction

Cancer is the second most common cause of death after heart-related illnesses. It is estimated that during the twenty-first century cancer will become the main cause of death in developed countries. However, there has also been an increase

H. González-Díaz (✉)
Department of Microbiology & Parasitology, University of Santiago de Compostela, 15782, Santiago de Compostela, Spain
e-mail: humberto.gonzalez@usc.es

G. Ferino
Dipartimento Farmaco Chimico Tecnologico, Universitá degli Studi di Cagliari, 09124, Cagliari, Italia

C.R. Munteanu and A. Pazos
Department of Information and Communication Technologies, University of A Coruña, 15071 A Coruña, Spain
e-mail: muntisa@gmail.com

S. Vilar, E. Uriarte, and F.J. Prado-Prado
Department of Organic Chemistry, University of Santiago de Compostela, 15782 Santiago de Compostela, Spain

W.C.S. Cho (ed.), *An Omics Perspective on Cancer Research*,
DOI 10.1007/978-90-481-2675-0_7, © Springer Science+Business Media B.V. 2010

in the survival rate of cancer patients. The five main causes of cancer deaths in order of prevalence are: lung cancer, colorectal cancer, breast cancer, prostate cancer and pancreatic cancer (Mayer et al. 1993; Welsh et al. 2001; Rivera and Stover 2004; Del Chiaro et al. 2007; Kirkegaard et al. 2007). Oncoproteomics is the application of proteomic technologies in cancer. Considerable progress has been made during the past decade in the refinement of proteomic technologies and their application, so that the pathological mechanisms and the discovery of biomarkers and diagnosis of the disease could be better understood (Jain 2007). With the advent of new and improved proteomics technologies, such as the development of quantitative proteomic methods, high-resolution, -speed and -sensitivity mass spectrometry and protein arrays, as well as advanced bioinformatics for data handling and interpretation, it is now possible to discover biomarkers that can reliably and accurately predict outcomes during cancer management and treatment (Cho and Cheng 2007).

Omics is a general term for a broad discipline of science and engineering that analyses the interactions of biological information objects in various *omes* (the Greek term for "all", "every", "whole" or "complete"): Genome (Coghlan et al. 2005; Notebaart et al. 2008), Proteome (the totality of proteins in an organism, tissue type or cell) (Cruz-Monteagudo et al. 2008; Klose 1989), Transcriptome (an mRNA complement to an entire organism, tissue type, or cell) (Hu et al. 2004; Latha and Venkatesh 2004), Metabolome (the totality of metabolites in an organism) (Wishart et al. 2008; Zhu and Qin 2005), Lipidome (the totality of lipids) (Bougnoux et al. 2006, 2008; Ding et al. 2008), Localizome (a whole set of localization information of protein domains and proteins) (Lee et al. 2006), Glycome (the total list of sugar/carbohydrate molecules in an organism) (Freeze 2006; Morelle et al. 2006), Expressome (a whole set of gene expression in a cell, tissue, organ, organisms, and species) (Borges et al. 2007) and Interactome (a whole set of molecular interactions in cells) (Altaf-Ul-Amin et al. 2006; Bader and Hogue 2003; Chen et al. 2006). Our aim is to map the information objects, such as genes, proteins and ligands, to find the interaction relationships between the objects and to engineer the networks and objects in order to get to understand and manipulate the regulatory mechanisms.

We give a solution for two types of problems related to Oncoproteomics at different levels of chemical matter organization. The first problem is to predict whether a specific protein is involved in Human Breast Cancer (BC) or in Human Colon Cancer (CC), given the protein sequence. The second problem is the use of the information provided by Mass Spectra (MS) analysis of human serum proteome and the level of prostate specific antigen (PSA) in the blood; this would improve the predictive power of the PSA test to detect prostate cancer. We established mathematical relationships between the structure/activity of the proteins/proteome mass spectra and the type of cancer by using a graphical method, the graph theory, in order to solve these problems. The graphic approaches applied to the biological systems can provide useful insights, as indicated by several previous studies on a series of important biological topics, such as enzyme-catalyzed reactions (Andraos 2008; Chou 1989), protein folding kinetics (Chou 1990), inhibition kinetics of processive nucleic acid polymerases and nucleases (Althaus et al. 1996; Chou et al. 1994), analysis of codon usage (Chou and Zhang 1992), base frequencies in the anti-sense strands (Chou et al. 1996), analysis of DNA sequence (Qi et al. 2007).

The graph theory can be used to obtain macromolecular descriptors named topological indices (TIs) and connectivity indices (CIs). The branch of mathematical chemistry dedicated to encode the DNA/protein information in graph representations by the use of the TIs has become an intense research area that led to the interesting works made by Liao and Wang (2004), Liao et al. (2006), Randic, Nandy, Balaban, Basak, and Vracko (Randic 2000; Randic and Balaban 2003), Bielinska-Waz et al. (2007) or our group (Perez et al. 2004; Aguero-Chapin et al. 2006). The TIs and CIs are parameters that describe numerically the patterns of interconnections (edges or arcs) between the parts of a system represented as a graph or network. These graphs or networks have become a flexible and general method to describe biological systems. First, we split the system in parts (nodes) and studied the presence of any kind of geometrical, physical, functional, dynamic or other classes of relationships between all pairs of nodes. In the case of the protein sequence, the nodes are amino acids and, regarding the mass spectra, the nodes are represented by contracted/averaged signals (Fig. 7.1). We can associate these graphs or networks with different classes of numerical matrices and with the invariant calculated parameters of the graph (the TIs and the CIs). In the present work, we did not use simple molecular graphs but two types of simple graphs, the so-called Lattice networks and Star graphs. Depending on the geometrical shape of the graphs, there are several types of representations such as the spiral, circular or random. The advantages of using the Lattice and Star graphs are the simplicity of the calculations and the efficacy of these graphical representations to encode the complex information in indices. The protein sequences and the mass spectrum signals have no simple property directly linked to cancer. Thus, the graph method transforms the protein sequence and the proteome mass spectrum signal in a unique discrete series of indices (in the case of the MS proteome, the exact nature of all compounds is unknown; only the signal intensities are used). These parameters can be used as inputs to seek new models, which are in essence equations connecting the structure with the properties of the system.

One of the widely-used models for the prediction of protein properties is the Quantitative Structure Activity Relationship, QSAR (Devillers and Balaban 1999). We used the actual graphs to represent protein sequences or the outcomes of the MS analysis of blood proteomes and to connect them to diseases (different types of cancer). Consequently, the models reported here may be seen as Quantitative Protein (or Proteome)-Disease Relationships (QPDRs) (Estrada and Uriarte 2001; Barabasi and Bonabeau 2003; Balaban et al. 2004; Barabasi and Oltvai 2004; Barabasi 2005; González-Díaz et al. 2007; Randic et al. 2007; Ferino et al. 2008; González-Díaz et al. 2008). The most used methods to obtain models in proteomics/genetics are the Linear Discriminant Analysis (LDA), the machine-learning classifiers, such as the artificial neural networks (ANN), the Bayesian belief network (BBN), the support vector machine (SVM), the radial basis Gaussian kernel function (RBF) neural network or the Rand-Tree genetic programming.

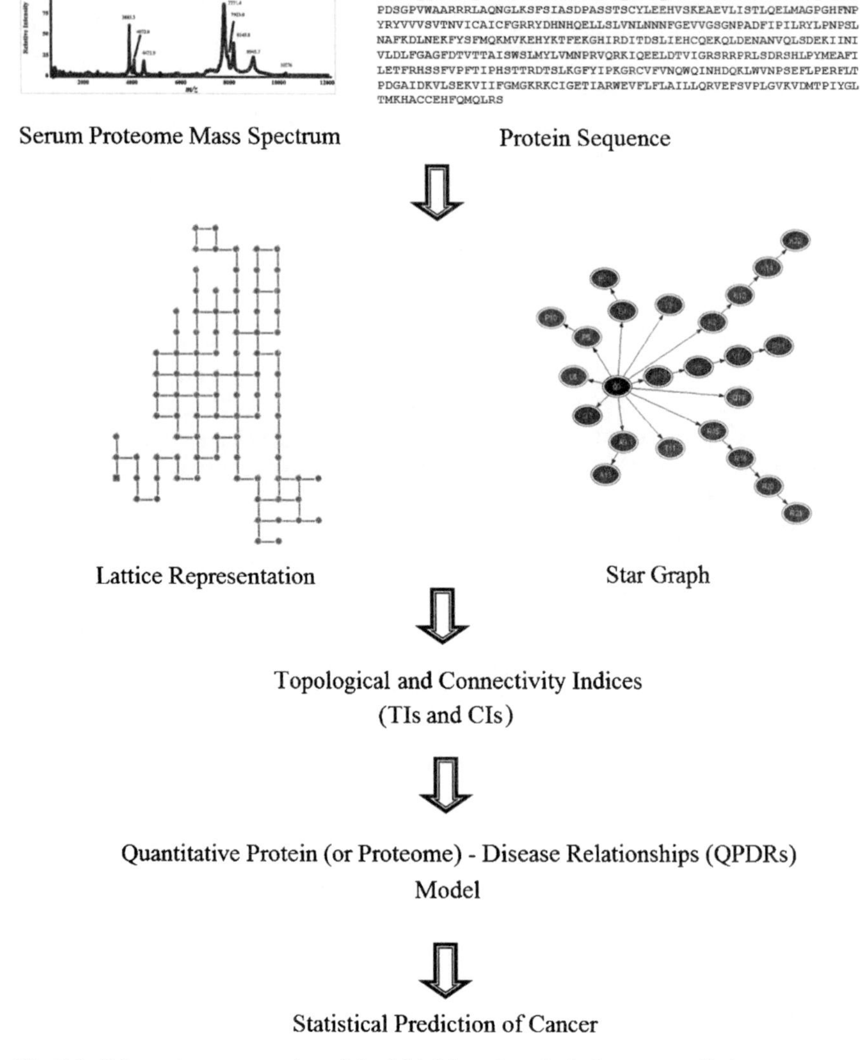

Fig. 7.1 Schematic representation of the QSAR-based method of cancer prediction

For instance, Ward et al. (2006) studied the colorectal cancer (CRC) which is often diagnosed at a late stage with concomitant poor prognosis. Early detection greatly improves prognosis; however, the invasive, unpleasant and inconvenient nature of current diagnostic procedures limits their applicability. No serum-based test is currently of sufficient sensitivity or specificity for widespread use. In the best currently available blood test, carcinoembryonic antigen exhibits low sensitivity and specificity particularly in the setting of early disease. Hence, there is a great need for new biomarkers that lead to an early detection of CRC. Surface-enhanced laser desorption/ionisation (SELDI) has been used to investigate the serum proteome of 62 CRC patients and 31 non-cancer subjects. Proteins with diagnostic potential have been identified. Thus, the ANN trained was able to classify the patients in this study (95% sensitivity and 91% specificity), using only the intensities of the SELDI peaks corresponding to the identified proteins.

Green and Karp (2004) have developed a method that efficiently combines homology and pathway-based evidence to identify candidates for filling pathway holes in Pathway/Genome databases. Not only does the application identify potential candidate sequences for pathway holes, but also combines data from multiple, heterogeneous sources to assess the likelihood that a candidate could have the required function. The new algorithm emulates the manual sequence annotation process, considering not only evidence from homology searches, but also considering evidence from genomic context (i.e. Is the gene part of an operon?) and functional context (e.g. Are there functionally-related genes nearby in the genome?) to determine the posterior belief that a candidate has the required function. The method can be applied across an entire metabolic pathway network and it is generally applicable to any pathway database. The programme uses a set of sequences that encode the required activity in other genomes to identify candidate proteins in the genome of interest, and then evaluates each candidate by using a simple Bayes classifier to determine the probability that the candidate has the desired function.

Wang et al. (2008) explored the diagnostic value for ovarian cancer, using proteomic pattern established by SELDI-time-of-flight mass spectrometry (TOF-MS) profiling of plasma proteins coupled with SVM data analysis, and investigated whether the proteomic pattern established by advanced ovarian cancer could be used for diagnosis of early-stage ovarian cancer patients. The study included 44 ovarian cancer patients (11 early-stage and 33 advanced ovarian cancer patients) and 31 age-matched non-cancer controls. SELDI-TOF-MS coupled with SVM analysis was performed to establish a proteomic pattern to discriminate 33 advanced ovarian cancer patients from 31 non-cancer controls. A blind test, including 11 early-stage ovarian cancer cases, was carried out to investigate whether proteomic pattern established by advanced ovarian cancer could be used for diagnosis of early-stage ovarian cancer patients. A 7-peak proteomic pattern was established, which discriminated 33 advanced ovarian cancer patients from 31 non-cancer controls effectively.

In present work, the QSAR/QPDR models were obtained with the simplest and fastest linear discriminant analysis (LDA) method (Van Waterbeemd 1995).

7.2 Materials and Methods

7.2.1 Protein Database

It is important that the database used to develop the discriminant equation is varied and representative enough in order to obtain a high quality statistical model with good predictive power. The database carried out by Sjoblom et al. (2006) was compiled and it included a series of 122 genes responsible for the formation of BC and another series of 69 genes responsible for the development of CC. The genetic sequence enabled information to be obtained on the codified protein sequence. Taking into consideration the high diversity of biological functions of the proteins represented in the dataset, it is obviously expected that these proteins were related to cancer by different mechanisms. The set of proteins not related to BC or CC (non-BCp or non-CCp) was also experimentally confirmed as negative by the same authors and it was used to provide additional examples of useful human proteins in other QSAR studies by Dobson and Doig (2003, 2005).

7.2.2 MS Database

The database of Petricoin et al. (2002) has been used. It contains the MS spectra of 322 patients, classified into 4 groups according to their PSA level. The PSA is a protein produced by both normal and cancerous prostate cells. A high level of PSA can be a sign of cancer, but the PSA level can also be raised in prostate conditions that are not cancer (are benign), or in case of infection. One group was made up of 190 patients with a PSA level greater than 4 (group A), a second group consisted of 63 patients with a PSA less than 1 (group B), 26 patients with a PSA level in the 4–10 ng/ml range made up the third group (group C) and 43 patients with a PSA level greater than 10, the fourth group (group D). The clinical analyses have demonstrated that the healthy patients are those belonging to the groups A and B, the patients of the groups C and D are PCa (Petricoin et al. 2002; Ferino et al. 2008; González-Díaz et al. 2008).

7.2.3 MS Data Coding

More than 15,000 signals are present in each spectrum, each consisting of the m/z value and the corresponding *I* value; considering the successive ranges of 500 values of this parameter, 31 regions have been obtained for each MS. For the Randic Star, the *I* value of all 31 regions was expressed as a percentage and coded into 8 branches (B, C, D, E, F, G, H, I) depending on the weight of the region in each MS. In the Lattice Network representation we applied the same coding for the Cartesian coordinates used by Randic to describe the 3D primary sequence of DNA (Randic et al. 2000). For the sake of convenience, we have also used the 4 Nucleic Bases to encode the MS of each patient. A 25% cut-off value is chosen to codify each data point according to their respective average *I* values. For example, all regions of the MS with an *I* value between 0% and 25% are classified with the letter A. The letter T includes regions with an *I* value between 25% and 50%, and the same label was used for the letter G and C, respectively.

7.2.4 *MARCH-INSIDE Software*

Almost all graphs developed are created with the programme MARCH-INSIDE 2.0 (Ferino et al. 2008; González-Díaz et al. 2008). This application can be obtained at request from the authors. The sequences of different proteins and the proteome MS of all patients were initially introduced into the programme. This programme was used to generate both Lattice-type networks (for protein sequence and proteome MS) and the Randic Star graph for each proteome MS.

7.2.5 *Lattice Network Representations*

The methodology used to construct lattice networks has been previously described in detail in the case of proteins (Aguero-Chapin et al. 2006) or MS outcomes of blood proteome (Cruz-Monteagudo et al. 2008). The representation was constructed in a similar way to the DNA representations introduced by Nandy and adapted to proteins according to a recently reported protocol (Arteca and Mezey 1990; Nandy and Basak 2000; Randic and Vracko 2000; Estrada 2002; Randic and Balaban 2003; González-Díaz et al. 2007). The method is based on splitting 20 types of amino acids into 4 groups according to the Hydrophobic (H) or Polar (P) nature of the different amino acids. Thus, four groups of amino acids can be generated in order to characterize the physicochemical nature of amino acids as polar, non-polar, acidic or basic. The 2D Cartesian representation and the Stochastic Matrix for a given protein sequence are shown in Table 7.1. The classification as acidic or basic prevails over the polar/non-polar classification in such a way that the four groups do not overlap each other. Subsequently, each amino acid in the sequence is placed in a 2D Lattice defined by a Cartesian space with its centre at the (0, 0) coordinates (see Table 7.1). The coordinates of the successive bases are calculated as follows, to form a Lattice Network with a step equal to 1:

For the protein sequence:

a. The abscissa axis increases by +1 for an acid amino acid (rightwards-step).
b. The abscissa axis decreases by –1 for a basic amino acid (leftwards-step).
c. The ordinate axis increases by +1 for a polar amino acid (upwards-step).
d. The ordinate axis decreases by –1 for a non-polar amino acid (downwards-step).

For proteome MS:

a. The abscissa axis increases by +1 for $0 < IR < 25$ (rightwards-step).
b. The abscissa axis decreases by –1 for $25 < IR < 50$ (leftwards-step).
c. The ordinate axis increases by +1 for $50 < IR < 75$ (upwards-step).
d. The ordinate axis decreases by –1 for $75 < IR < 100$ (downwards-step).

where I_R is the averaged Intensity value of proteome MS expressed in percentage terms. The three different non-equivalent coordinates were taken into consideration to create the 3D Lattice Network for proteome MS.

Table 7.1 Graph representation and associate stochastic matrix for protein sequence

N	Amino acids	X	Y
a	Tyr8	0	1
	Ala1	0	0
b	His7	0	0
	Glu2	1	0
	Lys4	1	0
c	Gln6	1	0
d	Asp3	2	0
e	Arg5	1	−1

HP lattice network

$$\begin{bmatrix} {}^1p_{aa} & {}^1p_{ab} & 0 & 0 & 0 \\ {}^1p_{ba} & {}^1p_{bb} & {}^1p_{bc} & 0 & 0 \\ 0 & {}^1p_{cb} & {}^1p_{cc} & {}^1p_{cd} & {}^1p_{ce} \\ 0 & 0 & {}^1p_{dc} & {}^1p_{dd} & 0 \\ 0 & 0 & {}^1p_{ec} & 0 & {}^1p_{ee} \end{bmatrix}$$

Star graph example

Star graph matrix

	*A*1	*C*2	*A*3	*D*4	*C*5	*E*6	*F*7	*D*8	*G*9	00
*A*1	0	0	1	0	0	0	0	0	0	1
*C*2	0	0	0	0	1	0	0	0	0	1
*A*3	1	0	0	0	0	0	0	0	0	0
*D*4	0	0	0	0	0	0	0	1	0	1
*C*5	0	1	0	0	0	0	0	0	0	0
*E*6	0	0	0	0	0	0	0	0	0	1
*F*7	0	0	0	0	0	0	0	0	0	1
*D*8	0	0	0	1	0	0	0	0	0	0
*G*9	0	0	0	0	0	0	0	0	0	1
00	1	1	0	1	0	1	1	0	1	0

7.2.6 MS Star Graph Representation

Regarding the second graphical representation, we opted for the Shining Star graph (Randic et al. 2007). In the same way as the Lattice Network, the Star graph was also constructed for DNA or protein sequences, as previously described, but in this case the nodes are regions of mass spectra (Nandy and Basak 2000). The starting point was the classification of each region into more categories according to the degree of average intensity (B, C, D, E, etc.). Afterwards, the number of regions of each type was counted and the Star graph was drawn by adding one node per region. The first node is the PSA value, represented in branch A, and 0, 1 or 2 nodes were also added in this branch depending on the PSA value = 0, 1 or 2.

7.2.7 *Protein Star Graph Representation*

In order to construct the protein Star graph, we have proceeded in the same manner as previously described in the case of MS Star. The only difference here is that the nodes of the star are not areas in the MS region but amino acids of the protein. The first node does not correspond to any amino acid; it is just one dummy node core of the star. If the vertices do not carry a label, the sequence information will be lost; for that reason, the best method is to construct a standard star graph where each amino acid/vertex holds the position in the original sequence and the branches are labelled by alphabetical order of the three-letter amino acid code (Randic et al. 2007). In the present study we use the alphabetical order of one-letter amino acid code. The standard star graph and the Stochastic Matrix for a random virtual nonapeptide (ACADCEFDG) are illustrated in Table 7.1. Thus, there are 20 groups corresponding to the possible star branches for any standard amino acid type. Starting from the beginning of the sequence, the amino acids are placed in the correspondent branch; the initial connectivity from the protein sequence is transformed into modified branch connectivity. If the initial sequence connections are added to the star graph, it will become an embedded graph.

7.2.8 *Entropy Measurements*

The protein chain sequences and the proteome MS are transformed into Network representations and subsequently characterised by the Shannon Entropy indices. We use our free S2SNet – Sequence to Star Networks application for the protein sequence Star graph (Munteanu et al. 2008). MARCH-INSIDE 2.0 was used for both Lattice Network representation and MS Star graph. In particular, the calculations presented here are characterized by Markov normalization and the power of matrices/indices (n) up to 5.

Shannon Entropy of the k powered Markov Matrices (θ_n) is described by the sum of i products between the p_i probability and its logarithm, where p_i are the n_i elements of the p vector, resulted from the matrix multiplication of the powered Markov normalized matrix ($n_i \times n_i$) and a vector ($n_i \times 1$) with each element equal to $1/n_i$.

7.2.9 *Linear Discriminant Analysis (LDA)*

Once the different Entropies were calculated, a LDA (Van Waterbeemd 1995) was performed using the STATISTICA package (Brzezinska 2003) to develop different classification functions. We have applied the same procedure, so that data processing could be placed in the software STATISTICA for both databases; the procedure applied to the database of Petricoin et al. (2002) is described below. The data set was randomly divided into two series, a training series (for model construction) and a cross-validation series (for model validation). The training series consisted of 230 patients (175 control group – healthy patients and 55 PCa); while the cross-validation series was made up of 92 patients (78 control group – healthy patients and 14 PCa). The activity or any given biological property (P) is expressed as a function of the entropy $\theta_k(R_i)$ of k order for every MS region R_i, generated in each case. Thus, in the classification function, P is the linear combination of the θ_k entropies plus the PSA level multiplied by the coefficient calculated for PSA region. The discriminant function was obtained by using the *Forward-stepwise* method with a minimum tolerance value of 0.01. In the forward stepwise model selection procedure, variables are sequentially added to an "empty" (intercept only) model, in contrast to the backward procedures that start with all the variables in the model, and proceed by removing them.

7.3 Results and Discussion

7.3.1 *Classification Function*

The best models developed using Lattice Network and Randic Star Graph is presented in Table 7.2. The statistical parameters show the quality of the model and are the standard results obtained from the STATISTICA analysis (Van Waterbeemd 1995). The most important values of a model are accuracy, specificity and sensitivity percentage that measure the ratio of the total number, cancer or non-cancer sequences/spectra correctly classified by the model with respect to the real classification. The results are presented in Table 7.2 and shows different values for the created models. Regarding the database Sjöblom et al. the best models were developed using Lattice Network representation. The protein Lattice Network model has reached a total accuracy amounted to 87%, an 87.2% specificity and an 86.9% sensitivity.

The initial assumptions have to be checked for any statistical model. The parametrical assumptions such as normality, homocedasticity (homogeneity of variances) and non-colinearity have the same importance in the application of multivariate statistic techniques to QSAR (Bisquerra Alzina 1989; Stewart and Gill 1998) as the correct specification of the mathematical form. The validity and statistical signification of any model are conditioned by the above mentioned

Table 7.2 Summary results for the QSAR models

Protein sequence				
Parameter	**Lattice network**		**Star graph**	
Coeff. of	θ(canc)	202.66	θ(canc)	191.61
Coeff. of	θ_0(prot)	−0.39	θ_0(prot)	−0.59
Coeff. of	θ_0(prot/canc)	2.56	θ_0(prot/canc)	0.86
Independent Coeff.		−30.24		−26.04
N	191		191	
R_c	0.76		0.76	
λ	0.41		0.41	
χ^2	247.66		246.35	
p	<0.001		<0.001	
Accuracy	87.0%		87.0%	
Specificity	87.2%		87.2%	
Sensitivity	86.9%		86.9%	
Classification matrices	**nCprot**	**Cprot**	**nCprot**	**Cprot**
nCprot	**163**	24	**163**	24
Cprot	25	**166**	25	**166**
	Proteome MS for patients			
Parameter	**Lattice network**		**Star graph**	
Coeff. of	θ_2(C1)	5.66	θ_0(total)	−938.80
Coeff. of	θ_3(C2)	−3.60	θ_1(total)	−158.80
Coeff. of	θ_0(C3)	13.06	θ_2(total)	161.60
Independent Coeff.		−22.92		1443.40
N	322		322	
R_c	0.34		0.91	
λ	0.88		0.17	
χ^2	28.49		402.09	
p	<0.001		<0.001	
Accuracy	73.9%		98.1%	
Specificity	78.3%		100%	
Sensitivity	58.0%		91.3%	
Classification matrices	**nC-patient**	**C-patient**	**nC-patient**	**C-patient**
nC-patient	**198**	55	**253**	0
C-patient	29	**40**	6	**63**

The bold characters refer to the number of cases classified correctly; Cprot is the cancer proteins, nCprot is the non-cancer proteins, C-patient is the cancer patients, nC-patient is the non-cancer patients; θ_0(prot/canc) = p(canc)*θ_0(prot); (C1), (C2) and (C3) are the non-equivalent coordinate used to develop the model; N is the number of patients included in the discriminant analysis calculation, R_c is the canonical regression coefficient, Wilk's λ is the standard statistic used to denote the statistical significance of the discriminatory power of the current model, χ^2 is the Chi-squared statistic, and p is the level of error.

factors. The simple linear mathematical form of the model in our work has been chosen in the absence of prior information. The basic assumption of LDA for this model was confirmed (Van Waterbeemd 1995). The distribution of the residuals for all patients (raw residuals *vs.* case number) shows no pattern.

The residual is the difference between the observed and the predicted values. A better threshold for the *a priori* classification probability can be estimated by means of the Receiver Operating Characteristic (ROC) curve (Hanley and McNeil 1982). The ROC curve presented a pronounced curvature (convexity) with respect to the y = x line for both training and predicting series. This result confirms that the present model is a significant classifier that has an area under the ROC curve (about 1.0 higher than 0.5), which is the value for a random classifier (Swets 1988).

The validity of the LDA models depends on the normal distribution of the sample used as well as on the homogeneity of their variances. Thus, we carry out two significant tests of normality: Chi-Square and Kolmogorov–Smirnov tests. In statistics, the Kolmogorov–Smirnov test (often called the KS test) is used to determine whether two underlying one-dimensional probability distributions differ, or whether an underlying probability distribution differs from a hypothesized distribution, in either case based on finite samples. The one-sample KS test compares the empirical distribution function with the cumulative distribution function specified by the null hypothesis. The main applications test goodness of fit with the normal and uniform distributions. The two-sample KS test is one of the most useful and general non-parametric methods to compare two samples, as it is sensitive to differences in both location and shape of the empirical cumulative distribution functions of the two samples. The Kolmogorov–Smirnov test for normality is based on the maximum difference between the sample cumulative distribution and the hypothesized cumulative distribution. If the d statistic were significant, consequently the hypothesis that the respective distribution would be normal should be rejected. The Chi-square test for independence examines whether knowing the value of a variable helps to estimate the value of another variable. The chi-square test for homogeneity examines whether two populations have the same proportion of observations with a common characteristic. We tested the normal distribution of residual using various fitting tests; the results were as follows: Kolmogorov–Smirnov d = 0.20 with $p < 0.01$; Chi-squared = 88.0, with $p = 0.00$. The tests have confirmed that it is very close to a normal distribution. Similar results were obtained with the Star Graph representation. The protein Star Graph model has indeed statistical parameters similar to those of the protein Lattice network model. In this case then, the goodness of the model does not seem to depend on the graphic representation but on the protein sequence. Different results were obtained using the database of Petricoin et al. In this case, the best models were developed from the Star Graph representation. The model 3.4 shows a 98.1% total accuracy, 100% specificity and 91.3% sensitivity.

These good results are confirmed by different statistical tests to which the model was subjected. The value of canonical regression coefficient R_c and Wilk's λ are, respectively, of 0.91 and 0.17. The scatter plots of the square standardized residual *vs.* the respective θ_k included in the MS Star Graph model reveal an adequate scatter on the points without any consistent pattern, which validates the assumption of homocedasticity (i.e. homogeneity of variance of variables) (Stewart and Gill 1998).

7.4 Conclusions

The actual work proposes several QPDR/QSAR models based on TIs and/or CIs of Star graphs or Lattice networks of protein sequence or MS outcomes of blood proteome in order to predict the proteins related to breast and colon cancer and to improve the diagnostic potential of the PSA biomarker for prostate cancer. We hope these results will help in the earlier detection of the BC and CC and in the discovering of new cancer-related proteins. The advantages of this method, such as simplicity (there are no complicated experiments needed), fast calculations and few necessary resources (MARCH-INSIDE and S2SNet are free software programmes and the calculation can run in normal computers), make it an ideal theoretical scheme that can be extended easily to other types of diseases or other *omics*, such as genomics, transcriptomics, and metabolomics.

Acknowledgements We thank for the grant (2007/127 and 2007/144) from the General Directorate of Scientific and Technologic Promotion of the Galician University System of the Xunta de Galicia, for the grant (PIO52048 and RD07/0067/0005) funded by the Carlos III Health Institute, and for the grant (File 2006/60, 2007/127 and 2007/144) from the General Directorate of Scientific and Technologic Promotion of the Galician University System of the Xunta de Galicia (Spain). González-Díaz and Munteanu acknowledge for the contract/grant sponsorships of Isidro Parga Pondal Programme, Xunta de Galicia (Spain). Ferino gratefully acknowledges a PhD scholarship from the University of Cagliari, actually developing a Bioinformatics research under Supervision of Professor Uriarte and Dr. González-Díaz at the University of Santiago de Compostela (Spain).

References

Aguero-Chapin G, González-Díaz H et al (2006) Novel 2D maps and coupling numbers for protein sequences. The first QSAR study of polygalacturonases; isolation and prediction of a novel sequence from Psidium guajava L. FEBS Lett 580:723–730

Altaf-Ul-Amin M, Shinbo Y, Mihara K, Kurokawa K et al (2006) Development and implementation of an algorithm for detection of protein complexes in large interaction networks. BMC Bioinformatics 7:207

Althaus IW, Chou KC, Franks KM et al (1996) The benzylthio-pyrididine U-31, 355 is a potent inhibitor of HIV-1 reverse transcriptase. Biochem Pharmacol 51:743–750

Andraos J (2008) Kinetic plasticity and the determination of product ratios for kinetic schemes leading to multiple products without rate laws: new methods based on directed graphs. Can J Chem 86:342–357

Arteca GA, Mezey PG (1990) A method for the characterization of foldings in protein ribbon models. J Mol Graph 8:66–80

Bader GD, Hogue CW (2003) An automated method for finding molecular complexes in large protein interaction networks. BMC Bioinformatics 4:2

Balaban AT, Basak SC, Beteringhe A et al (2004) QSAR study using topological indices for inhibition of carbonic anhydrase II by sulfanilamides and Schiff bases. Mol Divers 8:401–412

Barabasi AL (2005) Sociology. Network theory – the emergence of the creative enterprise. Science 308:639–641

Barabasi AL, Bonabeau E (2003) Scale-free networks. Sci Am 288:60–69

Barabasi AL, Oltvai ZN (2004) Network biology: understanding the cell's functional organization. Nat Rev Genet 5:101–113

Bielinska-Waz D, Nowak W, Waz P et al (2007) Distribution moments of 2D-graphs as descriptors of DNA sequences. Chem Phys Lett 443:408–413

Bisquerra Alzina R (1989) Introducción conceptual al análisis multivariante: Un enfoque informático con los paquetes SPSS-X, BMDP. LISREL y SPAD, PPU, Barcelona

Borges JC, Cagliari TC, Ramos CH (2007) Expression and variability of molecular chaperones in the sugarcane expressome. J Plant Physiol 164:505–13

Bougnoux P, Giraudeau B, Couet C (2006) Diet, cancer, and the lipidome. Cancer Epidemiol Biomarkers Prev 15:416–421

Bougnoux P, Hajjaji N, Couet C (2008) The lipidome as a composite biomarker of the modifiable part of the risk of breast cancer. Prostaglandins Leukot Essent Fatty Acids 79:93–96

Brzezinska E (2003) The QSAR analysis of tricyclic non-nucleoside inhibitors of HIV-1 reverse transcriptase. Acta Pol Pharm 60:3–13

Chen J, Chua HN, Hsu W et al (2006) Increasing confidence of protein-protein interactomes. Genome Inform 17:284–297

Chou KC (1989) Graphical rules in steady and non-steady enzyme kinetics. J Biol Chem 264:12074–12079

Chou KC (1990) Review: applications of graph theory to enzyme kinetics and protein folding kinetics. Steady and non-steady state systems. Biophys Chem 35:1–24

Chou KC, Kezdy FJ, Reusser F (1994) Review: steady-state inhibition kinetics of processive nucleic acid polymerases and nucleases. Anal Biochem 221:217–230

Chou KC, Zhang CT, Elrod DW (1996) Do antisense proteins exist? J Protein Chem 15:59–61

Chou KC, Zhang CT (1992) Diagrammatization of codon usage in 339 HIV proteins and its biological implication. AIDS Res Hum Retroviruses 8:1967–1976

Cho WC, Cheng CH (2007) Oncoproteomics: current trends and future perspectives. Expert Rev Proteomics 4:401–410

Coghlan A, Eichler EE, Oliver SG et al (2005) Chromosome evolution in eukaryotes: a multi-kingdom perspective. Trends Genet 21:673–682

Cruz-Monteagudo M, Gonzalez-Diaz H, Borges F et al (2008) 3D-MEDNEs: An alternative "*in silico*" technique for chemical research in toxicology. 2. Quantitative proteome-toxicity relationships (QPTR) based on mass spectrum spiral entropy. Chem Res Toxicol 21:619–632

Del Chiaro M, Zerbi A, Falconi M et al (2007) Cancer risk among the relatives of patients with pancreatic ductal adenocarcinoma. Pancreatology 7:459–469

Devillers J, Balaban AT (1999) Topological indices and related descriptors in QSAR and QSPR. Gordon & Breach, The Netherlands

Ding J, Sorensen CM, Jaitly N et al (2008) Application of the accurate mass and time tag approach in studies of the human blood lipidome. J Chromatogr B Analyt Technol Biomed Life Sci 871:243–252

Dobson PD, Doig AJ (2003) Distinguishing enzyme structures from non-enzymes without alignments. J Mol Biol 330:771–783

Dobson PD, Doig AJ (2005) Predicting enzyme class from protein structure without alignments. J Mol Biol 345:187–199

Estrada E (2002) Characterization of the folding degree of proteins. Bioinformatics 18:697–704

Estrada E, Uriarte E (2001) Recent advances on the role of topological indices in drug discovery research. Curr Med Chem 8:1573–1588

Ferino G, Gonzalez-Diaz H, Delogu G et al (2008) Using spectral moments of spiral networks based on PSA/mass spectra outcomes to derive quantitative proteome-disease relationships (QPDRs) and predicting prostate cancer. Biochem Biophys Res Commun 372: 320–325

Freeze HH (2006) Genetic defects in the human glycome. Nat Rev Genet 7:537–551

González-Díaz H, González-Díaz Y, Santana L et al (2008) Proteomics, networks and connectivity indices. Proteomics 8:750–778

González-Díaz H, Vilar S, Santana L et al (2007) Medicinal chemistry and bioinformatics – current trends in drugs discovery with networks topological indices. Curr Top Med Chem 7:1025–1039

Green ML, Karp PD (2004) A Bayesian method for identifying missing enzymes in predicted metabolic pathway databases. BMC Bioinformatics 5:76

Jain KK (2007) Recent advances in clinical oncoproteomics. J Buon 12:S31–38

Hanley JA, McNeil BJ (1982) The meaning and use of the area under a receiver operating characteristic (ROC) curve. Radiology 143:29–36

Klose J (1989) Systematic analysis of the total proteins of a mammalian organism: principles, problems and implications for sequencing the human genome. Electrophoresis 10:140–152

Kirkegaard T, McGlynn LM, Campbell FM et al (2007) Amplified in breast cancer 1 in human epidermal growth factor receptor – positive tumors of tamoxifen-treated breast cancer patients. Clin Cancer Res 13:1405–1411

Latha B, Venkatesh B (2004) Granulometric analysis of spots in DNA microarray images. Genomics Proteomics Bioinformtics 2:222–236

Lee S, Lee B, Jang I et al (2006) Localizome: a server for identifying transmembrane topologies and TM helices of eukaryotic proteins utilizing domain information. Nucleic Acids Res 34:W99–103

Liao B, Wang TM (2004) Analysis of similarity/dissimilarity of DNA sequences based on non-overlapping triplets of nucleotide bases. J Chem Inf Comput Sci 44:1666–1670

Liao B, Xiang X, Zhu W (2006) Coronavirus phylogeny based on 2D graphical representation of DNA sequence. J Comput Chem 27:1196–1202

Hu Z, Mellor J, Wu J et al (2004) VisANT: an online visualization and analysis tool for biological interaction data. BMC Bioinformatics 5:17

Mayer A, Takimoto M, Fritz E et al (1993) The prognostic significance of proliferating cell nuclear antigen, epidermal growth factor receptor, and mdr gene expression in colorectal cancer. Cancer 71:2454–2460

Morelle W, Flahaut C, Michalski JC et al (2006) Mass spectrometric approach for screening modifications of total serum N-glycome in human diseases: application to cirrhosis. Glycobiology 16:281–293

Munteanu CR, Gonzalez-Diaz H, Magalhães AL (2008) Enzymes/non-enzymes classification model complexity based on composition, sequence, 3D and topological indices. J Theor Biol 254:476–482

Nandy A, Basak SC (2000) Simple numerical descriptor for quantifying effect of toxic substances on DNA sequences. J Chem Inf Comput Sci 40:915–919

Notebaart RA, Teusink B, Siezen RJ et al (2008) Co-regulation of metabolic genes is better explained by flux coupling than by network distance. PLoS Comput Biol 4:e26

Perez MA, Sanz MB, Torres LR et al (2004) A topological sub-structural approach for predicting human intestinal absorption of drugs. Eur J Med Chem 39:905–916

Petricoin EF, Ornstein DK, Paweletz CP et al (2002) Serum proteomic patterns for detection of prostate cancer. J Natl Cancer Inst 95:489–490

Qi XQ, Wen J, Qi ZH (2007) New 3D graphical representation of DNA sequence based on dual nucleotides. J Theor Biol 249:681–690

Randic M (2000) Condensed representation of DNA primary sequences. J Chem Inf Comput Sci 40:50–56

Randic M, Balaban AT (2003) On a four-dimensional representation of DNA primary sequences. J Chem Inf Comput Sci 43:532–539

Randic M, Vracko M (2000) On the similarity of DNA primary sequences. J Chem Inf Comput Sci 40:599–606

Randic M, Vracko M, Nandy A et al (2000) On 3-D graphical representation of DNA primary sequences and their numerical characterization. J Chem Inf Comput Sci 40:1235–1244

Randic M, Zupan J, Vikic-Topic D (2007) On representation of proteins by star-like graphs. J Mol Graph Model 26:290–305

Rivera MP, Stover DE (2004) Gender and lung cancer. Clin Chest Med 25:391–400

Sjoblom T, Jones S, Wood LD et al (2006) The consensus coding sequences of human breast and colorectal cancers. Science 314:268–274

Stewart J, Gill L (1998) Econometrics. Prentice-Hall, London

Swets JA (1988) Measuring the accuracy of diagnostic systems. Science 240:1285–1293

Van Waterbeemd H (1995a) Chemometric methods in molecular design. Wiley-VCH, New York

Van Waterbeemd H (1995b) Discriminant analysis for activity prediction. In: Van Waterbeemd H (ed) Chemometric methods in molecular design, 2nd edn. Wiley-VCH, New York

Wang J, Zhang X, Ge X et al (2008) Proteomic studies of early-stage and advanced ovarian cancer patients. Gynecol Oncol 111:111–119

Ward DG, Suggett N, Cheng Y et al (2006) Identification of serum biomarkers for colon cancer by proteomic analysis. Br J Cancer 94:1898–905

Welsh JB, Sapinoso LM, Su AI et al (2001) Analysis of gene expression identifies candidate markers and pharmacological targets in prostate cancer. Cancer Res 61:5974–5978

Wishart DS, Knox C, Guo AC et al (2008) HMDB: a knowledgebase for the human metabolome. Nucleic Acids Res 37:D603–610

Zhu D, Qin ZS (2005) Structural comparison of metabolic networks in selected single cell organisms. BMC Bioinformatics 6:8

Chapter 8
The Use of Metabolomics in Cancer Research

B. van Ravenzwaay, G.C. Cunha, E. Fabian, M. Herold, H. Kamp, G. Krennrich, A. Krotzky, E. Leibold, R. Looser, W. Mellert, A. Prokoudine, V. Strauss, R. Trethewey, T. Walk, and J. Wiemer

Abstract The use of metabolite profiling techniques (metabonomics or metabolomics) in toxicology is a relatively new branch of this science. Due to their unique biochemical properties, cancer cells should, in principle, be an ideal field of application for metabolite profiling. However, due to technical and study design limitations there are only a few reliably metabolite profiles for human tumors. This chapter provides examples for the recognition of metabolic changes in animals induced by exposure to (carcinogenic) chemicals. In two major projects (COMET and MetaMapTox), data bases have been developed which are sufficiently large to evaluate the full potential of metabolite profiling in toxicology and cancer research. In both projects blood and urine were used as matrices which can be easily obtained with minimally-invasive methods. Based on a high degree of standardization and a large-scale controlled data collection, consistent patterns of metabolite changes have been identified which are associated with different toxicological modes of action, some of which are known to enhance tumor development in rodents.

B. van Ravenzwaay (✉), G.C. Cunha, E. Fabian, H. Kamp, W. Mellert, and V. Strauss
BASF SE, Experimental Toxicology and Ecology, Z 470, D-67056, Ludwigshafen, Germany
e-mail: bennard.ravenzwaay@basf.com

G. Krennrich
BASF SE, Computational Chemistry, B9, D-67056, Ludwigshafen, Germany

M. Herold, A. Krotzky, E. Leibold, R. Looser, A. Prokoudine, R. Trethewey, T. Walk, and J. Wiemer
Metanomics, GmbH, Tegeler Weg 33, 10589, Berlin, Germany

W.C.S. Cho (ed.), *An Omics Perspective on Cancer Research*,
DOI 10.1007/978-90-481-2675-0_8, © Springer Science+Business Media B.V. 2010

8.1 Introduction

8.1.1 General

The first section provides an overview of the potential of the metabolite profiling techniques (referred to as metabonomics if NMR-based and metabolomics if MS-based) in toxicology and its potential for cancer research. In the second section the two commonly used techniques for metabolite profiling (NMR- and MS-based technologies) are introduced. In the third section the current use of these technologies and their future potential for cancer diagnostics is discussed. In addition, the main challenges to progress metabolite profiling to an essential tool in cancer diagnostics are presented. In the forth section results obtained from metabolite profiling with NMR- and MS-based technique for several compounds affecting the liver (with modes of action related to tumor promotion) are presented and discussed. In the final part perspectives are given on how metabolite profiling can be improved and how it may be used in the early detection of (liver) tumor formation.

Tumors are thought to be the end result of a multi-stage (initiation, promotion, progression) process. The first phase in the process of tumor formation (initiation) is a consequence of the exposure to genotox compounds but can also result from pre-existing genetic conditions. The second phase (promotion) involves a great number of potential processes, which ultimately result in a proliferative stimulus, providing the "right" environment for the initiated cells to express their growth advantages compared to normal cells. The progression stage is again thought to involve genetic changes, either induced from external sources (genotoxic agents) or through genetic instability of the (preneoplastic) tumorigenic lesion (Pitot 1986).

Mechanisms involved in liver tumor promotion include: (1) liver toxicity: resulting in a hepatocyte loss and a subsequent regenerative cell proliferation stimulus (Schulte-Hermann, 1974); (2) liver enzyme induction: resulting in an adaptive cell proliferation stimulus, in rodents often accompanied by nuclear polyploidisation as well as a reduction of apoptosis (van Ravenzwaay et al. 1987; Gamer et al. 2002); (3) receptor mediated stimulus of cell proliferation, e.g. PPR-alpha receptor and Ah-Receptor. (Kota et al. 2005; Peters et al. 2005).

8.1.2 Metabolite Profiling of Liver Tumor Promoters

Due to the diversity of possible actions of liver tumor promoters, the correct identification and assessment is a rather complex process. Current methods comprise of clinical- and pathological-investigations and biochemical analysis. The main biochemical markers which are currently analyzed are alanine transaminase, aspartate transaminase, alkaline phosphatase and bilirubin. However, these biomarkers are subject to complex regulation and changes may sometimes occur only at rather progressed stages. A drawback of the histopathological assessments is that they are invasive

(Gunawan and Kaplowitz 2004). Taking into account the fact that most, if not all liver tumor promotion processes induce particular biochemical changes, the sensitive and selective analysis of these changes by metabolite profiling offers the possibility of a relatively easy, non-invasive and highly discriminative identification.

8.2 Methods

8.2.1 Metabolite Profiling – General

Metabolite profiling (analysis of endogenous low molecular compounds such as carbohydrates, amino acids, lipids, organic acids, etc.) has a long history of application in the plant sciences (Trethewey et al. 1999; Sauter et al. 1991) but it is a relatively new technology in toxicology studies to elucidate changes in biochemical pathways (Lindon et al. 2004b). The analysis is performed routinely by using blood or urine of animals. Two technologies are mostly applied: (1) the profiling by NMR is called "metabonomics" (Griffin et al. 2004; Lindon et al. 2004a); (2) the use of the LC-MS and GC-MS techniques is referred to as "metabolomics" (van Ravenzwaay et al. 2007; Looser et al. 2005; Weckwerth and Morgenthal, 2005; Wilson et al. 2005; Fernie et al. 2004). Most of the metabolite profiling data published so far have been developed using NMR metabonomics from urine of rats treated with compounds, which were toxic to the liver (Maddox et al. 2006; Clayton et al. 2003; Schoonen et al. 2007b). Although the focus on urinary metabolites is understandable as an easy, non invasive matrix, this does limit the potential number and nature of metabolites and will create a certain distortion of the information. Particularly with chemicals which would cause kidney toxicity and toxicity in any other organ at the same time, the effects on the kidney may greatly bias the overall metabolite profile. A number of researchers have used blood as a matrix for metabolite profiling (Lindon et al. 2003; Kleno et al. 2004; Heijne et al. 2005). Most of these studies have been conducted using NMR metabonomics. MS-based metabolite profiling, especially when coupled to a chromatographic preseparation (e.g. GC or HPLC), is several orders of magnitude more sensitive than NMR-based metabolomics so more metabolites (e.g. hormones) can be detected at lower concentrations. In a typical ^{1}H-NMR spectrum about 30-40 known metabolites can be detected and quantified from a biological sample of some 100 μL. MS-based methods are able to measure hundreds of metabolites from even smaller sample volumes. For most of the NMR applications only little sample preparation is necessary and the measurement is usually fast and non-destructive whereas for MS methods, extraction and derivatization steps may be necessary before the samples can be injected into the mass spectrometric system. The ability to run different ionization techniques in MS allow for detection of different classes of molecules either very broad or very specific. Especially for GC-MS the use of mass spectral libraries facilitates the identification of unknown metabolites even at low concentrations. For a detailed review of the advantages and disadvantages of NMR- and MS-based metabolite profiling see Lindon and Nicholson (2008).

8.2.2 NMR Metabonomics

^{1}H-NMR produces a spectrum containing a number of peaks. The heights and positions of these peaks enable researchers to accurately determine the carbon-hydrogen framework of an organic molecule. NMR spectroscopy combines high-resolution nuclear magnetic resonance techniques with statistical data analysis methods to evaluate the metabolic status of an organism. When a molecule is placed in a magnetic field, the magnetic momentum of the nuclei of these atoms tends to assume specific orientation with respect to the field. A pulse of electromagnetic energy at the resonant frequency is then used to tip the magnetic spins of the nuclei away from their orientation along the magnetic field lines. When the perturbing radio-frequency is switched off, the magnetic momentum of the nuclei returns to their original lowest energy orientations. During this process they transmit energy which can be picked up by a radiofrequency receiver and transformed with the aid of a computer into a spectrum (Claudino et al. 2007; Lindon et al. 2003; Heijne et al. 2005).

The chemical shift of a nucleus is the difference between the resonance frequency of the magnetic spins of the examined nucleus and a nucleus of the standard divided by the operation frequency of the magnet. In ^{1}H-NMR spectroscopy, this standard is often tetramethylsilane-Si $(CH_3)_4$. Different biologic molecules contain different arrangements of ^{1}H, ^{13}C, ^{15}N, and ^{31}P atoms and have distinct chemical shifts in the NMR spectra. A signal spectrum for a specific element can be obtained, and assignments of these signals enable chemical analysis of the samples. Hence, ^{1}H-NMR is a qualitative method. An array of data is produced, and chemoinformatic tools are used to clarify and interpret the results. Signal resolution has improved steadily for the past 2 decades as the field strength of magnets has increased while cost has dropped. Currently, the 600 MHz ^{1}H-NMR, is the mostly used spectroscopy hardware (Claudino et al. 2007; Ishihara et al. 2006; Connor et al. 2004).

8.2.3 GC-MS/LC-MS Metabolomics

The GC-MS/LS-MS metabolomics technology used by BASF metanomics is summarized below and the condition for the development of the MetaMapTox data base are described. For more detailed information see van Ravenzwaay et al. 2007 and Looser et al. 2005.

Wistar rats were maintained in an air-conditioned room under standardized environmental conditions. For all compounds tested blood samples were taken after 7,

14 and 28 days at the same period of time in order to avoid changes related to circadian rhythms. The study design can be best compared to a highly standardized OECD 407 guideline design with two dose levels. For mass spectrometry-based metabolite profiling analyses, plasma samples were extracted and a polar and a non-polar fraction prepared. For GC-MS analysis, the non-polar fraction was treated with methanol under acidic conditions to yield the fatty acid methyl esters. Both fractions were further derivatized with an O-methyl-oxime-generating reagent and subsequently with a trimethylsilylating reagent before analysis. For LC-MS analysis, both fractions were reconstituted in appropriate solvent mixtures. HPLC was performed by gradient elution on reversed phase separation columns. For mass spectrometric detection methods are applied which allows target and high sensitivity MRM (Multiple Reaction Monitoring) measurement of selected target metabolites in parallel to a full screen metabolite profiling analysis. Following comprehensive analytical validation steps, the data for each analyte were normalized against data from pool samples. These samples were run in parallel through the whole process to account for process variability.

The data generated were analyzed by univariate and multivariate statistical methods and a sex- and day-stratified heteroscedastic *t*-test ("Welch test") was applied to compare treated groups with respective controls. *P*-values and ratios of corresponding group medians were collected as metabolic profiles and fed into a database. Using the above mentioned MS based metabolomics technology, BASF has established a large metabolomics data base for chemicals, agrochemicals and drugs (MetaMap®Tox).

8.3 Metabolite Profiling and Cancer

Due to their unique biochemical properties, cancer cells should, in principle, be an ideal field of application for metabolite profiling. The biochemistry of neoplasia is characterized by a high glucose demand as well as elevated glycolytic activity. In addition to these features, noted by Warburg in the first half of the last century, there are other typical features of cancer cells such as decreased mitochondrial activity (due to the generally reduced availability of oxygen in many tumors), increased choline-phospholipid metabolism and increased lactate concentrations (Glunde and Serkova 2006; Serkova et al. 2007). These general features provide an opportunity for cancer diagnosis, however, lack of organ specificity when working with accessible body fluids such as blood or urine may limit the practical use of metabolite profiling. Nevertheless, for some tumors specific metabolite profiles have been observed in the tumor tissue. In brain tumors N-acetyl-aspartate has been shown to be decreased, whereas alanine was found increased according to several authors (Griffin and Shockcor 2004; Gillies and Morse 2005). In prostate cancer a decrease of citrate was noted (Griffin and Shockcor 2004; Costello and Franklin 2006) and in gliomas an increase of myo-inositol has been reported (Griffin and Shockcor 2004). These examples demonstrate definitive changes in the biochemistry of tumors, however, it is likely that changes in single (or few) metabolites (particularly if these involve common metabolites such as glucose, lactate or alanine) are not going to be unique enough for accurate diagnosis of cancer. Too many other causes may be related to changes in these metabolites: e.g. diseases, exposure to xenobiotica, and nutritional status. Therefore, the identification of more extensive patterns of metabolite changes is a prerequisite to accurately detect specific tumors.

One of the earliest reports describing the successful identification of tumors using a pattern of metabolite change was from Tate et al. (2000). Samples of malignant renal cortex tissue could be clearly separated from normal tissue using NMR metabonomics. With the aid of partial least square discriminant analysis (PLS-DA) they were able to distinguish with 100% accuracy between normal and malignant tissue, demonstrating the potential of metabolite profile recognition.

A step towards easier cancer diagnosis was reported by Odunsi et al. (2005) using serum samples of ovarian cancer patients and NMR based metabonomics pattern recognition. With principal component analysis (PCA) the study demonstrated a 100% identification of patients with epithelial ovarian cancer *versus* premenopausal healthy women.

A similar success was obtained for the detection of bladder cancer in human urine using MS-based metabolomics (Issaq et al. 2008). Urine collected from 48 healthy individuals was compared to that of 41 patients diagnosed with kidney transitional cell carcinoma. Using PLS-DA a correct prediction of all healthy and all tumor-bearing people involved in this study was possible. With PCA, 46 of the 48 healthy and 40 of the 41 bladder cancer urine samples were correctly identified. Despite the successful recognition of tumors with patterns of changes, in both studies, the metabolites which were responsible for the recognition pattern could not be exactly identified or quantified. Pattern recognition using PCA analysis was also successful in correctly identifying liver cancer patients. In this study reported by Yang et al. (2004), an important further step was achieved for practical cancer diagnosis. This study compared the urine of 50 healthy adults, 77 patients with liver diseases (27 hepatocirrhosis patients, 30 acute hepatitis patients and 20 chronic hepatitis patients) with those of 48 liver cancer patients using MS-based and NMR-based metabolite profiling. Starting with an initial set of 15 known urinary nucleosides a differentiation was obtained between patients with acute hepatitis, chronic hepatitis and hepatocirrhosis using quantitative analysis of 7 specific metabolites. Unfortunately, even with the quantitation of all 15 metabolites, liver cancer patients could not be separated from hepatocirrhosis patients. When PCA analysis was applied on a metabolite pattern of 113 urinary metabolite peaks, however, correct separation was achieved. This study, however, also suffered from the problem that most of the peaks could not be identified.

Metabolite profiling of normal human colon tissue and that of patients with colorectal cancer resulted in a total of 82 (out of 206) metabolites being different at a level of $p < 0.01$ following supervised analysis between normal and neoplastic tissue (Denkert et al. 2008). Twenty-five metabolites were up-regulated in neoplastic tissue (mainly metabolites related to the urea cycle, i.e. purines, pyrimidines and amino acids), while 57 were down-regulated (particularly TCA cycle metabolites and lipids).

The high percentage of changes between normal and neoplastic tissue was considered to be related to the biologically and biochemically highly different states of these tissues, and can be explained to a great extend by the metabolic disregulation of cancer cells. Thus, the findings as such cannot be immediately linked to a tissue specific neoplastic process. To help to interpret their findings, they connected the metabolite profiles with the existing knowledge on metabolic pathways as described in the Kyoto encyclopedia of genes and genomes (KEGG).

The importance of pattern recognition was also shown by Denkert et al. (2006). In their studies they analyzed 66 invasive ovarian carcinomas and nine borderline ovarian tumors using MS-based metabolomics. A total of 291 metabolites were detected in the tissues, out of which 114 were annotated as known compounds. *T*-test statistics ($p < 0.01$) showed that 51 metabolites were significantly different between borderline tumors and carcinomas. PCA analysis of the data allowed for an 88% separation of the borderline tumors from the carcinomas. The most prominent differences were seen as an up-regulation (>3 fold) of the following metabolites: alpha-glycerolphosphate (5.3 fold), uracil (4.2 fold) fold, hypoxantine (3.8 fold), pyrazine-2,5-bishydroxy (3.4 fold), inositol-2-phosphate (3.3 fold), phosphoric acid (3.2 fold). The most prominent down regulations were seen for nonadecanoic acid (1.2 fold), stearic acid (1.4 fold), heptadecanoic acid (14 fold), benzoic acid (1.6 fold) and lactic acid (2.2 fold). The ability to distinguish between different tumor characteristics within one organ is a major step towards the correct diagnosis of the disease. Griffin and Kauppinen (2007) provide an overview of metabolome differences between normal and neoplasia tissue from brain, connective tissue, lymphomas, liver, colon and prostate based on metabolite profiling in humans, animals and cell lines. These observations indicate, as predicted, that metabolic profiles in neoplastic tissue is fundamentally different from that of normal tissue.

In conclusion, the work described above demonstrates that: (1) tumors have a general metabolite profile that is different from normal cells; (2) most tumors will have a unique profile of their own; (3) the profile of at least some tumors is different during early stage and late stage (carcinoma) of the cancer process; and (4) advances are being made in the detection of tumors in patients using non invasive methods such as urine or blood sampling.

8.3.1 Challenges

There are still significant challenges which need to be addressed before metabolite profiling can progress to an essential tool in cancer diagnostics. The overwhelming amount of data accumulated in the omics sciences make it difficult to see the forest through the trees, i.e. to identify those changes which are particularly relevant to carcinogenesis. A further challenge is to integrate all of the relevant information in a transparent and systematical manner. These necessities are not new and have been raised several years ago for plant metabolomics. Bino et al. (2004) noted that for the maturation of metabolomics three objectives need to be achieved: (1) improvement in the comprehensive coverage of [plant] metabolomics; (2) facilitation of comparison of results between laboratories and experiments; and (3) enhanced integration of metabolomics with other functional information. No doubt that these objectives are universal and apply also to human and environmental health metabolomics. ECETOC (2008) concluded that quality standards need to be defined for all omics technologies to improve confidence and (regulatory) acceptance of data and conclusions. In addition to more general features of standardization, one additional quality standard, which was used in the MetaMapTox project, is the regular inclusion of "positive" controls, (compounds with known effects on the metabolome), during the data base development. These positive control substances help to monitor the quality of both the biological as well as the analytical procedures.

Visualization of genetic and metabolic pathways using the KEGG diagrams is a useful tool for the interpretation of large-scale data sets (Denkert et al. 2008). However, as metabolite changes may result from different pathways and not all metabolites from a single pathway can be detected there are some limitations to the general use of this tool. Moreover, the various cross-links between the different pathways that may be relevant are difficult to visualize in a complete manner.

The clinical utility of metabolite profiles relies on the establishment of correlations between metabolite data and clinical measurements. Because metabolite profiling is a quantitative study, the assessment of tumors based on visual (histo-) pathological assessment introduces potential biases which may limit the reliability of such correlations. Burns et al. (2004) have shown that this potential problem can be overcome by the introduction of computer aided image analysis in the case of prostate pathology slides. Their studies showed a two fold difference between human visual assessment and computer aided assessment of 28 samples of prostatectomy cases. Positive linear correlations were found between metabolites being indicative for normal epithelium (polyamines and citrate) and metabolites indicative of prostate cancer (choline and the sum of phosphocholine and choline) using the computer aided diagnostics.

An additional challenge is the lack of correlation between *in vitro* (cell line or fresh culture) data and *in vivo* (human) tumors. Part of this problem is related to the limited amount of data comparing the metabolite profile of compounds tested *in vitro* and *in vivo*. Even less data exist for tumor cell lines and *in vivo* tumors.

A relatively simple but nevertheless important challenge is the concept of biomarker(s). A biomarker is defined as a characteristic that is measured and evaluated as an indicator of normal biological processes, pathogenic processes or a pharmacological response to a therapeutic intervention. The potential problem is the term biomarker implicates that one specific parameter will be the sole solution provider. In reality, however, patterns of change contain far more information and a more robust and reliable than single markers (Christians et al. 2005). Specifically for cancer, which is a polygenetic disease, most diagnostic assessments will probably have to rely on multiple markers (Jain 2007). A further complicating factor is that specific pharmacological information may only be noted at low dose levels and that this information is lost or at least distorted at higher, toxic, dose levels due to competing mechanisms. In the MetaMap®Tox project, we have come across several of such cases. This indicates that dose selection is a potential confounding factor in metabolome research and needs to be taken into account during subsequent assessment of findings. The identification of specific biomarkers or patterns of change, however, is only one step towards clinical application of such findings. Brennan et al. (2007) note that whilst great advance have been made in discovery of putative biomarkers in DNA microarrays, few have been transformed into clinical applicability. Metabolite profiling may have some advantages: (1) metabolite changes are the product of several processes with profound feedback mechanisms; (2) first attempts to integrate omics information into systems biology suggest that metabolite profile is quite robust; and (3) metabolites have been traditionally used for clinical assessments, there may be reason to believe that metabolite profiling will be more successfully transferred into clinical applications.

8.4 Metabolite Profiles

In this chapter an overview will be given of the most salient results obtained so far using either NMR metabolite profiling of mostly urinary samples or GC-MS/LC-MS metabolite profiling of serum samples with compounds affecting the liver.

8.4.1 Control Animals

8.4.1.1 NMR Metabonomics

There are relatively few data available on the reproducibility of metabolite profiles following NMR metabolomics. This lack of such data can at least partly be explained by the fact that there have been few attempts to develop and publish metabolic profiles of control animals obtained in a standardized data base development. One exception, however, is the COMET project. In this project all animal studies (and the subsequent metabonomics) were carried out according to a standardized protocol.

8.4.1.2 GC-MS/LC-MS Metabolomics

In the MetaMap®Tox project, a great number of studies with untreated control rats were performed over a time period of approximately 3 years and the data were used to establish base line values and to investigate if stable and reproducible values could be obtained. In an analysis of control samples over nearly 3 years no seasonal fluctuation was observed (ECETOC 2008). Currently the method employed allows for the detection of almost 300 blood metabolites, which fulfilled the criteria for quantification, stability and reproducibility. These metabolites were used for the detection of biomarkers and specific patterns of change in the subsequent studies in which animals were exposed to test compounds.

Using PCA analysis of the metabolite profile of untreated control rats it can be seen that the major component driving the differences in the collective control group is the sex (see Fig. 8.1).

Using blood samples obtained from control animals over the course of nearly 3 years the variance of the parameters over time was analyzed. Figure 8.2 shows

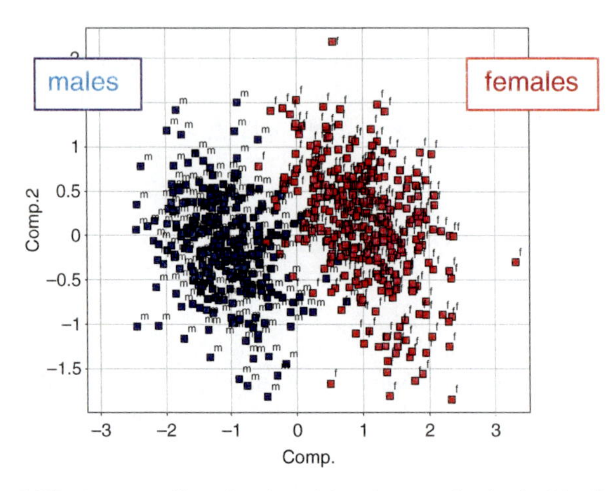

Fig. 8.1 PCA of 670 plasma profiles of male and female control animals. Distribution of profiles in space of first two principal components (first component horizontal – 23% variance, second component vertical – 6% variance)

Fig. 8.2 Technical and biological variation during the MetaMap Tox project development. Metabolite specific coefficient of variation (CV) was determined for reference samples (technical pool – *green line*) as well as for male and female control samples (male controls – *blue line* and female controls – *red line*). The plots shows median values (*circles*) and first and third quartiles (*error bars*) of the CV distribution of 233 standard high quality metabolites

the analytical variance and the biological variance in rats. From these data several conclusions can be drawn: (1) the biological variance is far larger than the analytical variance; (2) the variance within males and females is quite similar; (3) There is no seasonal component in the variance. Over time both analytical and biological variance was reduced slightly through increased standardization and experience gained through the course of the project.

The variance found during metabolite profiling using HPLC coupled with electrochemical array detection was reported by Shurubor et al. (2007). They observed that the median coefficient of variation (CV) for analytical parameters (total of 66) was 12%. This value is quite in line with the analytical CV shown in Fig. 8.2 (for 233 metabolites) which is approximately 10%. The work of Shurubor et al. (2007) also shows that biological variation is by far the dominating factor for the total variability of metabolite profiling. Using duplicate and triplicate measurements of human plasma samples, they conclude that the total CV (median value) is about 50%. Given the analytical CV of 10% this would result in a biological CV of approximately 40%. The biological CV shown in Fig. 8.2 is approximately 20%. The fact that the latter biological CV is much lower than the former, can probably be explained by the fact that these were obtained from highly standardized conditions (van Ravenzwaay 2007) and from the fact that human genetic background is more diverse than that of laboratory rats.

8.4.2 *Studies with Liver Enzyme Inducers*

8.4.2.1 NMR Metabonomics

There are relatively few studies reported with "classical" liver enzyme inducers, which do not cause severe liver toxicity. Schoonen et al. (2007a), reported on the effects of phenobarbital and tetracycline on the urinary metabolite profiling in rats. These changes consist of the following metabolic profile, which was identical for both compounds: succinate, citrate, ketoglutarate, allantoin and hippurate were all up-regulated. Both compounds did not induce liver toxicity in the study of Schoonen et al. (2007a), but are known to cause fatty changes (steatosis) in the liver.

8.4.2.2 GC-MS/LC-MS Metabolomics

Using the metabolic profiles of a number of typical compounds known to induce liver enzymes (reference compounds) we investigated which of the metabolites were regulated in a similar fashion. The reference compounds which were used to develop this pattern include phenobarbital sodium, Aroclor 1254, pentachlorobenzene ethyl-benzene and vinclozoline (see Table 8.1).

The metabolite patterns for both male and female rats treated with liver enzyme inducers are quite similar, with a predominant up regulation of cholesterol and,

Table 8.1 MS-based analysis: liver enzyme inducers, male and female rats, plasma

Metabolites	Direction of regulation	Strength of regulation (× fold)
Males		
Stearic acid	Up	1.2 –1.7
Lignoceric acid	Up	1.2–1.5
Behenic acid	Up	1.1–1.7
Cholesterol	Up	1.1–1.7
Eicosapentanoic acid	Up	1.4–2.1
Females		
Glycerol	Up	1.3–2.9
Palmitic acid	Up	1.2–2.4
Linoleic acid	Up	1.3–3.0
Stearic acid	Up	1.3–2.1
Arachidonic acid	Up	1.3–2.9
Cholesterol	Up	1.3–2.4
Lignoceric acid	Up	1.1–1.9
Eicosanoic acid	Up	1.2–2.2
Behenic acid	Up	1.2–2.1

metabolites belonging to the triacylglycerides pathway. Comparing the results obtained with urinary NMR based metabonomics (8.4.2.1), with plasma MS-based metabolomics (Table 8.1), it can be seen that the metabolites investigated were not similar.

8.4.3 Studies with Hepatotoxic Compounds

8.4.3.1 NMR Metabonomics

Particularly hydrazine has been frequently used to study metabolite profile changes. In the following Table 8.2, these changes have been summarized.

Alanine, citrulline and 2-amino-adipic acid are consistently up-regulated. For other metabolites no consistent pattern was noted. Comparing rat urinary metabolites with rat liver metabolites it can be seen that a similar regulation was obtained for citrulline, creatinine, and alanine. For glucose, there was a difference in the direction of regulation, up-regulated in the urine, down-regulated in the liver. Comparing the metabolite profile of hydrazine in the urine of rats and mice, a remarkable match can be observed. Nearly all metabolites found to be regulated in mouse urine following hydrazine administration are also regulated in the same manner in rats. The match of both species is an important positive finding for an across species application of metabolite profiling.

Other hepatotoxic compounds are shown in Table 8.3.

There is a relatively good match between the metabolic profile of bromobenzene in rat urine and rat plasma (five out of six plasma metabolites were similarly regu-

Table 8.2 Hepatotoxic compounds: hydrazine

Compound	Matrix	Metabolite	Direction	References
Hydrazine	Rat, urine	β-Alanine	Up	Bollard et al. (2001)
		2-Aminoadipic acid	Up	
		3-D-hydroxy-butyrate	Up	
		Citrate	Down	
		Citruline	Up	
		Nα-acetyl-citrulline	Up	
		Creatinine	Up	
		Glucose	Up	
		Hippurate	Down	
		Hypotaurine	Up	
		Lactate	Up	
		2-Oxoglutarate	Down	
		Succinate	Down	
		Taurine	Up	
		Trimethylamine-N-oxide	Down	
Hydrazine	Rat, urine	Citruline	Up	Kleno et al. (2004)
		2-Aminoadipic acid	Up	
		Amino acids	Up	
Hydrazine	Mouse, urine	2-Aminoadipic acid	Up	Bollard et al. (2001)
		Citrate	Down	
		Creatinine	Up	
		Guanidinoacetic acid	Up	
		Hippurate	Down	
		Lactate	Up	
		Succinate	Down	
		Trimethylamine	Down	
Hydrazine	Rat, liver tissue	Choline	Down	Kleno et al. (2004)
		Glucose	Down	
		Lipoproteine (VLDL, LDL)	Down	
		Lipids	Down	
		Amino acids (Alanine, valine, tyrosine)	Up	
		2-Amino adipic acid	Up	
		Citruline	Up	
		Creatinine	Up	
Hydrazine	Rat, liver tissue	Glycogen	Down	Garrod et al. (2005)
		Alanine	Up	
		Glucose oligomers	Down	
		Unsaturated fatty acids	Up	
		ω3 type fatty acids	Up	

lated). The metabolite 5-oxoproline, which is up-regulated in the urine of rats treated with bromobenzene as well as in mouse urine following treatment with acetaminophen (Farkas and Tannenbaum 2005) could be a potential common metabolite of oxidative liver toxicity. 1-naphthyl-isothiocyanate appears to induce a metabolic profile that is dominated by down regulation. Unfortunately, there are very few matches between the metabolites found to be regulated. Only citrate was found to be consistently lower in rat urine. The apparent heterogeneity of metabolic profiles of compounds causing liver toxicity, may be resolved if the actual mode of

Table 8.3 Other hepatotoxic compounds

Compound	Matrix	Metabolite	Direction	References
Bromobenzene	Rat, urine	Citrate	down	Waters et al. 2006
		2-Oxoglutarate	down	
		Creatinine	up	
		Glucose	up	
		Choline	up	
		Alanine	up	
		5-Oxoproline	up	
Bromobenzene	Rat, plasma	Choline	up	Waters et al. 2006
		Creatinine	up	
		Acetoacetate	up	
		Amino acids	up	
		5-Oxoproline	up	
3Z-3[(H-pyrrol-2-yl)-methylidene]-1-(1-peperidinylmethyl)-1,3,2H-indol-2-one	Rat, urine	Citrate	up	Wang et al. 2006a
		Lactate	up	
		2-Oxo-glutarate	up	
		Succinate	up	
		Creatinine	down	
		TMAO	down	
		Acetate	up	
		Succinate	up	
		2-Oxo-glutamate	up	
3Z-3[(H-pyrrol-2-yl)-methylidene]-1-(1-Peperidinylmethyl)-1,3,2H-indol-2-one	Rat, plasma	TMAO	up	Wang et al. 2006
		Lipids	up	
		Glucose	down	
		Phosphatidylcholine	down	
Ibuprofen	Rat, urine	Dimethylglycine	down	Schoonen et al. 2007a
		α-ketoglutarate	down	
		Creatinine	down	
		Allantoin	down	
		Uridine	down	
		Formate	down	
1-Naphthyl-Isothiocyanate (ANIT)	Rat, urine	α-ketoglutarate	Down	Schoonen et al. 2007a
		Citrate	Down	
		Dimethylamine	Down	
		Taurine	Down	
		Creatinine	Down	
		Allantoin	Down	
		Uridine	Down	
		Formate	Down	

(continued)

Table 8.3 (continued)

Compound	Matrix	Metabolite	Direction	References
ANIT	Rat, urine	Citrate	Down	Azmi et al. 2005
		Succinate	Down	
		2-oxo-glutarate	Down	
		Choline	Up	
		Glucose	Up	
		Lactate	Up	
		Creatinine	Up	
		Acetate	Up	
		Taurine	Up	
		Bile acids	Up	
		Hippurate	Down	
ANIT	Rat, plasma	Glucose	Up	Azmi et al. 2005
		TMAO	Up	
		LDL	Down	
		VLDL	Down	
		Choline	Down	
		Lactate	Down	

toxicological action resulting in liver toxicity is taken into consideration, knowing that different modes of action may cause different metabolic profiles. In this context it is interesting to see that Ibuprofen and 1-naphthyl-isothiocyanate both induce a profound down-regulation of metabolites. As both compounds exert their liver toxicity through a cholestatic mode of action, this mode of action may in fact be responsible for the metabolic changes noted in the urine, rather than the liver toxicity *per se*.

8.4.3.2 MS/LC-MS Metabolomics

Typical compounds known to induce liver toxicity were investigated for their common profile changes. The reference compounds which were used to develop this pattern, include tetrahydrofurane, cyproterone acetate, dimethylformamide and toxaphene (see Table 8.4).

There is a reasonable similarity in the profiles of male and female rats treated with compounds known to be toxic to the liver. As these compounds are also known to be potent inducers of liver metabolism, it is not entirely surprising that this profile is somewhat comparable to that found with MS based metabolomics for liver enzyme induction. The extent of regulation of those metabolites which are similar (belonging to the triacylglycerides), however, is far stronger in profiles from liver toxic compounds, than in liver enzyme inducers. It is also striking that there is no down regulation noted in these patterns.

Table 8.4 Metabolite pattern for hepatotoxic compounds

	Direction of regulation	Strength of regulation (× fold)
Male		
Palmitic acid	Up	1.4–1.6
Linoleic acid	Up	1.5–1.7
Arachidonic acid	Up	1.4–2.2
Docosahexaenoic acid	Up	2.9–3.6
Eicosatrienoic acid	Up	1.6–5.0
Behenic acid	Up	1.3–2.2
Lignoceric acid	Up	1.5–2.5
Cholesterol	Up	1.4–2.2
Nervonic acid	Up	1.7–3.6
myo-Inositol-2-monophosphate	Up	1.7–3.3
Glycerol-phosphate-lipid fraction	Up	1.5–2.6
Female		
Palmitic acid	Up	1.4–2.2
Stearic acid	Up	1.2–1.7
Arachidonic acid	Up	1.4–2.0
Docosahexaenoic acid	Up	1.4–3.2
Cholesterol	Up	1.6–2.4
Lignoceric acid	Up	1.4–2.5
Behenic acid	Up	1.5–2.5
Linoleic acid	Up	1.2–3.0
Threonic acid	Up	1.4–2.4
Nervonic acid	Up	1.3–5.1

8.4.4 Studies with Peroxisome Proliferators

8.4.4.1 NMR Metabonomics

The metabolic profile of mouse urine following administration of the peroxisome proliferator Wy-14,643 was studied by Zhen et al. (2007) (see Table 8.5).

8.4.4.2 GC-MS/LC-MS Metabolomics

Typical compounds known to induce peroxisome proliferation were investigated for their common profile changes. The reference compounds which were used to develop this pattern, include fenofibrate, clofibrate, diethylhexyl-phthalate and Wy14643 (see Table 8.6).

For rats treated with peroxisome proliferators there is a predominant down regulation of metabolites. The results appear to suggest a reduction in the presence of fatty acids in the rat serum. This would be in line with the lipid reduction effects of these compounds. Overall, the pattern of peroxisome proliferators is distinctly different from that of liver enzyme inducers and liver toxic compounds.

Table 8.5 Studies with peroxisome proliferators

Compound	Matrix	Metabolite	Direction	Reference
Wy-14,643	Mouse, urine	11β-Hydroxy-dioxopren-4en-21-oic acid	Up	Zhen et al. (2007)
		11β,20-dihydrxy-3-oxopregn-4-en-21-oic acid	Up	
		Nicotinamide	Up	
		1-Methylnicotinamide	Up	
		Hippuric acid	Up	
		2,8-Dihydroxyquinoline-β-d-glucuronide	Up	
		Xanturenic acid	Down	
		Hexanoylglycine	Down	
		Phenylpropyonyl glycine	Down	
		Cinnamoylglycine	Down	

Table 8.6 Metabolite pattern for peroxisome proliferators

	Direction of regulation	Strength of regulation (× fold)
Males		
Coenzyme Q10	Down	1.4–2.3
16-Methylheptadecanoic acid	Down	2.3–4.8
17-Methyloctadecanoic acid	Down	2.3–4.8
Eicosatrienoic acid	Up	1.4–3.0
trans-4-hydroxyproline	Down	1.3–2.1
Females		
Pantothenic acid	Up	1.8–3.3
Glycerol	Up	1.1–5.0
Linoleic acid	Down	1.3–1.8
16-Methylheptadecanoic acid	Down	2.5–4.2
Cytosine	Down	1.1–1.5
Phosphatidylcholine	Down	1.1–1.9

8.5 Perspectives

Metabolomics is considered by many to be complementary to genomics, transcriptomics, and proteomics (Griffin and Bollard 2004; Lindon et al. 2004a). This technology, however, does have a number of potential advantages over the other omics-technologies that could help to overcome many of the problems now encountered with genomics/transcriptomics and proteomics (Bilello 2005). In the events following a toxic insult, genomics/transcriptomics analyses the earliest change which may or may not result in changes at the protein level (as analyzed by proteomics). Protein changes (e.g. enzyme activities) in their turn again may or may not result in changes in metabolites. As both gene expression and protein changes are subject to complex homeostatic control and feedback mechanisms, the end result of changes in this sequence are often alterations in metabolite levels. Consequently, metabolite profiling should provide information on a higher level of integration than the other omics-technologies and is closer to everyday toxicology (van Ravenzwaay et al. 2007). Moreover, the knowledge concerning structure and function of metabolites is significantly greater than that of genes and their corresponding proteins. In addition, the number of metabolites is lower than those of genes and proteins. Consequently, the chances of finding meaningful changes (i.e. changes that can be interpreted both biochemically as well as in terms of effect) are greater. A further advantage of metabolite profiling is that changes are detectable in accessible body fluids such as urine and blood. If the analysis of a great number of individual organs can be replaced by such matrices, then this will provide several significant practical advantages (less invasive method, no need to sacrifice animals, time course analysis possible).

The (early) recognition of toxicological mode of action through metabolite profiling is also a promising tool for the discovery of processes which may be involved in tumor development.

The profiles established for the three liver specific mode of action (liver enzyme induction, liver toxicity and liver peroxisome proliferation) within the MetaMapTox project are sufficiently different from each other to allow for a clear identification of each of them, not only with reference compounds, but also with compounds subsequently tested. PLS-DA visualizations of profiles for specific compounds are shown in Figs. 8.3 and 8.4.

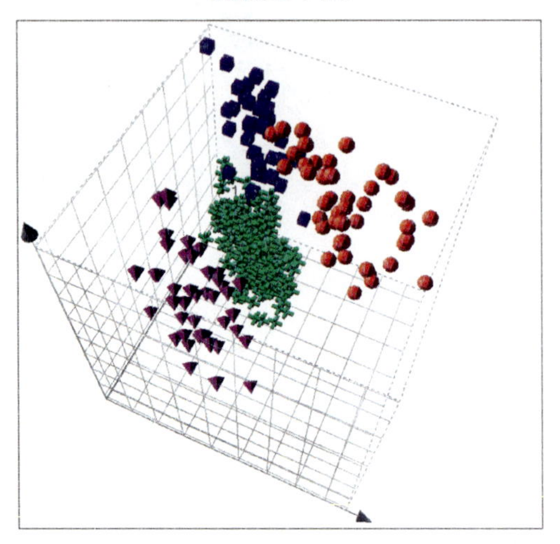

Fig. 8.3 PLS-DA analysis of high-dimensional metabolite profiling (256 standard high quality metabolites) of samples from male rats treated with compounds assigned to three liver specific modes of action: liver enzyme induction (*orange spheres*), liver toxicity (*blue cubes*) and peroxisome proliferation (*violet tetrahedrons*) relative to untreated male controls (*green crosses*). The plot shows the first three PLS-DA scores (cross-validation score: Q^2(cum) = 0.80, Q^2(cum, scores 1–3) = 0.71). Each mode of action comprises three different compounds. Separate clusters can be seen for the three modes of action and the control group

Scatter Plot

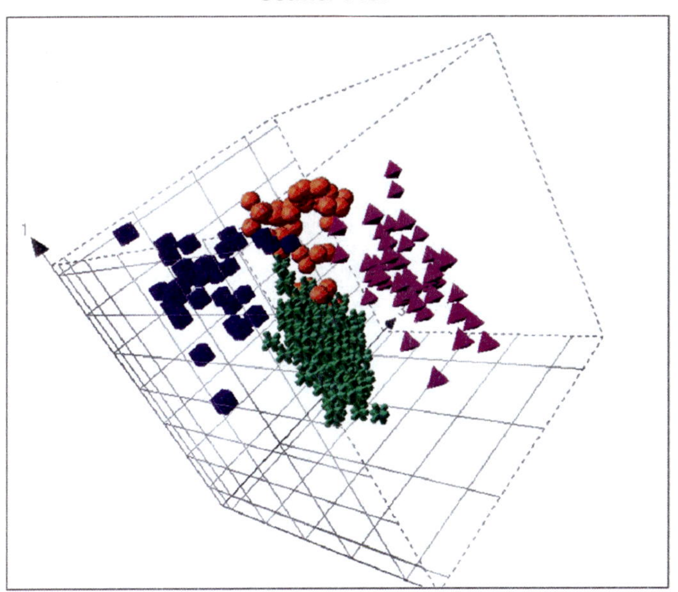

Fig. 8.4 PLS-DA analysis of high-dimensional metabolite profiling (256 standard high quality metabolites) of samples from female rats treated with compounds assigned to three liver specific modes of action: liver enzyme induction (*orange spheres*), liver toxicity (*blue cubes*) and liver peroxisome proliferation (*violet tetrahedrons*) relative to untreated female controls (*green crosses*). The plot shows the first three PLS-DA scores

Recognition of toxicological modes of action which contribute to tumor promotion would greatly enhance the quality of selection processes for drug or active ingredient development. The finding that blood metabolome analysis within a 28 day study can provide data which allow for the identification of liver enzyme inducers, liver toxicants and peroxisome proliferators is clearly a step towards this goal. Several of the enzyme inducers correctly identified in the data base are known to enhance liver tumor development in species and strains known to be susceptible for this effect, e.g. phenobarbital, Arochlor. Similarly some of the hepatotoxic compounds also are known to enhance liver tumor development in rodents, e.g. tetrahydrofurane. It should be noted that not all of the liver enzyme inducers, nor all of the hepatotoxic compounds, are necessarily enhancers of liver tumor formation and that this is not directly related to the primary mode of action, but rather to the response of the liver to these effects, i.e. cell proliferation. The association between these modes of action and the extent of liver tumor promotion will vary depending on the potential of the chemical and the susceptibility of the species tested. The importance of metabolite patterns as markers of a carcinogenic process was indicated in the work reported by Thomas et al. (2007). In their analysis of metabolite changes of several compounds in 90 day and 2 year rodent studies they conclude that individual endogenous metabolites make relatively poor biomarkers, but the metabolite profile as a whole gives a more accurate prediction of the bioassay results. The results obtained in the MetaMap Tox project points in a similar direction; very rarely are there individual metabolites which can be used as biomarkers of effect (mode of action). It is the consistent pattern of change that allows for a reliable identification of effect.

Further developments in metabolite profiling may increase the specificity of the data obtained and can provide a deeper insight into biochemical processes in cancer cells. The stable isotope dynamic metabolic profiling (SiDMAP) measures the flow of molecules through metabolic pathways. Thus, SiDMAP measures the activity of pathways, in terms of molecular flux over time in intact systems. With this technology, metabolism was demonstrated to be involved as a cancer cell mechanism in drug resistance (Maguire et al. 2006). The primary metabolic difference between drug induced apoptotic-sensitive and – resistant cells is the more rapid rate at which resistant cells synthesis medium – and long-chain fatty acids. In addition, indirect acetyl-CoA formation through high fatty acid chain desaturase activity allows tumor cells to synthesize a membrane, which is independent of dietary fatty acids. SiDMAP investigations have also revealed the mechanism through which resistance to imatinib mesilate occurs. This drug controls glucose metabolizing enzymes such as hexokinase II and glucose-6-phosphate dehydrogenase. Resistance occurs when cells use the non-oxidative branch of the pentose cycle for DNA- and RNA-synthesis, which is not controlled by the drug.

8.6 Integrated Approaches

A challenge and opportunity, is the integration of omics data. An integrated functional genomics and metabolomics approach was reported by Ippolito et al. 2005, in their investigations to identify features of human neuroendocrine cancers associated with poor outcome. In their analysis of gene-chip data sets of primary prostate tumors, as well as lymph node and liver metastases from neuroendocrine tumors, they identified 446 genes whose expression was enriched in neoplastic cells. This gene signature was used for *in silico* metabolic reconstruction of neuroendocrine cell metabolism and metabolite profiling of cell lines and human neuroendocrine tumors with good or poor prognosis. A distinguishing feature of poor prognosis is the GABA production through a glutamic acid decarboxylase independent pathway and the production of imidazole-4-acetate through a dopa decarboxylase pathway. The difficulty, but also the power of combining data from genomics, proteomics and metabolite profiling has been demonstrated by Craig et al. 2005. Following high dose administration of methapyrilene for 3 days to rat, causing periportal liver cell necrosis, they sampled liver, whole blood and urine for omics analysis. Their results show a great number of gene changes, a good amount of protein changes and some metabolite changes. Of all the changes noted, only two pathways were consistently affected in all three areas; glucose metabolism and choline metabolism. These findings are in line with the general notion that the number and extend of changes in genome activity and protein concentrations are higher than that of metabolites (ECETOC 2008). As metabolite profiling resides at the end of the biological road that commences with DNA, it is the place were variations in genome expression and protein formation become integrated. As the omics technologies provide quantitative data, robust computational methods for judging and interpreting omics data are needed (Willard et al. 2005). Given the advantages of metabolite profiling, its integration in such a concept will be essential but will require the development of bioinformatics and accessible data bases. A possible tool to achieve this objective was proposed by Toyoda and Wada (2004) and consists of the concept of an omics space comprising of a series of layers of information (from genomics to metabolomics and ultimately whole system observations called "phenomics"). The connections between these layers of information are enhanced by taking into account the dynamics of the interaction as well as its probability. With this tool the authors identified the oncogene *cMyc* and the cell division cycle homolog *Cdc25A* as the pair of genes which had the strongest connection to polyoma Middle-T induced mammary tumors.

Thomas and Ganji (2006) question, in a review of on the integration of genomics and metabolomics, "are we there yet?" They note that the search for correlations between genes and metabolites is largely limited to linking specific genes with defined metabolic events, and, that analysis appears to be heavily reliant upon an empirical search for correlations. Although valuable, this deductive approach is not likely to elucidate entirely new pathways. The approach taken by Griffin et al. (2004), in which hepatic extracts, intact liver tissue and plasma were used for metabolite profiling and gene analysis followed by a comprehensive multivariate statistical analysis appears to be a promising route. Particularly, the attempt to model time progression of the data sets using PLS seems to be an important step forward. As a result of this type of analysis, the authors were able to conclude that orotic acid-induced fatty liver is related with a decreased transcriptional activation of sterol regulatory elements.

The question of the influence of time progression and the dynamics of (early) responses deserves attention. Thomas and Ganji (2006) note that it is likely that data collected in steady-state situations are difficult to correlate with dynamic trends in levels of transcripts, protein, metabolite and phenotype. In the strive to elucidate the earliest, and ideally key events, in toxicological or pharmacological effects of a compound, most of the omics investigations have been performed within days, sometimes even hours after the administration of a compound. The search for very early information to derive mode-of-action knowledge from these data is one of the paradigms in safety testing of new active ingredients. Given the highly dynamic processes, with numerous feedback mechanisms occurring after a compound has started its interaction with an organism, it may be questioned whether this is an appropriate approach for all omics technologies. During the initial phases of this interaction, time and place of sampling of the material to be investigated will have a tremendous effect on the outcome of the results. In addition, inter-individual and inter-laboratory differences as well as the normal biological variation will further increase fluctuations of results. The sum of all these factors is the most likely cause of the current comparability problem. Long-term investigations of highly standardized metabolite profiling may offer a chance to tackle this problem. The combination of the higher level of integration of metabolite proofing, relative to other omics technologies (van Ravenzwaay et al. 2007) with sampling during a steady-state phase is highly likely to solve the problem of the comparability problem of omics data. In the MetaMap®Tox project, in which blood sampling takes place after 7, 14 and 28 days of administration, it was noted that the vast majority of all profiles established shows a stable qualitative pattern of change, irrespective of the sampling time point. Quantitative differences can be noted, but these are either minor (which is a general property of metabolomics relative to other omics technologies – ECETOC 2008), or can be explained by bio-accumulation of the compound over time, or the cumulative toxic effect on an organ.

Metabolite profiling is starting to demonstrate its potential in toxicology and its application in cancer diagnosis appears to be a matter of time. With metabolite profiling still being a new field, there are only a few data bases available that offer comprehensive pathway contents. However, with the current pace of progression within metabolomic research and with the data bases that are currently being developed, it is highly likely that metabolite profiling will result in an enhanced recognition and assessment of toxicological modes of action, as well as cancer diagnosis. Plotting of metabolite profiles in biochemical pathways may lead to the discovery of key events and may also allow for a further distinction of compounds within a mode of action. Moreover, the quantitative nature of the observed changes is also likely to allow for a potency ranking and could enhance the recognition of compounds associated with tumor promoting properties. For some years to come metabolite profiling, as any of the omics sciences is likely to retain its "research" status, as indicated by EPA's framework for the use of genomics data (Dix et al. 2006). EPA's policy states that genomics data alone are currently insufficient as a basis for risk assessment and management decisions. A workshop organised by ECETOC in 2007 derived the same conclusion (ECETOC 2008). With respect to the question if a meaningful NOAEL (no observed adverse effect level) can be derived from omics data the following consensus was reached in this workshop: changes of individual parameters (genome, proteome or metabolome) cannot be used to derive a NOAEL. It is necessary to identify patterns of change which need to be correlated to observable changes at the microscopic and macroscopic level. This correlation can be established by comparing patterns of change with those of known reference compounds. Moreover, it must be assured that the pathway identified is related to an adverse effect. With this proposed framework, with the generation of tools that connect omics information with observable changes in whole biological systems and with the increasing availability of data, it should not be long before the omics sciences will become an integrated part of the toxicological, pharmacological, medical and regulatory communities.

Acknowledgements The authors are grateful to Gertrud Skawran and Gunter Rank for their skilful technical assistance performing the animal study as well as Irmgard Weber for doing the clinical pathology work and the lab and data analysis teams for performing the extensive metabolite profiling analyses. Meyasse Bugay and Marina Herbst are acknowledged for their assistance in the preparation of this chapter.

References

Azmi J, Griffin JL, Shore RF et al (2005) Chemometric analysis of biofluids following toxicant induced hepatotoxicity: a metabonomic approach to distinguish the effects of 1-naphthylisothiocyanate from its products. Xenobiotica 35:839–852

Bilello JA (2005) The agony and ecstasy of omic technologies in drug development. Curr Mol Med 5:39–52

Bino RJ, Hall RD, Fiehn O et al (2004) Potential of metabolomics as a functional genomics tool. Trends Plant Sci 9:418–425

Bollard ME, Holmes E, Lindon JC et al (2001) Investigations into biochemical changes due to diurnal variation and estrus cycle in female rats using high-resolution ^{1}H-NMR spectroscopy of urine and pattern recognition. Anal Biochem 295:194–202

Brennan DJ, Kelly C, Rexhepaj E et al (2007) Contribution of DNA and tissue microarray technology to the identification and validation of biomarkers and personalised medicine in breast cancer. Cancer Genomics Proteomics 4:121–134

Burns MA, He W, Wu CL et al (2004) Quantitative pathology in tissue MR spectroscopy based human prostate metabolomics. Technol Cancer Res Treat 3:591–598

Christians U, Reisdorph N, Klawitter J et al (2005) Biomarkers of immunosuppressive drug toxicity. Curr Opin Organ Transplant 10:284–294

Claudino WM, Quattrone A, Biganzoli L et al (2007) Metabolomics: available results, current research projects in breast cancer, and future applications. J Clin Oncol 25:2840–2846

Clayton TA, Lindon JC, Everett JR et al (2003) An hypothesis for a mechanism underlying hepatotoxin-induced hypercreatinuria. Arch Toxocol 77:208–217

Clarke J, Haselden JN (2008) Metabolic profiling as a tool for understanding mechanisms of toxicity. Toxicol Pathol 36:140–147

Connor SC, Hodson MP, Ringeissen S et al (2004) Development of a multivariate statistical model to predict peroxisome proliferation in the rat, based on urinary ^{1}H-NMR spectral patterns. Biomarkers 9:364–385

Costello LC, Franklin RB (2006) The clinical relevance of the metabolism of prostate cancer, zinc and tumor suppression: connecting the dots. Mol Cancer 5:10–16

Craig A, Sidaway J, Holmes E et al (2005) Systems toxicology: integrated genomic, proteomic and metabonomic analysis of methapyrilene induced hepatotoxicity in the rat. J Proteome Res 5:1586–1601

Denkert C, Budczies J, Kind T et al (2006) Mass spectrometry-based metabolic profiling reveals different metabolite patterns in invasive ovarian carcinomas and ovarian borderline tumors. Cancer Res 66:10795–10804

Dix DJ, Gallagher K, Benson WH et al (2006) A framework for the use of genomics data at the EPA. Nat Biotechnol 24:1108–1111

ECETOC (2008) Workshop on the application of omics technologies in toxicology and ecotoxicology: case studies and risk assessment 6-7 December 2007, Malaga. ECETOC, Brussels, Workshop Report No. 11

Farkas D, Tannenbaum SR (2005) *In vitro* methods to study chemically-induced hepatotoxicity: a literature review. Curr Drug Metab 6:111–125

Gamer AO, Jaeckh R, Leibold E et al (2002) Investigations on cell proliferation and enzyme induction in male rat kidney and female mouse liver caused by tetrahydrofurane. Toxicol Sci 70:140–149

Garrod S, Bollard ME, Nicholls AW et al (2005) Integrated metabonomic analysis of the multiorgan effects of hydrazine toxicity in the rat. Chem Res Toxicol 18:115–122

Gavaghan CL, Holmes E, Lenz E et al (2000) An NMR-based metabonomic approach to investigate the biochemical consequences of genetic strain differences: application to the C57BL10J and Alpk:ApfCD mouse. FEBS Lett 484:169–174

Gavaghan CL, Nicholson JK, Connor SC et al (2001) Directly coupled high-performance liquid chromatography and nuclear magnetic resonance spectroscopic with chemometric studies on metabolic variation in Sprague-Dawley rats. Anal Biochem 291:245–252

Gillies RJ, Morse DL (2005) *In vivo* magnetic resonance spectroscopy in cancer. Annu Rev Biomed Eng 7:287–326
Glunde K, Serkova N (2006) Therapeutic targets and biomarkers identified in cancer choline phospholipid metabolism. Pharmacogenomics 7:1109–1123
Griffin JL, Shockcor JP (2004) Metabolic profiles of cancer cells. Nat Rev Cancer 4:551–561
Griffin JL, Bollard ME (2004) Metabonomics: its potential as a tool in toxicology for safety assessment and data integration. Curr Drug Metab 5:389–398
Griffin JL, Bonney SA, Mann C et al (2004) An integrated reverse functional genomic and metabolomic approach to understanding orotic acid-induced fatty liver. Physiol Genomics 17:140–149
Griffin JL, Kauppinen RA (2007) Tumor metabolomics in animal models of human cancer. Proteome Res 6:498–505
Gunawan B, Kaplowitz N (2004) Clinical perspectives on xenobiotica hepatotoxicity. Drug Metab Rev 36:301–312
Heijne WH, Lambers RJ, van Bladeren PJ et al (2005) Profiles of metabolites and gene expression in rats with chemically induced hepatic necrosis. Toxicol Pathol 33:425–433
Holmes E, Nicholson JK, Tranter G (2001) Metabonomic characterization of genetic variations in toxicological and metabolic responses using probabilistic neural networks. Chem Res Toxicol 14:182–191
Ippolito JE, Xu J, Jain S et al (2005) An integrated functional genomics and metabolomics approach for defining poor prognosis in human neuroendocrine cancers. Proc Natl Acad Sci USA 102:9901–9906
Ishihara K, Katsutani N, Aoki T et al (2006) A metabonomics study of the hepatotoxicants galactosamine, methylene dianiline and clofibrate in rats. Basic Clin Pharmacol Toxicol 99:251–260
Issaq HJ, Nativ O, Waybright T et al (2008) Detection of bladder cancer in human urine by metabolomic profiling using high performance liquid chromatography/mass spectrometry. J Urol 170:2422–2426
Jain KK (2007) Cancer biomarkers: current issues and future directions. Curr Opin Mol Ther 9:563–571
Jaeschke H (2002) Mechanisms of hepatotoxicity. Toxicol Sci 65:166–176
Kleno TG, Kiehr B, Baunsgaard D et al (2004) Combination of omics' data on investigate the mechanism(s) of hydrazine-induced hepatotoxicity in rats and to identify potential biomarkers. Biomarkers 4:116–138
Kota BP, Huang TH, Roufogalis BD (2005) An overview on biological mechanisms of PPARs. Pharmacol Res 51:85–94
Li H, Ni Y, Su M et al (2007) Pharmacometabonomic phenotyping reveals different responses to xenobiotic intervention in rats. J Proteome Res 6:1364–1370
Lindon JC, Nicholson JK, Holmes E et al (2003) Contemporary issues in toxicology. The role of metabonomics in toxicology and its evaluation by the COMET project. Toxicol Appl Pharmacol 187:137–146
Lindon JC, Holmes E, Bollard ME et al (2004a) Metabonomics technologies and their applications in physiological monitoring, drug safety assessment and disease diagnosis. Biomarkers 9:1–31
Lindon JC, Holmes E, Nicholson JK (2004b) Toxicological applications of magnetic resonance. Prog Nucl Magn Reson Spectrosc 45:109–143
Lindon JC, Nicholson J (2008) Spectroscopic and statistical techniques for information recovery in metabonomics and metabolomics. Annu Rev Anal Chem 1:45–69
Looser R, Krotzky AJ, Trethewey RN (2005) Metabolite profiling with GC-MS and LC-MS - a key tool for contemporary biology. In: Vaidyanathan S, Harrigan GG, Goodacre R (eds) Metabolome analyses – strategies for systems biology. Springer, New York, pp 103–118
Maddox JF, Luyendyk JP, Cosma GN et al (2006) Metabonomic evaluation of idiosyncrasy-like liver injury in rats cotreated with ranitidine and lipopolysaccharide. Toxicol Appl Pharmacol 212:35–44
Maguire G, Lee P, Manheim D et al (2006) SiDMAP: a metabolomics approach to assess the effects of drug candidates on the dynamic properties of biochemical pathways. Exp Opin Drug Discov Devel 1:351–359

Mandard S, Muller M, Kersten S (2004) Peroxisome proliferator-activated receptor α target genes. Cell Mol Life Sci 61:393–416

Noritaka I, Hidenobu Y, Yasuji A et al (1991) Effects of ethyl 4-chloro-2-methylphenoxyacetate on hepatic peroxisomal enzymes in rats. Tohoku J Exp Med 165:59–61

O'Donoghue JL (1986) Subchronic oral toxicology of 4-chloro-3-nitroaniline in the rat. Fundam Appl Toxicol 6:551–558

Odunsi K, Wollman RM, Ambrosone CB et al (2005) Detection of epithelial ovarian cancer using ^{1}H-NMR-based metabonomics. Int J Cancer 113:782–788

Peters JM, Cheung C, Gonzalez FJ (2005) Peroxisome proliferator-activated receptor-α and liver cancer: where do we stand? J Mol Med 83:774–785

Plumb RS, Granger JH, Stumpf CL et al (2005) A rapid screening approach to metabonomics using UPLC and oa-TOF mass spectrometry: application to age, gender and diurnal variation in normal/Zucker obese rats and black, white and nude mice. Analyst 130:844–849

Pitot HC (1986) Fundamentals of oncology (3rd ed). Marcel Decker, New York

Robosky LC, Wells DF, Egnash LA et al (2005) Metabonomic identification of two distinct phenotypes in Sprague-Dawley (Crl:CF(SD)) rats. Toxicol Sci 87:277–284

Russel ST, O'Connell TM, Pluta L et al (2007) A comparison of transcriptomic and metabonomic technologies for identifying biomarkers predictive of two-year rodent cancer bioassays 2006. Toxicol Sci 96:40–46

Sauter H, Lauer M, Fritsch H et al (1991) Metabolic profiling of plants–A new diagnostic technique. In: Baker DR, Fenyes JG, Moberg WK, (Hrsg.) Synthesis and chemistry of agrochemicals II. ACS Symposium Series 443 American Chemical Society Washington, D.C., 288–299

Schoonen WG, Kloks CP, Ploemen JP et al (2007a) Uniform procedure of ^{1}H NMR analysis of rat urine and toxicometabonomics part II: comparison of NMR profiles for classification of hepatotoxicity. Toxicol Sci 98:286–297

Schoonen WG, Kloks CP, Ploemen JP et al (2007b) Sensitivity of ^{1}H NMR analysis of rat urine in relation to toxicometabonomics part I: comparison of NMR profiles for classification of hepatotoxicity. Toxicol Sci 98:271–285

Schulte-Hermann R (1974) Induction of liver growth by xenobiotic compounds and other stimuli. Crit Rev Toxicol 3:97–158

Shurubor YI, Matson WR, Willett WC et al (2007) Biological variability dominates and influences analytical variance in HPLC-ECD studies of the human plasma metabolome 2007. BMC Clin Pathol 7:9

Serkova NJ, Spratlin JL, Eckhardt GS (2007) NMR-based metabolomics: translational application and treatment of cancer. Curr Opin Mol Ther 9:572–585

Stanley EG, Bailey NJ, Bollard ME et al (2005) Sexual dimorphism in urinary metabolite profiles of Han Wistar rats revealed by nuclear-magnetic-resonance-based metabonomics. Anal Biochem 343:195–202

Tate AR, Foxall PJ, Holmes E et al (2000) Distinction between normal and renal cell carcinomakidney cortical biopsy samples using pattern recognition of (1)H magic angle spinning (MAS) NMR spectra. NMR Biomed:13, 64–71

Teague CR, Dhabbar FS, Barton RH et al (2007) Metabonomic studies on the physiological effects of acute and chronic psychological stress in Sprague-Dawley rats. J Proteome Res 6:2080–2093

Thomas CE, Ganji G (2006) Integration of genomic and metabonomic data in systems biology – Are we "there" yet? Curr Opin Drug Discov Devel 9:92–100

Thomas RS, O'Conell TM, Pluta L et al (2007) A comparison of transcriptomic and metabonomic technologies for identifying biomarkers predictive of two-year rodent cancer bioassays. Toxicol Sci 96:40–46

Toyoda T, Wada A (2004) Omic space: coordinate-based integration and analysis of genomic phenomic interactions. Bioinformatics 20:1759–1765

Trethewey RN, Krotzky A, Willmitzer L et al (1999) Metabolic profiling: a Rosetta Stone for genomics? Curr Opin Plant Biol 2:83–85

Vainio H, Linnainmaa K, Kähönen M et al (1983) Hypolipidemia and peroxisome proliferation induced by phenoxyacetic acid herpicides in rats. Biochem Pharmacol 32:2775–2779
van Ravenzwaay B, Tennekes HA, Stöhr M et al (1987) The kinetics of nuclear polyploidization and tumor formation in livers of CF-1 mice exposed to dieldrin. Carcinogenesis 8:265–269
van Ravenzwaay B, Mellert W, Deckardt K et al (2005) The comparative toxicology of 4-chloro-2-methylphenoxyacetic acid and its plant metabolite 4-chloro-2-methylcarboxyphenoxyacetic acid in rats. Reg Toxicol Pharmacol 42:47–54
van Ravenzwaay B, Coelho-Palermo CG, Leibold E et al (2007) The use of metabolomics for the discovery of new biomarkers of effect. Toxicol Lett 172:21–28
Wang Q, Jiang Y, Wu C et al (2006a) Study of a novel indolin-2-ketone compound Z24 induced hepatotoxicity by NMR-spectroscopy-based metabonomics of rat urine, blood plasma, and liver extracts. Toxicol Appl Pharmacol 215:71–82
Wang Y, Holmes E, Tang H et al (2006b) Experimental metabonomic model of dietary variation and stress interactions. J Proteome Res 5:1535–1542
Waters NJ, Waterfield CJ, Farrant RD et al (2006) Integrated metabonomic analysis of bromobenzene-induced hepatotoxicity: novel induction of 5-oxoprolinosis. J Proteome Res 5:1448–1459
Watkins SM, Reifsnyder PR, Pan H et al (2002) Lipid metabolome-wide effects of the PPARγ agonist rosiglitazone. J Lipid Res 43:1809–1817
Weckwerth W, Morgenthal K (2005) Metabolomics: from patter recognition to biological interpretation. Drug Discov Today 10:1551–1558
Wheelock CE, Goto S, Hammock BD et al (2007) Clofibrate-induced changes in the liver, heart, brain and white adipose lipid metabolome of Swiss-Webster mice. Metabolomics 3:137–145
Wilson ID, Plumb R, Granger J et al (2005) HPLC-MS-based methods for the study of metabonomics. J Chromatogr B Analyt Technol Biomed Life Sci 817:67–76
Willard HF, Angrist M, Ginsburg GS (2005) Genomic medicine: genetic variation and its impact on the future of health care. Philos Trans R Soc Lond B Biol Sci 360:1543–1550
Yamamoto T, Tomizawa K, Fujikawa M et al (2007) Evaluation of human hepatocyte chimeric mice as a model for toxicological investigation using panomic approaches – effect of acetaminophen on the expression profiles of proteins and endogenous metabolites in liver, plasma and urine. Toxicol Sci 32:205–215
Yang J, Xu G, Zheng Y et al (2004) Diagnosis of liver cancer using HPLC-based metabonomics avoiding false-positive result from hepatitis and hepatocirrhosis diseases. J Chromatogr B Analyt Technol Biomed Life Sci 813:59–65
Zhen Y, Krausz KW, Chen C et al (2007) Metabolomic and genetic analysis of biomarkers for peroxisome proliferator-activated receptor alpha expression and activation. Mol Endocrinol 21:2136–2151

Chapter 9
Interactomics and Cancer

Gautam Chaurasia and Matthias E. Futschik

Abstract Cancer is a complex disease with a myriad of genes and molecular processes involved. To unravel its underlying mechanisms, the main approach to date has been the study of individual genes and their association with carcinogenesis. As a recently emerging new paradigm, systems biology has complemented this time-honoured concept by promoting a holistic view of cancer as a network-associated disease. This new strategy is reflected par excellence by the construction of genome- and proteome-wide interaction networks and their utilization. We give here an overview of the current status of the human interactome and report first successes in its application in cancer research. In particular, interactomics-based analyses have been successfully undertaken for the characterization and de novo prediction of cancer-associated genes and processes. Although considerable challenges are still to overcome, interactomics promises to become a cornerstone in the systems biology of cancer.

9.1 Introduction

Cancer is not a single uniform disease, but displays a striking heterogeneity in its cause, progression and prognosis. In fact, more than 100 distinct types of cancer have been identified in a variety of tissues over the last decades (Hanahan and Weinberg 2000). The recent progress in molecular profiling of cancer is likely to contribute to an even larger number of biologically and clinically distinct tumor sub-types (Alizadeh et al. 2000). Such observed heterogeneity is not only of interest for cancer researchers, but has also direct consequences in the clinical prognosis and medical treatment of cancer patients.

G. Chaurasia
Charité, Humboldt-University, Berlin, Germany

M.E. Futschik (✉)
Centre for Molecular and Structural Biomedicine, University of Algarve, Faro, Portugal
e-mail: mfutschik@ualg.pt

W.C.S. Cho (ed.), *An Omics Perspective on Cancer Research*,
DOI 10.1007/978-90-481-2675-0_9,

Where does the observed heterogeneity originate from? Intensive research has discovered a large number of genes involved in the development of cancer. Especially, the study of genetic mutations identified many cancer-associated genes and has led to the view of cancer as a primarily genetic disease. A recent census of human cancer genes showed that somatic and germline mutations in almost 400 genes have repeatedly been reported to contribute to oncogenesis (Futreal et al. 2004). Additionally, numerous epigenetic and transcriptional changes have been associated with cancer (see also chapters 4 and 5).

How can we cope with this complex and heterogeneous disease in which so many genes and processes are involved? For a long time, the main approach to unravel oncogenesis has been to identify single cancer-associated genes and to characterize them one at a time. Undoubtedly, this paradigm in cancer research has supplied us with an impressive catalogue of pathogenic changes on molecular level. Despite considerable success, however, it has not yet delivered the much anticipated "magic bullets" against this disease.

Recently, a new discipline has emerged with the advent of large-scale biological data sets: *Systems biology*. It can be viewed as a complementary – but not opposing – approach to the classical reductionistic strategy for the study of the biological processes. In contrast to reductionistic approaches based on the dissection of processes into their most elementary levels, systems biology is more holistically orientated. The guiding principle of systems biology is that the total system can be more than the sum of its parts and can acquire properties that are not implied in the single components.

Following this principle, we seek to study a biological system as a whole. The aim is to determine the rules governing its behaviour and eventually to generate qualitative and quantitative predictions concerning its response to perturbations and modifications. To achieve this, two requirements have to be fulfilled: (1) a sufficient amount of data and information describing the system has to be available, and (2) a computational model of the system has to be designed. Whereas the first requirement is increasingly met with the development of new high-throughput techniques, the second one still demands considerable efforts. For instance, when we aim to represent the whole system, we need to choose an adequate level of resolution. Finding this level is challenging, since there is usually a trade-off between computational feasibility and detailed representation of the molecular systems due to their mere size and complexity. The inclusion of too many components can lead to ill-determined models of the system with many parameters unknown, whereas a too severe restriction can results in an incomplete model with a lack of coherence. In fact, the choice of a suitable model depends not only on the research objective, but also, more practically, on the quality and quantity of data and information present.

In response to this difficulty, various methodologies for different levels of resolution have been brought forward in systems biology to date. A nowadays very popular approach is based on the representation of biological systems as mathematical graphs and has laid the ground for the blooming field called *network biology*. In the context of molecular systems, for instance, the molecules are typically

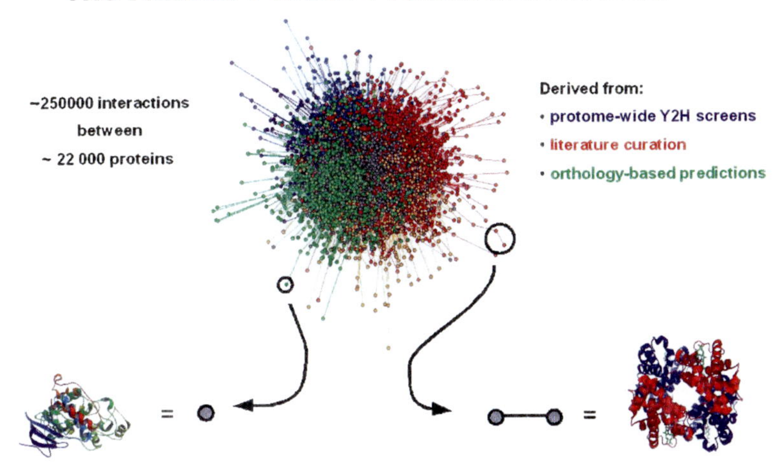

Fig. 9.1 Graphical representation of the current human protein–protein interactome as stored in the UniHI database (http://www.unihi.org). Altogether, it comprises over a quarter of a million of interactions derived from experimental resources and by computational prediction. Nodes and edges in the displayed graph represent proteins and their interactions. The different colours indicate the source of interactions: blue - Y2H screens, red - literature curation, green - orthology-based prediction, and grey - multiple evidence. Notably, distinct regions of the interactome are covered by different methods indicating the potential benefits of integration. The figure also illustrates the grade of simplification achieved by the graph-theoretical approach. The highlighted nodes symbolizing the shown proteins (*left*: mitogen activated protein kinase; *right*: haemoglobin complex consisting of alpha and beta chains) are depicted for illustration only; they do not represent the actual location of these proteins in the interactome. Displayed protein structures were taken from the Protein Data Bank

represented as nodes and their interactions as edges (Fig. 9.1). Although this type of representation is clearly a stark simplification of the underlying physical system, a major advantage of this approach is that the analysis of large networks becomes feasible. Also, the underlying graph-theory has been well developed and offers researchers a variety of tools. In fact, with its beginning dating back to Leonard Euler in 1736, graph theory has made profound impact in social, physical and computer sciences (Euler 1736). The application of graph-theory to biology seems to be well suited where large networks are involved in the process of interest. Thus, it is not surprising that the concepts of network biology have been especially applied to elucidate the complex processes during oncogenesis and to consolidate the hitherto divergent observations. A short introduction to graph theory and its basic concepts is presented in Box 1.

The reminder of this chapter is following: We first present an overview of current strategies to chart, to store and to analyse interaction maps. We focus here on protein–protein interaction data as many concepts of network biology have originally been demonstrated using protein interactions. Notably, we describe the generation of protein interaction maps in some details, as it can have a considerable

Box 1 Introduction to Graph Theory and Its Application to Network Biology

Graph-Theoretical Description of Molecular Networks

One of the most basic descriptions of molecular systems is given by their representation as mathematical graphs. For protein interaction networks, for instance, proteins are commonly represented as nodes and their physical interactions as undirected edges. For transcriptional regulatory networks, nodes symbolize both transcription factors and their target genes and are connected by direct edges. The resulting graphs can be analyzed using various graph-theoretical measures:

A fundamental characteristic of a node in a mathematical graph is its degree, i.e. the number of edges to other nodes. The *degree distribution P(k)* for a network can be defined as fraction of proteins with k interactions in the total network. It is an important feature of network to distinguish different network classes. Of special importance here is the power-law distribution ($P(k) \sim k^{-\gamma}$) which is characteristic for the class of scale-free networks. It has been shown that such network architecture is more robust against random failure of single components. A consequence of the scale-free topology is the emergence of so-called network *hubs*, i.e. highly connected nodes. Hubs are of particular importance for the network integrity and were associated with essential proteins (Jeong et al. 2001). Finally, the *shortest path* length between two nodes is defined as the minimum number of edges included in the (directed) path between the two nodes.

influence on the final maps. In fact, it is of critical importance for researchers to have a basic understanding how interaction maps were derived to avoid pitfalls in their usage. Subsequently, several studies and methodologies utilizing protein interaction networks to study cancer are reviewed. For sake of completeness, some references to the application of transcriptional networks to cancer research are given. Finally, we discuss future challenges and directions in the generation of human protein interaction maps and their applications.

9.2 The Human Protein–Protein Interactome: Generation and Analysis

In the last few years, we have witnessed the rapid increase in the large-scale protein–protein interaction maps for various model organisms. This striking rise is mainly due to advances in the high-throughput experimental techniques such as Yeast-two-Hybrid (Y2H), the coordinated efforts to systematically chart interactions by human experts as well as the progress in computational text-mining and prediction.

As all these methods can lead to considerably divergent protein interaction maps (von Mering et al. 2002; Futschik et al. 2007a, b), it is important to have a basic understanding of the applied methodologies. In the following sections, we therefore introduce several current methods, discuss their pros and cons and outline their application to the human interactome.

9.2.1 Yeast-Two Hybrid System

The Y2H method is based on a screening approach using a set of modified proteins. The experimental basis of Y2H is the reconstitution of a multi-domain transcription factor (such as GAL4). Specifically, a protein-encoding cDNA of interest is cloned into a bait vector, and fused with the DNA binding domain of the multi-domain transcription factor. A second cDNA encoding a potentially interacting protein is cloned into a prey vector and fused to the transcription factor's activation domain. Subsequently, the two yeast strains carrying the bait and prey hybrid proteins in plasmids are mated, resulting in yeast carrying both plasmids. If the bait and prey proteins interact, a functional transcription factor is reconstituted leading to the transcription of a reporter gene such as lacZ encoding for β-galactosidase. In the high-throughput mode, whole libraries of bait and prey vectors can be screened for interactions. Thus, the main advantage of this approach is that it provides a platform for the rapid generation of large-scale protein–protein interaction networks and it does not need to be biased towards known interactions. However, the false positive rate for Y2H screens can be considerable and can even exceed the estimated true positive rates (Hart et al. 2006).

Recently, the Y2H system was applied in two large-scale studies to screen human proteins identifying in total over ~5,500 new protein interactions, of which a selected sub-set was experimentally validated (Rual et al. 2005; Stelzl et al. 2005). Notably, the overlap between the two studies is small: Only 17% of interactions between common proteins were detected by both groups.

9.2.2 Literature Curation and Text-Mining

Besides high-throughput experimental approaches, the numerous small-scale experiments described in the literature can be exploited to create large-scale protein interaction maps. Tapping into the wealth of published experiments, information on protein interactions is systematically extracted from the literature either by human experts or text-mining algorithms. The advantages of such procedures are that it is not biased *a priori* towards a particular experimental technique and that the charted interactions are determined under a broad range of conditions and protocols. Characteristic disadvantages are the inherent difficulty to estimate the false positive rate and the bias towards highly studied proteins. Numerous research groups have followed this strategy to create large-scale human protein interaction maps (Bader et al. 2001; Salwinski et al. 2004; Pagel et al. 2005; Ramani et al. 2005; Mishra et al. 2006; Kerrien et al. 2007; Breitkreutz et al. 2008).

9.2.3 Computational Prediction of Human Protein Interactions

Alternative to the large-scale experimental and literature-curation, *in silico* prediction has been used to build large-scale protein–protein interaction maps (Lehner and Fraser 2004; Brown and Jurisica 2005; Persico et al. 2005). This strategy is based on the assumption that interactions are evolutionarily conserved for orthologous proteins and thus interactions detected between proteins in lower organisms can be extrapolated to their human orthologs. A main advantage of this method is that it is entirely computational and thus enables rapid and cost-effective construction of human protein–protein interaction maps. Disadvantages are that it is purely predictive in nature and false positives can arise through erroneous mapping to human orthologs or that interactions are simply lost during evolution.

9.2.4 Databases for Human Protein Interactions

Several human protein interaction databases have been established to help researchers find and analyze interaction partners of proteins of interest. These databases can generally be divided into two different categories: The first one is based on the manual-curation of published literature and includes the Human Protein Reference Database (HPRD), the Biological General Repository for Interaction Datasets (BioGRID), IntAct, the Database of Interacting Proteins (DIP), the Biomolecular Interaction Network Database (BIND) and the MIPS Mammalian Protein–Protein Interaction Database (MPPI) (Bader et al. 2001; Salwinski et al. 2004; Pagel et al. 2005; Mishra et al. 2006; Kerrien et al. 2007; Breitkreutz et al. 2008). The other category of databases also includes computationally predicted interactions; examples of such databases are the Online Predicted Human Interaction Database (OPHID) and HomoMINT (Brown and Jurisica 2005; Persico et al. 2005). Currently, HPRD is one of the major sources for human interaction data and – as the name implies – dedicated to human proteins. Besides interactions, it also provides information on domain architecture, post-translational modifications, disease association and biological pathways. Other databases e.g. BioGRID, IntAct, DIP and BIND are the repositories for a more diverse set of organisms and provide access to interaction data for other model organisms such as yeast, worm and fly.

Although these databases chart thousand of interactions from human proteins, their coverage in terms of the whole human interactome remains rudimentary. Comparative analysis revealed a very limited overlap between them (Chaurasia et al. 2006; Futschik et al. 2007a; Ramírez et al. 2007) (Fig. 9.1). Naturally, the question arises why these maps have such small degree of overlap. One reason is likely that current maps are highly unsaturated. Given an estimated total size of human interactome of ~650,000 interactions, even HPRD – as one of the largest sources – covers less than 10% of the total interactome (Stumpf et al. 2008). Additionally, current maps display a strong detection bias, i.e. they are enriched in characteristic types of proteins while depleted of other types (Futschik et al. 2007a). For example, literature-based maps show a significant enrichment in signalling proteins which is probably due to their popularity as biomedical research topic. Since currently available maps are incomplete and might contain complementary information, we and others reasoned that their integration can be beneficial. Therefore, several research groups have started to integrate the diverse protein interaction datasets available (Prieto and De Las Rivas 2006; Chaurasia et al. 2007). For instance, the Unified Human Interactome database (UniHI) including human interaction data from 14 different sources stores over ~250,000 interactions between ~22,000 proteins and thus constitutes one of the most comprehensive collections of human protein interactions at present (Chaurasia et al. 2009). Such centralized repositories liberate researchers from laborious and time-consuming integration of the diverse interaction data sets. An overview of several current resources for human protein interactions is provided in Table 9.1. A more complete list of protein interaction databases is compiled by the Pathguide project (Bader et al. 2006).

9.3 Application of Interactomics to Cancer Research

9.3.1 Network-Based Characterization of Cancer Genes

One of the first questions addressed by network-based approaches in cancer research is also one of the most intriguing: What makes a gene to a cancer gene? Although such naïve question may be somewhat puzzling at first, it makes naturally sense in network biology to ask whether cancer-associated genes have characteristic properties within interaction networks. To address this question, graph-based methods can be applied to study network properties of cancer genes. An important concept here is *centrality* which evaluates the location within a network. Centrality of a node can be defined simply by its degree, i.e. the number of interactions or, more elaborately, by the number of shortest paths passing through this node.

Several research groups have applied such concepts to reveal the graph-theoretical properties and the role of cancer genes in human protein interaction networks (Wachi et al. 2005; Jonsson and Bates 2006; Hernández et al. 2007; Platzer et al. 2007). For the analysis, the set of cancer-associated genes has first to be determined, for which commonly databases or microarray studies are used. As a second step, a disease network is created by integrating the cancer genes products (i.e. proteins encoded by cancer-associated genes) with available large-scale protein

Table 9.1 Resources for human protein–protein interactions described in the chapter. The size, the construction approach and additional information are given. For the calculation of the number of proteins and interactions in each dataset, proteins were mapped to their corresponding Entrez Gene identifiers

Resource	Proteins	Interactions	Method	References	Resource location
MDC-Y2H	1,703	3,186	Y2H SCREEN	Stelzl et al. (2005)	www.mdc-berlin.de/neuroprot
CCSB-Y2H	1,549	2,754	Y2H SCREEN	Rual et al. (2005)	www.vidal.dfci.harvard.edu (flat file only)
HPRD-BIN	8,788	38,800	LITERATURE	Peri et al. (2003)	www.hprd.org
DIP	1,085	1,397	LITERATURE	Salwinski et al. (2004)	www.dip.doe-mbi.ucla.edu
BIOGRID	7,953	24,624	LITERATURE	Breitkreutz et al. (2008)	www.thebiogrid.org
INTACT	7,273	19,404	LITERATURE	Hermjakob et al. (2004)	www.ebi.ac.uk/intact
BIND	5,286	7,394	LITERATURE	Bader et al. (2001)	www.bind.ca
COCIT	3,737	6,580	TEXT MINING	Ramani et al. (2005)	www.Bioinformatics.icmb.utexas.edu/idserve
REACTOME	1,554	37,332	LITERATURE	Joshi-Tope et al. (2005)	www.reactome.org
ORTHO	6,225	71,466	ORTHOLOGY	Lehner and Fraser, (2004)	www.sanger.ac.uk/PostGenomics/signaltransduction/interactionmap
HOMOMINT	4,127	10,174	ORTHOLOGY	Persico et al. (2005)	www.mint.bio.uniroma2.it
OPHID	4,785	24,991	ORTHOLOGY	Brown and Jurisica (2005)	www.ophid.utoronto.ca
UniHI	22,307	200,473	INTEGRATION	Chaurasia et al. (2009)	www.unihi.org

interaction networks. Finally, the topological properties (e.g. degree distribution, centrality) of cancer genes within this network are computed and compared to those of genes that have not been associated with cancer.

Wachi and co-worker applied the above outlined strategy to study the centrality of genes that are differentially expressed in cancer (Wachi et al. 2005). For their analysis, human interaction data was collected from OPHID. Microarray data were obtained from five patients with squamous cell lung cancer and compared to normal samples of the same patients. Using a paired *t*-test, differentially regulated genes were determined and mapped onto the protein interaction network. The subsequent analysis revealed that up-regulated genes tend to be highly connected and more centrally located in the network compared to randomly selected genes. Down-regulated genes tended to be also more highly connected but not significantly. Furthermore, they did not show an increased centrality. Based on their findings, the authors suggested that a core set of central genes has to be activated during the course of carcinogenesis.

Similar results were reported in a separate topological analysis performed by Jonsson and Bates (2006). In contrast to Wachi et al., this analysis did not depend on microarray experiments to define cancer-associated genes. To avoid a bias towards a particular cancer type, they selected a general set of cancer genes that were previously identified in a literature-based census (Futreal et al. 2004). The human interaction network was constructed using an orthology-based approach. After mapping of cancer genes onto the human protein interaction network, the connectivity of each protein in the integrated network was computed. Results indicate that the cancer proteins show higher degrees than non-cancer proteins. Cancer proteins also tend to function as central hubs, reflecting their role as a key player in protein–protein interaction network. Clustering analysis additionally showed that cancer proteins, on average, are more frequently located in the interfaces between clusters indicating an enhanced role in the coordination of different cellular processes.

Following the same strategy as Wachi et al., Hernández and colleagues reported somewhat contrasting results for the topological properties and organization of cancer gene products in the human interactome network (Hernández et al. 2007). They started their analysis by creating an integrated set of human interactome originated from five manually-curated literature-based dataset. Microarray data sets for prostate, lung and colorectal cancers were utilized and differential expression was calculated. Topological analysis of the integrated network revealed that down-regulated genes consistently tend to be more centrally located. In contrast, the centrality of up-regulated genes was dependent on the chosen cancer type. They also found that topological properties of down-regulated cancer genes are correlated with common biological processes and pathways that lead to cancer. However, both types of genes appear to be important for the organization and integrity of network structure. In particular, the elimination of cancer-associated genes from the network results in a faster breakage of the original network in smaller networks than those observed for elimination of randomly chosen genes.

Finally, the most comprehensive graph-theoretical study for cancer to date was conducted by Platzer et al. (2007). Altogether, they analysed 29 genome-wide cancer expression data sets using 22 individual graph-theoretical measures. For each study, differential gene expression was determined and sub-graphs of differentially regulated genes were constructed based on interaction data from OPHID. Various properties of the sub-graphs such as size, modularity and density were subsequently examined. The main result was that genes showing differential expression in cancer tend to interact and to form larger sub-networks than expected by chance. Strikingly, however, the prevalence of hub proteins was not increased in cancer-associated subgraphs. The authors speculated that extended graphs with low density indicate networks of high robustness against the failure of single genes. This is especially intriguing in the context of cancer, as such finding would demand for the simultaneous therapeutic targeting of multiple proteins.

In summary, the described network studies give a first overview about the structural role of cancer genes in protein interaction networks. Nevertheless, care has to be taken in interpretation as current interaction maps often show divergence in structure due to different methods used for their assembly (Futschik et al. 2007b).

9.3.2 Identification of New Cancer-Associated Genes and Processes Using Protein Interaction Networks

A second area in which protein interaction networks have been utilized in cancer research is the identification of new cancer-associated genes. The rationale behind these investigations is that interacting proteins are likely linked to the same or similar phenotype. A leading example is Fanconi anemia, a genetic disease, for which seven of the nine associated proteins form a physical complex involved in DNA repair. Although interaction data can provide a suitable first basis for *de novo* identification of disease-causing genes, additional information has commonly been utilized to improve specificity.

For many years, genetic linkage studies were the most potent approach to find new disease-causing genes. A major difficulty, however, is to pick the right gene within extended chromosomal regions that have been linked to a disease. Oti et al. showed that this task can be considerably facilitated using protein interaction data (Oti et al. 2006). For genetically homogenous diseases, they predicted new disease associations when genes fell within an identified susceptibility locus and have protein interactions with a gene known to cause this disease. This simple method of data integration led to a tenfold increased specificity compared to randomly selected candidate genes at the same locus. Notably, Oti et al. also deduced that protein interactions added as much information as localization to the prediction accuracy. In a similar study, Franke et al. extended the protein interaction network by including microarray and gene annotation to generate a functional interaction network (Franke et al. 2006). Also, new candidate genes were identified in the larger network neighbourhood of known disease genes, avoiding the restriction to direct interactors only.

One requirement of these studies is that we have to know already a set of genes associated with a certain disease. This set can be then used to "anchor" a disease in the human interactome. If however no such genes are known, this approach cannot be used. To overcome this limitation, Lage et al. catalogued human phenotypes in a computationally tractable manner (Lage et al. 2007). Their motivation was that similar diseases might share the same molecular basis. Having defined a score for the similarity of phenotypes, information for a specific disease can then be deduced from similar diseases. Thus, candidate genes can be predicted even if no other gene associated with the specific disease is known yet. For prediction, Lage et al. integrated human protein interaction with linkage data in a similar manner as Oti et al. and Franke et al. Using an *in silico* pull-down approach and the similarity of phenotypes, they extracted known and new complexes and predicted several novel candidate disease genes involved in disorders such as cancer, Alzheimer's, diabetes and coronary heart diseases. Detailed analysis for epithelial ovarian cancer lead to the identification of a new candidate gene, Fanconi anemia group D2 protein (*FANCD2*) placed in a complex with breast cancer type 1 susceptibility protein (*BRCA1*) and breast cancer type 2 susceptibility protein (*BRCA2*). This protein has been associated with different types of cancer, but not with epithelial ovarian cancer so far.

A conceptually similar network-based modelling approach was applied by Pujana et al. to predict new candidate genes involved in breast cancer (Pujana et al. 2007). They assumed that genes, which are functionally related or showed conserved co-expression across species, might cause a similar phenotype. To test their hypothesis, they created a cancer-specific network with four known breast cancer-associated genes: *BRCA1*, *BRCA2*, *ATM*, and *CHEK2*. Neighbours of each reference gene set were further ranked using a scoring system based on co-expression, phenotypic similarity, and genetic or physical interactions among orthologs of the proteins in other species. They identified a new gene (*HMMR*) that was found to be associated with an increased risk of breast cancer.

In addition to prediction of novel cancer-associated genes, interaction networks were also employed to unravel cancer-related molecular processes. As one example, Chuang et al. applied a network-based classifier to identify sub-networks as markers for breast cancer prognosis (Chuang et al. 2007). To find the sub-networks, they mapped the gene expression profiles of metastatic and non-metastatic patients on a human protein–protein interaction network. Subsequently, they computed activity scores of all associated members to rank the sub-network as a whole. Their finding showed that high scoring sub-networks were enriched in many cancer-related biological processes such as apoptosis, proliferation, tissue remodelling, signalling and survival. Their analysis also indicated that identified modules were more reproducible than individual genes selected without network information, and that they achieve a higher accuracy in the classification of metastatic *versus* non-metastatic tumors. Another advantage of this approach is that it also captures those genes which may have not been detected based on gene expression data alone. Such non-differentially expressed genes could be an integral part of a complex and be required for connecting high scoring proteins in a sub-network. In fact, Chuang et al. found that a large number of the identified network structures contained at least one protein that was not significantly expressed in metastasis while most of them served as a bridge between high scoring proteins in a sub-network. This integration provides the opportunity to analyze the relationships between members of the complexes, and also increases the accuracy of the overall prediction.

9.3.3 Analysis of Transcriptional Regulatory Networks in Cancer Research

Besides physical protein–protein interactions, transcriptional regulations have been analyzed in network biology to shed light on oncogenesis. The main building blocks of the constructed transcriptional regulatory networks are transcription factors and their target genes. In contrast to the protein interaction networks, the resulting graphs are directed, i.e. include edges directed from transcription factors to their target genes. Since transcription factors can be themselves target genes of other transcription factors, this wiring scheme can lead to large connected networks. The ultimate goal is to build models that can "explain" observed expression patterns in terms of the underlying regulatory networks. Such models would go beyond the simple description of expression changes and could eventually provide us with a causative framework. This has become particularly interesting in the context of microarray technologies that have enabled a rapid genome-wide monitoring of expression.

In particular for yeast, this line of investigation has proven to be fruitful in revealing regulatory principles that are not detectable from the mere analysis of expression data (Janga et al. 2008). Early studies, for instance, could link changes in the structure of regulatory networks to the type of external stimuli and the corresponding transcriptional response (Luscombe et al. 2004). Such impressive interrogations were made possible by the systematic experimental mapping of yeast transcription factor binding sites using Chromatin-Immunoprecipitation on chip (ChIP-chip) experiments. Unfortunately, the systematic experimental charting of human transcription factor binding sites is still at a very early phase with experiments being limited to a small number of transcription factors and cell types. At present, many collections of transcription factor binding sites for humans thus rely considerably on *in silico* matching between promoter regions and position weighted matrices describing the consensus binding sites of transcription factors. Further difficulties in the construction of comprehensive regulatory networks are (1) a high number of false positive predictions of transcription factor binding sites based on simple sequence matching, (2) the choice of an adequate size of human promoter regions, (3) the combinatorial action of transcription factors within *cis*-regulatory modules and (4) the influence of the – generally unknown – chromatin structure on the accessibility of binding sites.

Despite these challenges, first efforts have been undertaken to construct genome-wide regulatory networks for cancer research. Notably, Kluger and colleagues examined the topological properties of regulatory networks to characterize gene deregulation during tumorigenesis (Tuck et al. 2006). For construction of a regulatory network, they utilized a collection of transcription factors stored in the Transcription factor (TRANSFAC) database. Potential target genes were determined by position weighted matrices. The basal connectivity network was then intersected with co-expression of genes from different cancer microarray studies to obtain condition-specific regulatory networks. In the subsequent analysis, network features such as degree distributions were used to differentiate between diseased and healthy patent samples. Although no significant improvement of classification accuracy was achieved compared to conventional microarray analysis, the procedure offered some valuable insights in the potential causative mechanisms of gene deregulation. Most intriguingly, genes that discriminate best between disease conditions tend to be highly localized on the transcriptional network. It is important to note that the applied strategy implies that expression levels of transcription factors can be proxies for their activity states. However, this might neglect important post-translational modifications.

An impressive project, which can also serve as a prime example for integrative network biology, is the assembly and analysis of the B-cell interactome by Califano and co-workers. This model of the molecular network for B-cells not only includes transcriptional regulatory, but also protein–protein and modifying post-translational interactions derived from a variety of experimental and computational resources. In the study by Mani et al., a strategy was developed to scrutinize the B-cell interactome for dysregulated interactions in three distinct types of lymphoma (Mani et al. 2008). In contrast to conventional microarray analysis focusing on the differential regulation of genes, the loss or gain of correlation between interacting genes was analyzed. Remarkably, the examination of dysregulated interactions pointed more clearly to the set of known genetic lesions than simple differential gene expression did. Furthermore, potential downstream effectors could be identified which would have been missed using gene expression alone. Notably, these results probably would not have been derived without the construction of a cell type-specific network.

9.4 Summary and Outlook

Cancer shows a striking complexity in the cellular mechanisms involved and, despite all successes in cancer research, the untangling of these interwoven processes remains one of the most formidable tasks in molecular biology and medicine. For a long time, genes and their implications in cancer were studied one at a time. This time-honoured strategy has now been complemented with systems-wide studies of disease-associated mechanisms. A central position in the new paradigm has taken the uprising field of network biology. Applied to biomedicine, diseases represent particular states of the underlying molecular network; a perspective that was already brought forward several decades ago by S. Kauffman (Kauffman 1993). Following his influential ideas, cancer can be perceived as attractor states that might display remarkable robustness. Although based mostly on theoretical reasoning, we might argue to view cancer as a network-associated disease which requires complex intervention for its treatment (Kitano 2007).

A pivotal role in this new system biological strategy will be the study of protein interaction networks. Proteins and their aberrant interactions have long been known to be crucial in oncogenesis. With the construction of comprehensive interaction maps, we are now approaching a stage where the influence of dysfunctional proteins can systematically be dissected and potential interventions designed. Despite its early successes and rapidly growing popularity, the application of interactomics requires some caution.

Interaction maps of molecular processes are frequently highly rudimentary. This is also the case for human protein–protein maps in spite of their impressive size. At present, they are still scanty and are likely to include a considerable number of false positives. These shortcomings of current protein interaction networks – as well as of other types of molecular networks – underline the necessity of integrate complementary data and information. In fact, only by constructing multi-dimensional datasets, one can harvest the full potential of protein interaction maps. At present, this is mainly performed by simple mapping of expression changes onto generic interaction maps extracted from databases. Notably, such simple strategies account poorly for the complex spatial and temporal aspects of carcinogenesis. One step towards a more accurate representation can be the creation of tissue-specific networks. This might be especially relevant for cancer research where the examination of genes can lead to contradictory results depending on the used experimental model. For instance, *RAS*, a classical oncogene, has been shown to function in a tumor suppressing manner under certain conditions indicating the importance of the molecular context (Zhang et al. 2001). Also, the usefulness of streamlined interaction networks has already been demonstrated by the described study of the B-cell interactome. Future molecular maps reflecting this complexity will provide highly valuable tools for biomedical research. Indeed, the integration of independent information concerning expression and localization has already been used for the identification of dynamic as well as constitutive protein modules (Futschik et al. 2007c).

To conclude, early applications have indicated the large potential of network biology in cancer research. Progress in experimental techniques and computational methods will continue to improve the coverage and sensitivity of interaction networks. A focus of interactomics – especially in its application to cancer research – will be on the combination of different types of networks, such as protein-protein, transcriptional regulatory and metabolic networks, to enable the creation of detailed molecular models of oncogenesis. Furthermore, the integration of interactions networks with the rich datasets generated by ongoing cancer-related sequencing, microarray or imaging projects is likely to provide us with molecular maps of unprecedented detail for the human organism in health and disease. Thus, network biology promises to contribute substantially to a better understanding of the complexity of cancer and eventually to its cure.

Acknowledgements We would like to thank Paulo Martel and Nuno dos Santos for their important contributions to this chapter.

References

Alizadeh AA, Eisen MB, Davis RE et al (2000) Distinct types of diffuse large B-cell lymphoma identified by gene expression profiling. Nature 403:503–511

Bader GD, Donaldson I, Wolting C et al (2001) BIND – The biomolecular interaction network database. Nucleic Acids Res 29:242–245

Bader GD, Cary MP, Sander C (2006) Pathguide: a pathway resource list. Nucleic Acids Res Database 34:D504–506

Breitkreutz BJ, Stark C, Reguly T et al (2008) The BioGRID Interaction Database: 2008 update. Nucleic Acids Res Database 36:D637–640

Brown KR, Jurisica I (2005) Online predicted human interaction database. Bioinformatics 21:2076–2082

Chaurasia G, Herzel H, Wanker EE et al (2006) Systematic functional assessment of human protein–protein interaction maps. Genome Inform 17:36–45

Chaurasia G, Iqbal Y, Hänig C et al (2007) UniHI: an entry gate to the human protein interactome. Nucleic Acids Res Database 35:D590–594

Chaurasia G, Malhotra S, Russ J et al (2009) UniHI 4: new tools for query, analysis and visualization of the human protein–protein interactome. Nucleic Acids Res Database 37:D657–D660

Chuang HY, Lee E, Liu YT et al (2007) Network-based classification of breast cancer metastasis. Mol Syst Biol 3:140

Euler L (1736) Solutio problematis ad geometriam situs pertinentis, Commentarii academiae scientiarum Petropolitanae 8:128–140

Franke L, van Bakel H, Fokkens L et al (2006) Reconstruction of a functional human gene network, with an application for prioritizing positional candidate genes. Am J Hum Genet 78:1011–1025

Futreal PA, Coin L, Marshall M et al (2004) A census of human cancer genes. Nat Rev Cancer 4:177–183

Futschik ME, Chaurasia G, Herzel H (2007a) Comparison of human protein–protein interaction maps. Bioinformatics 23:605–611

Futschik ME, Tschaut A, Chaurasia G et al (2007b) Graph-theoretical comparison reveals structural divergence of human protein interaction networks. Genome Inform 18:141–151

Futschik ME, Chaurasia G, Tschaut A et al (2007c) Functional and transcriptional coherency of modules in the human protein interaction network. J Integr Bioinform 4:76

Hanahan D, Weinberg RA (2000) The hallmarks of cancer. Cell 100:57–70

Hart GT, Ramani A, Marcotte E (2006) How complete are current yeast and human protein-interaction networks? Genome Biol 7:120

Henning Hermjakob, Luisa Montecchi-Palazzi, Chris Lewington et al (2004) IntAct: an open source molecular interaction database. Nucleic Acids Res 32: D452–D455

Hernández P, Huerta-Cepas J, Montaner D et al (2007) Evidence for systems-level molecular mechanisms of tumorigenesis. BMC Genomics 8:185

Janga SC, Collado-Vides J, Babu MM (2008) Transcriptional regulation constrains the organization of genes on eukaryotic chromosomes. Proc Natl Acad Sci USA 41:15761–15766

Jeong H, Mason SP, Barabasi AL et al (2001) Lethality and centrality in protein networks. Nature 411:41–42

Jonsson PF, Bates PA (2006) Global topological features of cancer proteins in the human interactome. Bioinformatics 22:2291–2297

Joshi-Tope G, Gillespie M, Vastrik I, et al (2005) Reactome: a knowledgebase of biological pathways. Nucleic Acids Res., 33:D428–D443

Kauffman SA (1993) Origins of order: self-organization and selection in evolution. Oxford University Press, New York

Kerrien S, Alam-Faruque Y, Aranda B et al (2007) IntAct – open source resource for molecular interaction data. Nucleic Acids Res Database 35:D561–565

Kitano H (2007) Towards a theory of biological robustness. Mol Syst Biol 3:137

Lage K, Karlberg EO, Størling ZM et al (2007) A human phenome-interactome network of protein complexes implicated in genetic disorders. Nat Biotechnol 25:309–316

Lehner B, Fraser BA (2004) A first-draft human protein-interaction map. Genome Biol 5:R63

Luscombe NM, Babu MM, Yu H et al (2004) Genomic analysis of regulatory network dynamics reveals large topological changes. Nature 431:308–312

Mani KM, Lefebvre C, Wang K et al (2008) A systems biology approach to prediction of oncogenes and molecular perturbation targets in B-cell lymphomas. Mol Syst Biol 4:169

Mishra GR, Suresh M, Kumaran K et al (2006) Human protein reference database – 2006 update. Nucleic Acids Res Database 34:D411–414

Oti M, Snel B, Huynen MA et al (2006) Predicting disease genes using protein–protein interactions. J Med Genet 43:691–698
Pagel P, Kovac S, Oesterheld M et al (2005) The MIPS mammalian protein–protein interaction database. Bioinformatics 21:832–834
Peri S, Navarro JD, Amanchy R et al (2003) Development of human protein reference database as an initial platform for approaching 245 systems biology in humans. Genome Res., 13:2363–2371.
Persico M, Ceol A, Gavrila C et al (2005) HomoMINT: an inferred human network based on orthology mapping of protein interactions discovered in model organisms. BMC Bioinformatics 6:S21
Platzer A, Perco P, Lukas A et al (2007) Characterization of protein-interaction networks in tumors. BMC Bioinformatics 8:224
Prieto C, De Las Rivas J (2006) APID: Agile Protein Interaction DataAnalyzer. Nucleic Acids Res Web Server 34:W298–302
Pujana MA, Han JJ, Starita LM et al (2007) Network modeling links breast cancer susceptibility and centrosome dysfunction. Nat Genet 39:1338–1349
Ramani A, Bunescu R, Mooney RJ et al (2005) Consolidating the set of known human protein–protein interactions in preparation for large-scale mapping of the human interactome. Genome Biol 6:R40
Ramírez F, Schlicker A, Assenov Y et al (2007) Computational analysis of human protein interaction networks. Proteomics 7:2541–2552
Rual JF, Venkatesan K, Hao T et al (2005) Towards a proteome-scale map of the human protein–protein interaction network. Nature 437:1173–1178
Salwinski L, Miller CS, Smith AJ et al (2004) The database of interacting proteins: 2004 update. Nucleic Acids Res Database 32:D449–D451
Stelzl U, Worm U, Lalowski M et al (2005) A human protein–protein interaction network: a resource for annotating the proteome. Cell 122:957–968
Stumpf M, Thorne T, de Silva E et al (2008) Estimating the size of the human interactome. Proc Natl Acad Sci USA 105:6959–6964
Tuck DP, Kluger HM, Kluger Y (2006) Characterizing disease states from topological properties of transcriptional regulatory networks. BMC Bioinformatics 7:236
von Mering C, Krause R, Snel B et al (2002) Comparative assessment of large-scale data sets of protein–protein interactions. Nature 417:399–403
Wachi S, Yoneda K, Wu R (2005) Interactome–transcriptome analysis reveals the high centrality of genes differentially expressed in lung cancer tissues. Bioinformatics 21:4205–4208
Zhang Z, Wang Y, Vikis HG et al (2001) Wildtype Kras2 can inhibit lung carcinogenesis in mice. Nat Genet 29:25–33

Chapter 10
Cytomics and Predictive Medicine for Oncology

A.O.H. Gerstner and G. Valet

Abstract Cytomics combines the multimolecular cytometric analysis of cell and cell system (cytome, cytomes) heterogeneity on a single cell level with the exhaustive bioinformatic knowledge extraction from all analysis results (cytomics = system cytometry + bioinformatics). It therefore yields a maximum of information about the apparent molecular cell phenotype.

At present, in the typical *hypothesis driven* way the high amount of information collected by multiparameter single cell flow- or slide-based cytometry measurements is preferentially used to investigate the molecular behaviour of specific cell populations in the perspective of the hypothesis. The information outside the scope of the hypothesis remains frequently unused.

In contrast, under the *predictive medicine by cytomics* concept, the entire available information is processed ("sieved") in a *data driven* way under the general data mining hypothesis that such data may contain useful information for clinical diagnosis and especially for therapy related predictions about disease progress in individual patients.

The present experience from clinical data sets of various malignant and other diseases suggests that this is a promising concept for cancer patients since it has amongst others the potential to identify high risk patients prior to an anticipated therapy as being unsusceptible with accuracies of greater 95% or 99%. This opens the way for early decision on alternative therapies by objective and molecularly standardised criteria. This has been traditionally difficult by current prognosis evaluation according to the widely used Kaplan-Meier statistics for patient groups.

The cytomics concept is also useful for cancer research in general because it favours the enrichment of informative parameters concerning disease outcome in individual organisms or cell cultures from an essentially unlimited number of parameters.

A.O.H. Gerstner (✉)
Department for Otorhinolaryngology/Head and Neck Surgery, University of Bonn, Bonn, Germany
e-mail: gerstner@web.de

G. Valet
Max-Planck-Institut für Biochemie, Martinsried, Germany
e-mail: valet@biochem.mpg.de

W.C.S. Cho (ed.), *An Omics Perspective on Cancer Research*,
DOI 10.1007/978-90-481-2675-0_10,

The selected parameters are useful as a starting point for mathematical modelling in systems biology without requirement for detailed pre-existing knowledge about potential disease inducing mechanisms. It has therefore the potential for the discovery of new molecular cell pathways and for their subsequent molecular reverse engineering.

10.1 Background

Single cell measurements by flow or slide-based cytometry in cancer medicine and cancer research are typically performed in a multiparameter setup using hypothesis driven parameter selection in order to simultaneously gather a maximum of diagnostic or prognostic information by the accurate assessment of the molecular phenotype of patient or experimental cells (Table 10.1).

Table 10.1 Comparison of flow- *vs.* slide-based cytometry**.** Some principle advantages and specific features of flow- (FCM) and slide-based (SBC) cytometry are listed. Depending on the specific setting one approach alone or both in combination might be applied. Whereas FCM allows to measure large sets of parameters in unsurpassed speed, SBC offers the unique feature to keep cells in their natural environment, i.e. tissue context and to follow up "individual" cells at different time points of experimental settings

	FCM	SBC
Technical specifications	Rapid speed	Slow
	Low CVs	Broader CVs
		Higher background, bleaching, during measurement
	Standard 6 colours	Standard 6 colours
	Optional 17 colours	Optional 8-colour, theoretical n-colour
Logical features	Single-cell	Multi-cellular complexes: tissue sections, cell/tissue cultures
	In suspension: cell-network destroyed	On slide: topology kept intact
	High-content analysis	High-content analysis
	Limited structural resolution	Morphological re-evaluation
Clinical/practical aspects	Large specimens	Hypocellular specimens
	Consumptive unless cell sorting	Non-consumptive, no cell loss
	Detection of ultra-rare events	Analysis of cell interaction
	Bulk-sorting of specific cell-subtypes	Re-analysis on a single-cell basis
		Combination of data at different functional states (pre-/post-fixation, pre-/post-stimulation) at single-cell level

With cells as the elementary function units of organisms at the one hand and diseases being caused by molecular alterations in cells and cellular systems (cytomes) at the other hand (Valet 2002), the disease associated molecular cell phenotype is a correlate of the disease process and a result of genotype realisation and the lifetime history of cell exposure to external and internal influences. The molecular cell phenotype (Fig. 10.1) is of interest for disease diagnosis but also for predictions about the future disease progress in individual patients.

These considerations lead to the development of the *data driven* system cytometry (Valet 1997) and cytomics (Valet 2002) approaches with the aim to extract knowledge from the entire available information of *hypothesis driven* cytometry and other investigations. Both, system cytometry and cytomics, regard a single cell as being a single biochemical cuvette. The analysis of the utmost cellular complexity instead of cellular monosystems constitutes the central feature of system cytometry. This requires to combine both: (a) to collect as much biochemical information as possible in a maximum of potentially related but nevertheless different cell populations of complex cellular systems (blood, bone marrow, transplant biopsies etc); and (b) extract this enormous amount of information

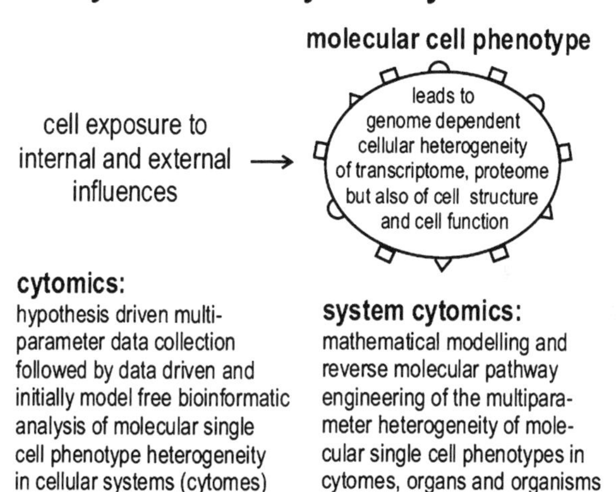

Fig. 10.1 Cytomics. Cytomics and system cytomics address the molecular heterogeneity of single cells in cell systems (cytomes). Genome expression in cytomes adapts to environmental influences and may lead to altered disease susceptibility in genetically identical organisms (Wirdefeldt et al. 2005). The resulting molecular cell phenotype represents a useful indicator of the actual balance between genetic setup and exposure in healthy and diseased individuals. It provides information for therapy dependent predictions on future disease course in individual patients (predictive medicine by cytomics) and is of interest for the molecular reverse engineering of disease pathways (Valet 2005b, 2002, 1997) by system cytomics as well as for purposes of drug discovery (Valet 2006)

efficiently by standardised multiparameter data classification (SMDC). This concept dates back to 1982 when a periodic system of cells was drafted for the first time (Schwemmler 1982).

Nevertheless, still multiparameter cytometry measurements are used in many instances to discriminate particular cell populations of interest while the information of the other cell populations remains unconsidered. This constitutes a significant waste of information since it is by no means certain that non-evaluated cells lack relevant diagnostic, prognostic, or predictive information.

Antigen as well as forward (FSC) and sideward (SSC) light scatter distributions of cells are typically broad with coefficients of variation in the 20–50% range. Cellular antigens are frequently expressed with little correlation to each other as evidenced by the presence of nearly round or spherical clusters in cytometric two or three parameter histograms displays (Fig. 10.2). Antigen expression, the correlations between different antigen expressions, and the spreads of value distributions

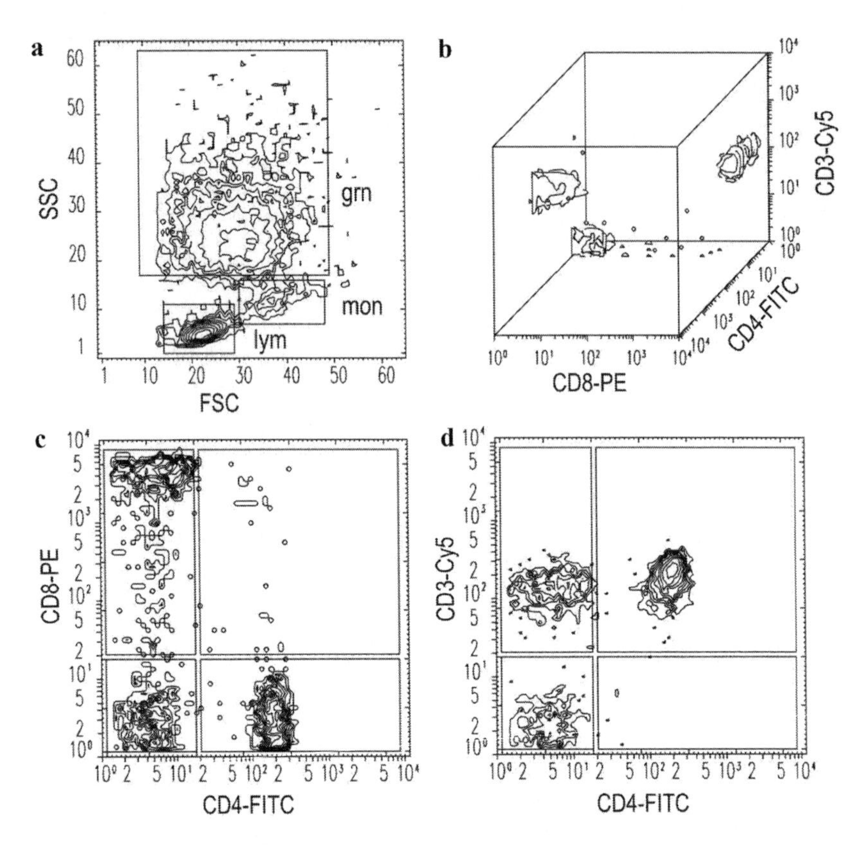

Fig. 10.2 CD4, CD8 and CD3 antigen expression on peripheral blood leukocytes of a healthy adult person by flow cytometry as determined by fluorescence labelled antibodies (data file from Valet et al. 2002). The spheroid and oblongated cell clusters show significant spread and little parameter correlation (**a–d**). The coefficients of variation (CVs) of the cell clusters are typically between one and two orders of magnitude higher than CVs from DNA cell cycle analysis where

may change during disease or experimental conditions. The observed cellular heterogeneity may furthermore contain information about the high adaptivity of the hemato- and immunopoietic systems. It seems worthwhile to study these phenomena by systematic, automated, and exhaustive information and knowledge extraction from all cells or biological particles in slide-based or flow cytometric measurements by bioinformatic data mining (Valet et al. 1986) under the general hypothesis that the thus obtained information may prove useful for disease diagnosis or especially for therapy related predictions on disease progress in individual patients. Data mining can be performed by statistical or algorithmic methods. Algorithmic data pattern classification (data sieving) (Valet et al. 1993) was found particularly efficient for this purpose.

10.2 Flow Cytometry

10.2.1 Clinically Oriented Studies

Initial efforts concerned cell functions with intracellular pH, esterase activity, and viable *E. coli* K12 phagocytosis as potential outcome predictors for intensive care patients with regard to recovery, development of sepsis, or posttraumatic shock (Rothe et al. 1990). Although at this stage not predictive, granulocyte function parameters showed a potential for individualised outcome predictions when granulocyte serine protease activity was flow cytometrically determined (Valet et al.

Fig. 10.2 (continued) G0/G1 phase CV values around and below 1% lead to the resolution of x- and y-spermatides by flow cytometry (Meistrich et al. 1978). The form and spread of the cell clusters indicates molecular heterogeneity of cell populations and not uncertainty of measurement. The uncorrelated heterogeneity of cell parameters is informative in view of knowledge extraction by data mining (Valet et al. 1993, 2003). Self-adjusting gates (Valet et al. 1993) separate peripheral blood leukocytes into lympho-, mono- and granulocytes (A). The CVs (CV = 100*standard deviation/mean) of the FSC/SSC clusters of lympho- (lym), mono- (mon) and granulocytes (grn) are 10.2/21.6%, 10.7/16.6%, 23.2/25.1% with correlation coefficients between FSC and SSC of r = 0.441, 0.269, 0.099. The selective display of cells within the lymphocyte gate (**a**) separate CD4/CD3 positive and CD8/CD3 positive T-lymphocyte clusters as well as a CD3/CD4/CD8 negative cell cluster together with some CD8 positive but CD3 negative cells (**b**). The CD4 and CD8 positive T-cell clusters (**c**) show coefficients of variation of 24.6% and 46.9% with CD4 to CD8 correlations of r = 0.183 and -0.053. The CVs for CD3 expression on CD3 positive/CD4 negative and CD3/CD4 positive lymphocytes (**d**) are 37.2% and 33.3% with 25.0% for the CD4 positive cells. The CD4 to CD3 correlations are r = 0.056 and 0.273 for the two lymphocyte clusters. The automated evaluation of the list mode file with the parameters forward (FSC) and sideward (SSC) light scatter, fluorescein isothiocyanate (FITC), phycoerythrin (PE) and phycoerythrin-cyanine 5 (Cy5) antibody labels, provides histograms containing 9789 (**a**), 2514 (**b**), 1872 (**c**) and 2384 (**d**) cells. Contour lines were drawn on a 3-decade logarithmic amplitude scale in 10% linear steps downwards from the respective maximum logarithmic channel content of each histogram (max = 253, 22, 70 cells) (**a, c, d**). The 10% contour line delimits the 3D clusters (max = 118 cells) (**b**). The lowest contour lines contour histogram channels containing minimally a single cell thus assuring the display of all cells in each histogram

1998). It was equally possible to identify risk patients for myocardial infarction from peripheral thrombocyte surface antigens CD62, CD63, and thrombospondin by algorithmic data pattern classification (Valet et al. 1986).

With this background, predictions in malignant diseases were addressed using clinical data collected outside of the own institution. It was possible from an initial data set of six parameters to prognosticate by meta-analysis individualised 10 year survival in melanoma patients at diagnosis (Valet et al. 2001) from the parameter triplet tumor diameter, tumor infiltration depth, and percentage of S-phase cells with negative and positive predictive values of around 80.3% for survivors and 79.8% for non-survivors (Valet et al. 1986).

Acute myeloid leukemia (AML) was investigated in the context of a multicenter study of the South German Hemoblastosis Group (SHG). Data on the expression of 23 of the most common cytogenetic abnormalities, of the frequency of positive cells for 36 antigen specificities, and of 9 clinical parameters at diagnosis were available (Valet et al. 2003). The predictive data pattern consisted of the following seven selected parameters: patient age, sample cellularity, percent CD2, CD4, CD13, CD36, and CD45 positive cells. This pattern identified with positive predictive values of 100% a subset of 49.8% and 91.5% of high risk non-survivors in the learning and unknown test sets. The result indicates an identifiable patient subgroup that will only have a survival chance upon immediate bone marrow stem cell transplantation. This clinically important conclusion cannot be derived from the concomitant prognostic analysis of the same data by Kaplan-Meier statistics (Repp et al. 2003).

Survivor AML patients, in contrast, are not well identified with 15.1% negative predictive value by the above data pattern (Valet et al. 2003). A more discriminatory data pattern may help. It may be important to analyse the "normal" cell populations by flow cytometry appearing during the first remission and to classify these results against final outcome to better understand the influence of non-malignant cells on final AML outcome.

Individualised outcome prediction by data pattern classification is not restricted to flow cytometry. The expression of mRNA in diffuse large B-cell lymphoma (DLBCL) was assessed on a Lymphochip cDNA array with 7,399 evaluable spots (Rosenwald et al. 2002). In the original study, this high amount of information produced a Kaplan-Meier prognosis graph, similar to the one for the international prognostic index (IPI) (The International Non-Hodgkin's Lymphoma Prognostic Factors Project 1993). The IPI is constituted of only five clinical parameters, namely age over 60 years, late-stage disease (stages 3 and 4), more than one extranodal site, high LDH, and poor general health. Since the Lymphochip evaluation heat maps did not allow individualised predictions it was of interest whether data pattern classification could do so.

High risk non-survivor patients were well identified at diagnosis by data pattern classification with 98.1% and 78.3% positive predictive values in the learning as well as in the unknown test set of patients (Valet and Höffkes 2004). Survivors, similarly as in the AML study, were less well recognised with negative predictive values of 67.3% and 45.3%.

In conclusion, data pattern classification has been shown to be of significant immediate clinical interest for the identification of high risk patients at diagnosis. The selected data patterns may be useful for research purposes and it may be promising to additionally classify "normal" remission cells against AML- or DLBCL-outcome.

10.2.2 Multiparameter Data Mining

Individualised predictions of therapy related disease progress in medicine are frequently considered impossible because in the majority of patients their future under therapy is evaluated by Kaplan–Meier statistics for patient groups (Bland and Douglas 1998). This approach is directed towards overall therapy optimisation but does not address the individual patient.

Parameter values of patients are typically introduced into value clusters in multiparameter data mining. At this point the concatenation, i.e. the coherence of the parameter values of a given patient, is lost. The value clusters for the various parameters are then correlated between the various classification categories and the parameters with the most discriminatory clusters are selected for classification. The lost coherence of patient parameters values leads typically to probabilistic conclusions with the inherent difficulty to generate individualised disease course predictions at accuracy levels greater than 95% or 99%.

Accurate individual predictions of therapy related disease progress are, however, important for patients and doctors but also for the public and private health care systems since potential adverse drug reactions in non-responders as well as higher health care costs due to inefficient therapies.

Data pattern classification according to the CLASSIF1 algorithm (Valet et al. 1993, 2001; Valet 2005a) performs always a concatenated parameter classification of the originally measured multiparameter data patterns of each patient to achieve individualised predictions. The most discriminatory parameters of the measured data patterns are selected during the iterative learning phase and kept whereas lesser discriminating parameters are eliminated from consideration after the end of the iteration process.

All parameter values of a data pattern are classified at the beginning of the learning phase, then one parameter is temporarily removed, followed by reclassification of the remaining learning set. If the removed parameter contained information, an effect of this removal on the classification result will be observed (deterioration or improvement); this result is retained. The temporarily removed parameter is then reinserted and the next parameter is temporarily removed for the next iteration, and so on until the last parameter. Only parameters that have improved the classification result are kept at the end.

The ensuing predictive classification patterns consist typically of between 5 and 20 parameters. The statistical probability that such patterns occur randomly decreases with 3^{-n} and is lower than 1% from five-parameter classification patterns onward. Statistical probabilities do therefore not significantly influence the classification result.

The robustness of the CLASSIF1 data patterns against random statistical fluctuations permits to further analyse them in order to discriminate subgroups of patients for example of particular genotype or exposure background as an additional data mining aspect. CLASSIF1 classifiers are typically standardised on a reference group of patients or on standard particles. The similarity of reference groups is verified by classifying them against each other. If they are indistinguishable classifiers can be compared between institutions and databases can be merged to build up standardised relational classification system of cells at the molecular level suitable for a human cytome project or for a periodical system of cells where normal and diseased cells can be compared in a standardised classification system for normal and abnormal cells (Valet 2005a, b).

10.3 Slide-Based Cytometry

Slide-based cytometry offers a different technological platform for the multiparametric analysis of cells. First instruments actually date back as early as 1955 (Caspersson et al. 1955; Göhde and Dittrich 1970): instead of passing the sample as a single-cell solution flowing through an analysis chamber as in flow cytometry, for slide-based cytometry the sample is prepared on an objective slide as for conventional microscopy or on some other kind of adequate solid support. The different commercially available instruments have in common that this slide with the fluorescent cells on it is moved by a motorised stage under the objective. Cells are illuminated by lasers or lamps (Xenon lamp or mercury arc lamp) and the fluorescence is detected by CCDs or photomultiplier tubes. This chapter does not aim to describe the different technical features in detail that have been realised in order to obtain stable illumination, spectral separation of different fluorochromes, detection of the fluorescence, or calculation of fluorescence intensity (for detailed comparison, see Mittag et al. 2009). We would rather concentrate on the potential of this technology in general in the context of clinical oncology.

The slide-based design opens several opportunities that have so far not been accessible by flow cytometry or image cytometry based on conventional chromatic cytological staining. First, fluorescence can be stoichiometrically detected so that the number of antigens bound by an antibody can be calculated (Rimm 2006), and modern fluorochromes allow combinations of up to 17 dyes that can be detected in parallel (Perfetto et al. 2004). This is a major advantage as compared to image cytometry. So far, for a slide-based system up to eight colours have been combined at a single run using two excition wavelengths (488 nm by an Argon- and 633 nm by a HeNe-laser) (Gerstner et al. 2002; Mittag et al. 2006). This can easily be increased by adding another wavelength (405 nm by a violet-dye laser) and adapting the filter systems. In image analysis there is in general only one channel available; taking the most popular application, which is the determination of DNA-ploidy by Feulgen-staining (Feulgen and Rossenbeck 1924), the staining protocol includes procedures that destroy most antigens keeping only cellular morphology intact.

Second, for image analysis in general the investigator selects just 150–300 single "typical" cells for analysis whereas the other cells on the slide are not analysed and therefore the information contained in these cells is lost. In slide-based cytometry, all cells on the slides can easily be included into the analysis.

The third advantage is that samples not necessarily have to be prepared as a single-cell solution. Whereas this separation yields to no loss of information in a tissue that constitutes of single cells physiologically, as it is the case in blood, in any other tissue crucial information is contained within the architecture of the tissue itself. A good example is the case of metastatic disease to a lymph node. Disrupting the architecture of the node for preparation of a single-cell solution only allows to detect the presence of metastatic cells among the lymphocytes; however, clinical even more relevant information such as extracapsular spread is lost. Detection of this topological information demands an intact anatomical specimen. This, however, would be inappropriate for flow cytometric assays.

In principle, any specimen that can be placed on a slide can be analysed in some way by slide-based cytometry. Modern instruments allow the analysis of chromatic dyes, too.

Since objects are immobilised on the slide at a fixed position they can be identified according to their x–y-co-ordinates. This puts the investigator into the position to rule out any artefact and to obtain proof that a given fluorescence signal is in fact generated by a cell. Also, cells are not lost; this has the advantage that sample too small for flow cytometry can be readily analysed by slide-based systems, such as hypocellular fine-needle aspirate biopsies or exfoliative swabs. For better visualisation of cell morphology cells can be re-stained by conventional chromatic dyes and inspected afterwards. Instead using chromatic dyes, specimens can be bleached and stained with another set of fluorochromes for another analyses. The data of the first analysis can be merged with those of the subsequent analyses to yield a data stack per cell. In principle, bleaching can even be omitted since the fluorescence of the prior staining can be subtracted from the subsequent staining (Mittag et al. 2006). Slides also can be stored as a conventional pathological specimen which is of utmost importance keeping medicolegal issues in mind.

The sum of these specific characteristics makes slide-based cytometry the superior technology for clinical applications in the oncology of solid tumors. Although the concept of predictive medicine by cytomics opens up n-dimensional data spaces so far only two or three parametric analyses have been established for the investigation of solid tumors and will be outlined below. Unlike in hematopoietic disorders, where clinical sample material is easily available, for solid tumors in general only minute samples are available which routinely are analysed by conventional histopathology. However, as will be shown by some non-oncological examples, the capacity of a slide-based technology in the analysis of tissue samples in principle has been established and soon will be transferred to oncological issues.

This specific clinical situation in solid tumors explains why the brilliant progress achieved by the cytomics approach in hematopoietic diseases has not yet been transferred to solid cancers.

10.3.1 Predictive Medicine in Solid Tumors

Strictly speaking, an assay that allows the prediction of the clinical course in a patient with a solid tumor has to be implemented as early as possible within the diagnostic work-up. From the clinician's point of view the first and most important issue is nothing more than to either verify or rule out the presence of a malignant tumor. This should be possible with samples available on an out-patient basis obtained by minimal- or non-invasive procedures. In theory there could be cases where the malignant phenotype is expressed only in a minority of cells which would be missed by a random sample such as a fine-needle aspiration biopsy. Keeping this in mind therefore a negative result hardly ever will be taken as a proof of absence of malignancy. However, in fact routine histopathology does not include an entire work-up of the whole specimen neither; instead, "representative" sections are screened. Actual, there is no proof that the sections in fact are representative other than the subjective judgement of the pathologist.

Exfoliative swabs on the other hand are confronted with the argument that the cells accessible by this method, i.e. the superficial layers of the lesion, are not the cells relevant for the course of the disease which rather is determined by the basal layers that infiltrate through the basilar membrane. However, from a clinical point of view just a stable surrogate marker of the disease would suffice. Besides, it is rather theoretical to argue that the superficial cells themselves do not infiltrate into the surrounding healthy tissue; in any way they serve as representatives of the basal cells. The only exception is tumors growing under intact mucosa hidden by intact epithelium as in sarcomas or some lymphomas. These tumors are hard to detect by any diagnostic procedure and make up only a minor part of malignancies.

In some anatomical regions such as the larynx the physiological function is tightly coupled to anatomical integrity. In tumors of the parotid gland incision biopsies are obsolete due to the risk of damaging the facial nerve and of spreading tumor cells in the surrounding tissue which would be hazardous even in the most common benign solid tumor, the pleomorphic adenoma. In cases like these minimal sampling is mandatory in order to avoid loss of quality of life due to diagnostic manoeuvres. "Deep" biopsies including basal layers in general induce a loss of function in these anatomical regions.

From a practical point of view predictive assays can be divided into assays that establish the diagnosis by minimal- or non-invasive ways on the one hand and into assays that predict the further clinical course of an already diagnosed disease on the other hand.

10.3.1.1 Diagnostic Cytomic Assays

As lined out above slide-based cytometry is an ideal tool to analyse hypocellular specimens. This has been exploited to determine the DNA-ploidy of tumor cells. Automated slide-based cytometry allows to analyse a multiple of cells that are

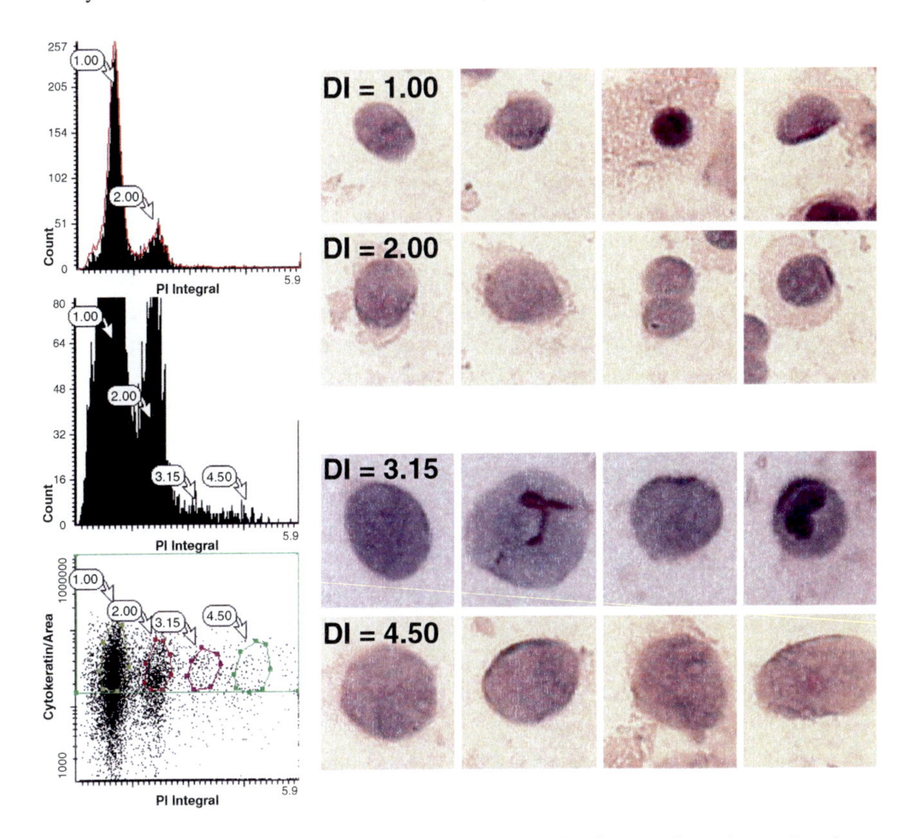

Fig. 10.3 Slide-based cytometry. Slide-based cytometry of a fine-needle aspirate taken from a solid parotid gland tumour. Cells were stained with anti-cytokeratin and PI as a stoichiometrical DNA-dye. DNA-ploidy of epithelial cells is determined taking leukocytes as an internal standard (DNA-index = 1.00). Note the minimal population of cells with a DNA-index of 3.15 and 4.50 making up less than 1% of all cells. Histology confirmed an adenocarcinoma. For detailed method see Gerstner et al. (2003) (reprinted with permission from Gerstner et al. 2008)

routinely selected by image cytometry, i.e. 20,000–100,000 *vs.* 150–300. This guarantees that also minor cell populations of less than 1% are included into the analysis that easily could be missed by manual screening with direct visualisation (Fig. 10.3). It has been shown that slide-based cytometry gives better prediction of the histopathological diagnosis than conventional cytology (Gerstner et al. 2003, 2005, 2006). In some cases, the correct diagnosis could be established weeks before an appropriate histopathological sample could be obtained (Remmerbach et al. 2001, 2004).

The combination of HPV-genotyping and determination of the DNA-ploidy with conventional cytology has been proven beneficial in the early detection of carcinomas in cervical smears (Bollmann et al. 2005a) and was shown to have the potential to predict the further clinical course (Bollmann et al. 2006).

In general, slide-based cytometry can yield quantitative and objective cytomics data where conventional cytology only can offer subjective judgements. The applications outlined above highlight only some peculiar examples.

10.3.1.2 Predictive Cytomic Assays

Prediction of the future clinical course of cancer patients by cytomic analyses date back to the 1990s (Hemmer and Schön 1993; Hemmer and Prinz 1997; Hemmer et al. 1997; Hemmer et al. 1999): in oral tumors it could be shown that DNA-aneuploid tumors have a higher rate of locoregional metastasis advocating resection of the locoregional lymph nodes in DNA-aneuploid tumors even in cases without clinical and radiological signs of metastatic disease. Since DNA-aneuploid tumors also have a higher rate of local recurrence closer follow-up is advised.

In patients with superficial bladder carcinoma slide-based cytometry has been successfully implemented in the follow-up using samples obtained from voided urine (Bollmann et al. 2005b).

10.3.2 Future Aspects

The more expertise with slide-based assays is gained the more complex samples are analysed. Several (non-)oncological applications are outlined below but it is expected that these assays soon will be transferred into oncological issues.

Slide-based cytometry allows to analyse tissue sections. This has been performed on sections from lymphoid organs: based on nuclear fluorescent staining the different microanatomical compartments of the lymphoid follicles, i.e. the mantel zone and the germinal centre, could be analysed separately and the relative content of specific subtypes (e.g. CD8+-lymphocytes) could be mapped to these compartments. The germinal centre could be further subdivided into the dark and the bright region containing the respective cell types (Gerstner et al. 2004). This concept of tissometry (Ecker and Steiner 2004; Ecker et al. 2006) has been further developed on brain sections – which have significant lower cellular density – aiming to analyse the three-dimensional interaction of different cell types within the tissue architecture (Mosch et al. 2006). This concept has recently been further pushed forward by developing a versatile hardware (Kim et al. 2007). Applications on lymphatic tissue are those most sophisticated (Harnett 2007). On prostate tissue sections the quantitative detection of α-methylacyl-CoA racemase has been exploited to allow automated classification of the tissue (Rubin et al. 2004).

Although these assays are still viewed sceptically by histopathologists they will in future be more and more relevant for clinicians who aim to base their decision about further diagnostic and therapeutic steps on objective parameters. This is not only the case in malignancy but is true for all somatic diseases.

For example, therapeutic intervention in organ transplantation is oriented on functional parameters. However, slide-based analysis is useful in thorough characterisation of graft-rejecting leukocytes in transplant rejection and therefore can help to better understand the underlying cellular mechanisms (Ecker and Steiner 2004). On this basis, therapy could be based on individual cytomic data.

The so far most detailed analysis of cellular interactions in the tissue has been achieved by a technology termed Multi-Epitop-Ligand-Kartographie MELK (Schubert 2003). A single section is repeatedly stained for an antigen with a fluorochrome, analysed, and bleached to remove the signal, followed by another cycle of staining-analysis-bleaching, and so on. Data of all analyses are combined according to the fixed x–y-co-ordinates of the slide. Since the natural micro-environment of the cells is not disrupted (as in flow cytometry in order to obtain a single-cell suspension) but the architecture of the tissue is kept this approach allows to generate hierarchical clusters of interacting proteins according to their spatial location, termed the toponome (Schubert 2006. The maximum number of antigens analysed on the same slide so far is 100 (Friedenberger et al. 2007).

It is expected that these assays will soon be applied to issues in the oncology of solid tumors: to characterise and quantitatively determine tumor-infiltrating leukocytes, to analyse the tumor invasion front where the tumor and its host have the crucial interaction, and to judge on residual tumor cells after primary radiochemotherapy.

The slide-based approach on cytomics analysis additionally allows a detailed absolute immunophenosubtyping, i.e. the measurement of absolute numbers per volume of a given specific cell type such as the number of CD8+ cytotoxic cells per liter blood (Laffers et al. 2007). In breast cancer, the quantification of circulating tumor cells after therapy has been established as a predictor for the outcome and therefore can be used in the decision about adjuvant therapy (Lobodasch et al. 2007). In future, detailed analysis of the functional capacity of these cells might develop significant input into the therapeutic regimen in other solid tumors as well.

However, quantitative data on cells can also be obtained by analysing the tissue in toto without taking a biopsy applying multispectral imaging. The idea of this technology has first been applied at the Landsat-program of the National Aeronautics and Space Administration (NASA) for earth imaging (Harris 2006). The molecular composition of a specimen interacts with the electromagnetic radiation. Therefore, the molecular phenotype of a cell is reflected by its specific spectral signature. Since the architecture of the tissue is composed by cells and their molecular products, the type of tissue can be classified by its spectral signature as well. This principle was drafted in 1998 (Farkas et al. 1998) and since then it has been developed to different biomedical applications: It has been applied to histological sections (Levenson and Mansfield 2006) where it distinguished metastasising cells in lymphoid tissue or infiltrating tumor cells in breast cancer. Without taking biopsies invasively it has been applied for non-invasive classification of pigmented naevi; in fact, this application was established even earlier. In order to verify multispectral classification the naevi were subsequently resected; histopathology confirmed exactly those regions within the naevi were malignant melanoma developed (Farkas and Becker 2001). This concept could develop tremendous impact on routine diagnostic work-up, especially in oncology: any inner or outer body surface accessible for visualisation could be imaged by multi- or hyperspectral imaging. The topological discrimination would be limited only by the optical resolution of the imaging tool and could in principle be scaled down to the single cell level. Beyond non-trivial problems concerning the hardware that so far have not been resolved the only limitations of this approach seem to be that the lesion should be at the inner or outer surface. However, novel imaging modalities could even break this barrier: for example, optical coherence tomography (OCT) gives optical sections non-invasively up to 2 mm deep into the tissue (Armstrong et al. 2006; Bibas et al. 2004; Kraft et al. 2007). Novel developments (μOCT) give subcellular resolution allowing to differentiate even single nuclei (Pan et al. 2007; Wang et al. 2007) without a cut or a single drop of blood.

10.4 Conclusions

The single cell approach in flow and slide-based cytometry with its currently rapid methodological progress has a significant clinical and research potential, especially when entire multiparameter data sets are subjected to outcome driven data mining.

The data pattern classification algorithm evaluates individual patient pattern against previously learned disease classifications patterns. This differs from statistical classifiers where the coherence of patient patterns is typically lost by introduction of the individual parameter values into data clusters resulting in lower resolution for pattern differences during the learning process. Statistical ambiguities concerning the best achievable classification as well as the limitation to prognostic conclusions for patient groups but not for individual patients make algorithmic data pattern classifiers attractive for clinical purposes. They provide by principle individualised predictions at the optimum discrimination conditions while the potential for subsequent statistical analysis of the selected parameter patterns is maintained.

Flow cytometry as the first single cell high-throughput technology has at present a certain lead with regard to predictive medicine by cytomics since a wealth of data sets and evaluations are available. With the fast progress of automated image segmentation in fluorescence microscopy the situation is likely to change. The difficulty to standardise molecular quantification in microscopy, is outweighed by the possibility to collect intracellular morphological as well as cellular 2D and 3D neighbourhood information as a further potential for knowledge extraction and mathematic modelling in system cytomics.

Molecule specific multiparameter fluorescence staining, high-throughput and high-content single cell measurements in conjunction with multiparameter data classification will open the way for generalized *disease course prediction* for patients for the practice of medicine, similarly as microscopic single cell observation of initially textile colour stained histological sections by Virchow (Virchow 1858) has enabled histopathological *disease diagnosis*.

References

Armstrong WB, Ridgway JM, Vokes DE et al (2006) Optical coherence tomography of laryngeal cancer. Laryngoscope 116:1107–1113

Bibas AG, Podoleanu AG, Cucu RG et al (2004) 3-D optical coherence tomography of the laryngeal mucosa. Clin Otolaryngol 29:713–720

Bland JM, Douglas GA (1998) Survival probabilities (the Kaplan–Meier method). BMJ 317: 1572–1580

Bollmann R, Bánkfalvi A, Griefingholt H et al (2005a) Validity of combined cytology and human papillomavirus (HPV) genotyping with adjuvant DNA-cytometry in routine cervical screening: results from 31.031 women from the Bonn-region in West Germany. Oncol Rep 13:915–922

Bollmann M, Heller H, Bánkfalvi A et al (2005b) Quantitative molecular urinary cytology by fluorescence *in situ* hybridisation: a tool for tailoring surveillance of patients with superficial bladder cancer? Br J Urol 95:1219–1225

Bollmann M, Várnai AD, Griefingholt H et al (2006) Predicting treatment outcome in cervical diseases using liquid-based cytology, dynamic HPV genotyping and DNA cytometry. Anticancer Res 26:1439–1446

Caspersson T, Lomakka G, Svensson G et al (1955) A versatile ultramicrospectrograph for multiple-line and surface scanning high resolution measurements employing automized data analysis. Exp Cell Res 3:40–51

Ecker RC, Steiner GE (2004) Microscopy-based multicolor tissue cytometry at the single-cell level. Cytometry A 59A:182–190

Ecker RC, Rogojanu R, Streit M et al (2006) An improved method for discrimination of cell populations in tissue sections using microscopy-based multicolor tissue cytometry. Cytometry A 69A:119–123

Farkas DL, Du C, Fisher GW et al (1998) Non-invasive image acquisition and advanced processing in optical bioimaging. Comput Med Imaging Graphics 22:89–102

Farkas DL, Becker D (2001) Applications of spectral imaging: detection and analysis of human melanoma and its precursors. Pigment Cell Res 14:2–8

Feulgen K, Rossenbeck H (1924) Mikroskopisch-chemischer Nachweis einer Nucleinsäure vom Typus der Thymonucleinsäure und die darauf beruhende elektive Färbung von Zellkernen in mikroskopischen Präparaten. Hoppe Seyler Z Physiolog Chem 135:203–248

Friedenberger M, Bode M, Krusche A et al (2007) Fluorescence detection of protein clusters in individual cells and tissue sections by using toponome imaging system: sample preparation and measuring procedures. Nat Protoc 2:2285–2294

Gerstner AOH, Lenz D, Laffers W et al (2002) Near-infrared dyes for six-color immunophenotyping by laser scanning cytometry. Cytometry 48:115–123

Gerstner AOH, Müller AK, Machlitt J et al (2003) Slide-based cytometry for predicting malignancy in solid salivary gland tumors by fine needle aspirate biopsies. Cytometry B (Clin Cytometry) 53B:20–25

Gerstner AOH, Trumpfheller C, Rasc P et al (2004) Quantitative histology by multicolor slide-based cytometry. Cytometry A 59A:210–219

Gerstner AOH, Thiele A, Tárnok A et al (2005) Preoperative detection of laryngeal cancer in mucosal swabs by slide-based cytometry. Eur J Cancer 41:445–452

Gerstner AOH, Thiele A, Tárnok A et al (2006) Prediction of upper aerodigestive tract cancer by slide-based cytometry. Cytometry A 69A:582–587

Göhde W, Dittrich W (1970) Simultane Impulsfluorimetrie des DNS- und Proteingehaltes von Tumorzellen. Z Anal Chem 252:328–330

Harnett MM (2007) Laser scanning cytometry: understanding the immune system *in situ*. Nat Rev Immunol 7:897–904

Harris AT (2006) Spectral mapping tools from the earth sciences applied to spectral microscopy data. Cytometry A 69A:872–879

Hemmer J, Schön E (1993) Cytogenetic progression and prognosis in oral carcinoma: a DNA flow cytometric study in 317 cases. Int J Oncol 3:635–640

Hemmer J, Prinz W (1997) Comparison of DNA flow cytometry and fluorescence *in situ* hybridization with a set of 10 chromosome-specific DNA probes in four head and neck carcinomas. Cancer Gent Cytogenet 97:35–38
Hemmer J, Thein T, van Heerden WFP (1997) The value of DNA flow cytometry in predicting the development of lymph node metastasis and survival in patients with locally recurrent oral squamous cell carcinoma. Cancer 79:2309–2313
Hemmer J, Nagel E, Kraft K (1999) DNA aneuploidy by flow cytometry is an independent prognostic factor in squamous cell carcinoma of the oral cavity. Anticancer Res 19:1419–1422
Kim KH, Ragan T, Previte MJR et al (2007) Three-dimensional tissue cytometer based on high-speed multiphoton microscopy. Cytometry A 71A:991–1002
Kraft M, Lerßen K, Lubatschowski H et al (2007) Technique of optical coherence tomography of the larynx during microlaryngoscopy. Laryngoscope 117:950–952
Laffers W, Schlenkhoff C, Pieper K et al (2007) Concepts for absolute immunophenosubtyping by slide-based cytometry. Transfus Med Hemother 34:188–195
Levenson RM, Mansfield JR (2006) Multispectral imaging in biology and medicine: slices of life. Cytometry A 69A:748–758
Lobodasch K, Fröhlich F, Rengsberger M et al (2007) Quantification of circulating tumor cells for the monitoring of adjuvant therapy in breast cancer: an increase in cell number at completion of therapy is a predictor of early relapse. Breast 16:211–218
Meistrich ML, Göhde W, White RA (1978) Resolution of x and y spermatids by pulse cytophotometry. Nature 274:821–823
Mittag A, Lenz D, Gerstner AOH et al (2006) Hyperchromatic cytometry principles for cytomics using slide based cytometry. Cytometry A 69A:691–703
Mittag A, Bocsi J, Laffers M et al (2009) Technical and methodological basics of slide-based cytometry. In: Sack U, Tárnok A, Rothe G (eds) Cellular diagnostics. Basics, methods and clinical applications of flow cytometry. Karger, Basel
Mosch B, Mittag A, Lenz D et al (2006) Laser scanning cytometry in human brain slices. Cytometry A 69A:135–138
Pan YT, Wu ZL, Yuan ZJ et al (2007) Subcellular imaging of epithelium with time-lapse optical coherence tomography. J Biomed Opt 12:050504
Perfetto SP, Chattopadhyay PK, Roederer M (2004) Seventeen-colour flow cytometry: unravelling the immune system. Nat Rev Immunol 4:648–655
Remmerbach TW, Weidenbach H, Pomjanski N et al (2001) Cytologic and DNA-cytometric early diagnosis of oral cancer. Anal Cell Pathol 22:211–221
Remmerbach TW, Weidenbach H, Hemprich A et al (2004) Earliest detection of oral cancer using non-invasive brush biopsy including DNA-image-cytometry: report of four cases. Anal Cell Pathol 25:159–166
Repp R, Schaekel U, Helm G et al (2003) Immunophenotyping is an independent factor for risk stratification in AML. Cytometry B (Clin Cytom) 53B:11–19
Rimm DL (2006) What brown cannot do for you. Nat Biotechnol 24:914–916
Rosenwald A, Wright G, Chan WC et al (2002) The use of molecular profiling to predict survival after chemotherapy for diffuse large-B-cell lymphoma. N Engl J Med 346:1937–1947
Rothe G, Kellermann W, Valet G (1990) Flow cytometric parameters of neutrophil function as early indicators of sepsis- or trauma-related pulmonary or cardiovascular organ failure. J Lab Clin Invest 115:52–61
Rubin MA, Zerkowski MP, Camp RL et al (2004) Quantitative determination of expression of the prostate cancer protein α-methylacyl-CoA racemase using automated quantitative analysis (AQUA). Am J Pathol 164:831–840
Schubert W (2003) Topological proteomics, toponomics, MELK-technology. Adv Biochem Eng Biotechnol 83:189–209
Schubert W (2006) Cytomics in characterizing toponomes: towards the biological code of the cell. Cytometry A 69A:209–211
Schwemmler W (1982) The endocytobiotic cell theory and the periodic system of cells. Acta Biotheor 31:45–68

The International Non-Hodgkin's Lymphoma Prognostic Factors Project (1993) A predictive model for aggressive non-Hodgkin's lymphoma. N Engl J Med 329:987–94

Valet G, Warnecke HH, Kahle H (1986) Automated diagnosis of malignant and other abnormal cells by flow-cytometry using the newly developed DIAGNOS1 program system. In: Burger G, Ploem B, Goerttler K (eds) Proceedings of the international symposium on histometry. Academic, London, pp 58–67

Valet G, Valet M, Tschöpe D et al (1993) White cell and thrombocyte disorders: Standardized, self-learning flow cytometric list mode data classification with the CLASSIF1 program system. Ann NY Acad Sci 677:233–251

Valet G (1997) Cytometry, a biomedical key discipline. 3. System cytometry, a new research strategy. In: Robinson JP (ed) Purdue cytometry CD series, vol. 4 (ISBN: 1-890473-03-0), http://www.cyto.purdue.edu/cdroms/flow/vol4/8_websit/valet/keyvirt1.htm#system

Valet GK, Roth G, Kellermann W (1998) Risk assessment for intensive care patients by automated classification of flow cytometric data. In: Robinson JP, Babcock GF (eds) Phagocyte function. Wiley-Liss, New York, pp 289–306

Valet G, Kahle H, Otto F et al (2001) Prediction and precise diagnosis of diseases by data pattern analysis in multiparameter flow cytometry: Melanoma, juvenile asthma and human immunodeficiency virus infection. Methods Cell Biol 64:487–508

Valet G (2002) Predictive medicine by cytomics: potential and challenges. J Biol Regul Homeost Agents 16:164–167

Valet G, Arland M, Franke A et al (2002) Discrimination of chronic lymphocytic leukemia of B-cell type by computerized analysis of 3-color flow cytometric immunophenotypes of bone marrow aspirates and peripheral blood. Lab Hematol 8:134–142

Valet G, Repp R, Link H et al (2003) Pretherapeutic identification of high-risk acute myeloid leukemia (AML) patients from immunophenotype, cytogenetic and clinical parameters. Cytometry B Clin Cytom 53B:4–10

Valet G, Höffkes HG (2004) Data pattern analysis for the individualised pretherapeutic identification of high-risk diffuse large B-cell lymphoma (DLBCL) patients by cytomics. Cytometry A 59A:232–236

Valet G (2005a) Human cytome project: a new potential for drug discovery. In: Real Academia Nacional de Farmacia (ed) Las Omicas genomica, proteomica, citomica y metabolomica: modernas tecnologias para desarrollo de farmacos. Madrid, pp 207–228

Valet G (2005b) Cytomics, human cytome project and systems biology: top-down resolution of the molecular biocomplexity of organisms by single cell analysis. Cell Prolif 38:171–174

Valet G (2006) Cytomics as a new potential for drug discovery. Drug Discov Today 11:785–791

Virchow R (1858) Die Cellularpathologie in ihrer Begründung auf physiologische und pathologische Gewebelehre. August Hirschwald, Berlin

Wang Z, Lee CS, Waltzer WC et al (2007) *In vivo* bladder imaging with microelectromechanical-systems-based endoscopic spectral domain optical coherence tomography. J Biomed Opt 12:034009

Wirdefeldt K, Gatz M, Pawitan Y et al (2005) Risk and protective factors for Parkinson's disease: a study in swedish twins. Ann Neurol 57:27–33

Chapter 11
The Frontiers of Computational Phenomics in Cancer Research

Eneida A. Mendonça and Yves A. Lussier

Abstract Understanding the molecular mechanisms underpinning prognosis and response to therapy of individuals suffering from cancer increasingly requires integrated and systematic approaches. Molecular-based strategies to more effectively prevent, diagnose, and treat cancer are seen as the future goal of oncology research. Although altered phenotypes can reliably be associated with altered gene functions, the systematic analysis of phenotypes relationships to study cancer biology remains nascent. The completion of the Human Genome Project has made possible high-throughput approaches such as the Cancer Genome Atlas to accelerate phenomics research. However, these approaches still face important challenges. In this chapter, we review these challenges, introduce current research efforts in the field, and highlight the importance of computational approaches to conduct large-scale phenomic studies.

11.1 Introduction

The foundations of the biology and genetics of cells and organisms were laid as early as 1869, when F. Miescher isolated the nuclein (later named nucleic acid). Almost a century later, the discovery of the DNA double helix by Watson and Crick in 1953 marked, for many, the beginning of molecular biology (Watson 1968). Since, molecular biology grew from this point delivering solutions to many biological questions. Current cancer research (including the understanding of how cancers occur) draws in most part from these discoveries, especially in biology and genetics. One example is the close relationship between genetic contents and phenotype expressions, which

E.A. Mendonça (✉)
Section of Hematology/Oncology and Stem Cell Transplantation, Department of Pediatrics, and Computation Institute, The University of Chicago, Chicago, IL, USA
e-mail: emendonca@uchicago.edu

Y.A. Lussier
Section of Genetic Medicine, Department of Medicine, Ludwig Center for Metastasis Research, Institute for Genomics and Systems Biology, Computation Institute, and Cancer Research Center, The University of Chicago, Chicago, IL, USA
e-mail: lussier@uchicago.edu

W.C.S. Cho (ed.), *An Omics Perspective on Cancer Research*,
DOI 10.1007/978-90-481-2675-0_11,

has been demonstrated for years. The recent Human Genome Project represented a landmark in science. The completion of the sequence of the human genome in 2001 (Lander et al. 2001; Venter et al. 2001) have generated a number of facts in science, including the beginning of the genomic era, making possible high-throughput approaches to accelerate the understanding of human diseases. More recently, The Cancer Genome Atlas further paves the way for unveiling the molecular underpinnings of specific cancers (http://cancergenome.nih.gov/).

Understanding normal and diseases states of any organism requires integrated and systematic approach. While the above-mentioned approaches allow for genome-wide unbiased assessments of molecular mechanisms, high-throughput assessments of supra cellular phenotypes are lacking. Although altered phenotypes are among the most reliable manifestations of altered "combinations" of gene functions, research using systematic analysis of phenotype relationships to study human biology is still in its infancy (Lussier and Liu 2007).

Phenomics is a field concerned with the characterization of phenotypes as a whole (phenome), which are characteristics of organisms that develop via the interaction of the genome with the environment. It is the systematic acquisition and objective documentation of phenotypic data at various levels, including clinical, molecular and cellular.

Advances in technology and reduced costs in the generation of genomic data have accelerated the amount of available genomic data for researchers. New technologies have also accelerated the pace of collecting phenotype data and the establishment of phenotype databases. Despite that, our ability to better understand and address the basis of human diseases continues to lag behind (Ball et al. 2004). Current research efforts in phenomics capitalize on novel high-throughput computation and informatics technologies to derive genome-wide molecular networks of genotypic-phenotype associations. These approaches involve the integration of multiple data repositories and the use of diverse statistical, machine learning, and data mining techniques.

In this chapter, we review the challenges facing high-throughput phenomics research and introduce research and development of phenomics databases.

11.2 High-Throughput Collection of Phenotypes: Challenges

The volumes of data available to researchers in life sciences can double every six months (Wong 2007). In addition, technologies to enable large scale, parallel, quantitative and inexpensive assessment of these data are increasingly becoming available. These advances are pushing an increase opportunity for more complex analyses and bringing different perspectives to research and to health care. However, several challenges, some well recognized in the development of health care systems, still limit the timely and effective use of these resources (Shortliffe and Sondik 2006; Pare and Trudel 2007).

Deriving useful knowledge from knowledge biomedical resources requires robust approaches in knowledge acquisition, representation, management, visualization, analysis and interpretation. High-throughput phenomics face major challenges, such as how to code and represent knowledge in databases and networks, in order to allow useful retrieval, visualization and proper analysis and interpretation of data (Lussier and Liu 2007; Chen et al. 2008a, b).

Phenotypic information can be found in many genomic databases. However, no single database or algorithm has successfully represented all information necessary to model the biological questions posed by the complex phenotypes in human diseases such as cancer (Juristica 2007). Very often the phenotypic information is coded in different formats, at different levels of granularity, and with different aims. These databases most likely are manually curated and have limited breath. These challenges have greatly limited the effectiveness of conducting combined phenotypic/genotypic analysis.

More recent studies have looked into mining the scientific literature, but the success of a great number of these efforts have been low, due to the lack of expressiveness of mining techniques (Lussier and Liu 2007). Granularity, synonyms, and ambiguity of biomedical entities, particularly of gene symbols, are a big challenge for text-mining systems (Xu et al. 2007; Spasic et al. 2005; Fan and Friedman 2008).

Another area that has been explored is the use of clinical data from electronic medical records. Researchers have studied automated methods for creating and updating knowledge bases from narrative reports of patient records (Mendonca et al. 2001; Rindflesch and Fiszman 2003; Rindflesch et al. 2000; Chen et al. 2008a, b; Wang et al. 2008). Although these studies have showed that natural language processing and statistical methods can generate meaningful relationships, similar limitations are described. In addition, still a challenge to accurately identify the nature of associations from patient records, especially to infer causal links such as between drug and diseases or symptoms.

In understanding human diseases, it is also important to consider the association of gene expression and proteomic patterns with phenotypes and other factors, such as clinical history, environmental exposures and experimental conditions (Butte and Kohane 2006). These relationships have been essential to medicine and could lead to new disease-associated genes. Gene expression microarray analysis has increasingly added additional forms of information about diseases and other biological processes (Chen et al. 2008a, b). Repositories of microarray data present similar challenges to the ones presented by phenotypic databases. Contextual annotations are, in general, represented by unstructured text, which make almost impossible to determine the phenotype and environment by manual processes (Butte and Kohane 2006).

11.3 Representation and Organization of Phenotypes for High-Throughput Analysis

11.3.1 Ontologies Related to Cancers

Ontologies are conceptual models that aim to support consistent and unambiguous knowledge sharing and that provide a framework for knowledge integration (Stevens et al. 2000). A comprehensive body of knowledge is currently stored in biomedical ontologies, which has led scientists to invest considerable efforts in establishing standards for the integration of phenotypes using ontologies. We present some examples of ontologies used in phenomics research below. Of particular interest for this chapter, ICD-O and SNOMED have been used for decades by pathologists to describe cancer morphologies.

11.3.1.1 Gene Ontology (GO)

The GO project is a collaborative effort to address the need for consistent descriptions of gene products in different databases. GO has succeeded in annotating genes with molecular functions, processes, and cellular locations. The ontology is organized in separate hierarchies and represented as a directed acyclic graph, in which nodes are GO terms and edges represent the GO relationships (GO terms may have synonyms comprise over 25,000 terms, with many specialized in processes involved in cancer biology, e.g. cell cycle, angiogenesis, etc.) (Ashburner et al. 2000).

11.3.1.2 The Medical Subject Headings (MeSH)

MeSH is the National Library of Medicine's (NLM) controlled vocabulary thesaurus. It is used by the NLM for indexing articles from biomedical journals for the MEDLINE/PubMed database. MeSH consists of sets of terms naming descriptors in a hierarchical structure that permits searching at various levels of specificity. A list of Descriptors and Supplementary Concept Records (SCRs) provides a more granular representation of biomedical entities, including chemicals and proteins (The Medical Subject Headings 2009).

11.3.1.3 The Systematized Nomenclature of Medicine (SNOMED)

Currently, SNOMED is probably the most comprehensive biomedical terminology and it comprises a specialized section on cellular morphologies of cancers, regional anatomies for description of primary or metastasis sites, and diverse types of cancers described as disease states. It contains over a half million biomedical concepts such as diseases, anatomy, morphology, functions, drugs, procedures, and treatments. Originally developed by the College of American Pathologists, SNOMED has since been owned and developed by the International Health Terminology Standard Development Organization, a not-for-profit association (http://www.ihtsdo.org/) (Spackman et al. 1997).

11.3.1.4 The Unified Medical Language System (UMLS)

The UMLS is a comprehensive knowledge source developed and distributed by the National Library of Medicine. It relates ontologies to one another. It has three major components: the UMLS Metathesaurus, the Semantic Network, and MetaMap. The Metathesaurus contains over 1 million of biomedical concepts, from more than 100 controlled vocabularies, classifications, and ontologies (including GO, SNOMED, NCI Thesaurus, MedDRA, and MeSH). The resource also presents relations among these concepts. Semantic types are assigned to the concepts and represented in the Semantic Network (Lindberg et al. 1993).

11.3.1.5 The Open Biomedical Ontologies (OBO)

The OBO is a consortium, which aims to establish a set of principles for ontology development with the goal of creating a suite of orthogonal interoperable reference ontologies in the biomedical domain. It covers several ontologies, including GO, Cell Ontology (CO), Sequence Ontology (SO), RNA Ontology (RnaO), and Protein Ontology (PRO). Data varies in granularity from organ and organism to cell and cellular component to molecule (Smith et al. 2007).

11.3.1.6 International Classification of Diseases for Oncology (ICD-O)

The ICD-O is developed by the World Health Organization and is used principally in tumor or cancer registries for coding the site (topography) and the histology (morphology) of neoplasms, usually obtained from a pathology report (Organization WH).

11.3.1.7 The Medical Dictionary for Regulatory Activities (MedDRA)

The MedDRA is an international medical terminology supported by the International Federation of Pharmaceutical Manufactures and Association (IFPMA) (http://meddramsso.com/MSSOWeb/index.htm) (Brown et al. 1999). It is commonly used to report adverse event data from clinical trials and for pharmacovigilance. Its use has also been reported in the analysis of associations between phenotypic and genomic data with drug safety adverse event data (Hernandez-Boussard et al. 2006).

11.3.2 Phenotypic Databases Related to Cancers

Data integration plays a key role in correlating heterogeneous phenotypic data with genomic data at different levels. Manual methods for developing and maintaining databases of phenotypes and their genomic information provide more accurate relations, but are time and labor consuming and, in general, more expensive. Automated high-throughput methods work in large scale, faster, but results are not as accurate when compare to the manual methods. In this section, we present some examples of efforts in both areas. A more detailed description and comparison of the databases can be found in Lussier and Liu (2007).

11.3.2.1 The Online Mendelian Inheritance in Man (OMIM)

OMIM is a comprehensive compendium of human genes and genetic phenotypes (Hamosh et al. 2000). Referenced overviews in OMIM contain information on all known mendelian disorders and over 12,000 genes of which about 700 are associated to cancer phenotypes. It focuses on the relationship between phenotype and genotype, and entries contain links to other genetic resources. Phenotypes are coded at different levels of granularity, in different formats and with different aims (Biesecker 2005).

11.3.2.2 The Online Mendelian Inheritance in Animals (OMIA)

OMIA is a compendium of genes, inherited disorders, and traits in more than 135 species other than human and mouse (Lenffer et al. 2006).

11.3.2.3 The Mouse Genome Informatics (MGI)

The Mouse Genome Informatics databases contain genes, phenotypic narratives, and references to the literature (Eppig et al. 2007). Its core database is the Mouse Genome Database (MGD), which contains genetic, genomic, and phenotypic data for laboratory mouse (Bult et al. 2008). MGI also contains the Mammalian Phenotype Ontology (MPO).

11.3.2.4 GeneCards

GeneCards is a web-based, integrated database of human genes. It provides concise genomic, proteomic, transcriptomic, genetic and functional information on all known and predicted human genes (Rebhan et al. 1997, 1998; Safran et al. 2002).

11.3.2.5 Gene2Disease (G2D)

G2D is a web application that allows researchers to inspect any region of the human genome to find candidate genes related to a genetic disease of their interest. It was build over OMIM and it prioritizes genes on a chromosomal region according to their possible relation to an inherited disease using a combination of data mining on biomedical databases and gene sequence analysis (Perez-Iratxeta et al. 2005).

11.3.2.6 PhenomicDB

PhenomicDB is a multi-species genotype-phenotype database developed by merging public data from a variety of model organisms and *Homo sapiens*. Current release includes data from several different sources: MGI, OMIM, FlyBase, WormBase, MAtDB, ZFIN, flyrnai.org, Phenobank, and CYGD (Kahraman et al. 2005).

11.3.2.7 PhenoGO

PhenoGO is a computed database designed for high-throughput mining that provides phenotypic and experimental context, such as the cell type, disease, tissue and organ to existing annotations between gene products and GO terms as specified in the Gene Ontology Annotations (GOA) for multiple model organisms. PhenoGO provides the broadest variety of binary and ternary relationships between genes, GO concepts, and phenotypes, including biological process of a specific gene in a particular phenotypic context (Sam et al. 2007, 2009).

11.4 Phenomic Analyses

Several researchers have used mining algorithms to automatically identify phenotype-genotype relationships from the scientific literature (Perez-Iratxeta et al. 2005; Hristovski et al. 2005; Korbel et al. 2005). These approaches have show limitations and moderate results.

Lussier and colleagues pioneered ontology-based phenomics using clinical databases. They integrated the Quick Medical Reference (QMR) with OMIM, generating relationships among genes, diseases, and traits of diseases (Miller et al. 1986; Lussier et al. 2002). In GeneTrace, they study an integrative approach between ontology-based phenotypes from the UMLS and their statistical and semantic relationships with GO and model organism databases (Cantor et al. 2005). These studies indicated the potential of exploiting existing curated databases to infer new gene-disease relationships.

Butte and Kohane (2006) have more recently developed and validated a system that identifies and represents phenotypic, environmental and experimental context for microarrays in the Gene Expression Omnibus (GEO) database by mapping annotations to biomedical concepts in the UMLS (Wheeler et al. 2004). This study provided a method for identifying genes related to phenotype and environment. In a subsequent study, Dudley and Butte (2008) described a method for the automated discovery of disease-related experiments within GEO using MeSH annotations derived from PubMED identifiers (Butte and Chen 2006). Their results showed that 62% of disease-related experiments contain sample subsets that could be automatically identified as normal controls. The work was important as an initial step to demonstrate that large-scale genomic data can be automatically mined for human disease categorization.

11.5 Future Challenges

"A deeper understanding of disease requires a database of human traits and disease states that is integrated with molecular biology (Butte 2008)."

Despite the advances in phenomic research, it is clear that we still face important challenges in integrating, organizing, and managing phenotypic databases across specifies, as well as enabling genome-wide analysis to associate phenotypic and genotypic data (Lussier and Liu 2007; Butte 2008). This need for advances in this area has driven several initiatives.

The Whole Genome Association studies launched by the National Institutes of Health aimed to link genetic data with phenotype datasets of large-scale clinical studies over several generations of patients. The NIH expects researchers to be able to identify variations in human DNA that underlie particular diseases or effects of medicines, genetic factors that influence health, disease and response to treatment.

The Human Phenome Project was proposed in 2003 as an international effort to create phenomic databases, and to develop new approaches for analyzing such phenotypic data (Freimer and Sabatti 2003). It focuses on establishing databases of phenotypes to establish their relation with genes and proteins. The Physiome Project is an international collaboration to define the physiome (description of the functional behavior of the physiological state of an individual or species) via the development of integrated quantitative and descriptive modeling. The computation methods will integrate biochemical, biophysical, and anatomical information about cells, tissues and organs (Hunter and Borg 2003). PhysioNet is a web public service of the PhysioNet Resource funded by the National Institutes of Health's NIBIB and NIGMS. The resource was intended to stimulate the study of complex biomedical and physiologic signals and already contains some complex human physiological traits (Physionet 2009).

The Cancer Genome Atlas (TCGA) (http://cancergenome.nih.gov/) is a comprehensive and coordinated effort of the USA National Cancer Institute to accelerate our understanding of the molecular basis of cancer through the application of genome analysis technologies, including large-scale genome sequencing. The pilot project in glioblastoma is currently conducted to assess the feasibility of a full-scale effort to systematically explore the entire spectrum of genomic changes involved in human cancer (Lussier et al. 2002).

The promise of these advances is "personalized medicine" where treatment strategies can be individualized based on a combination of factors, such as gene expression, protein expression. Personalized medicine has an extreme importance for cancer patients due to the with significant molecular differences in the expression and distribution of tumor cell markers among patients, the tendency for cellular mutation with disease progress, and the toxic effect of most therapies to normal cells. The emerging field of cancer phenomics is likely to focus on therapeutic predictions and gene-disease associations.

Acknowledgements This work was supported in part by the 1U54CA121852 (National Center for Multiscale Analyses of Genomic and Cellular Networks - MAGNET), the Cancer Research Foundation, the University of Chicago Cancer Research Center and the Ludwig Center for Metastasis Research.

References

Ashburner M, Ball CA, Blake JA et al (2000) Gene ontology: tool for the unification of biology. The Gene Ontology Consortium. Nat Genet 25:25–29

Ball CA, Sherlock G, Brazma A (2004) Funding high-throughput data sharing. Nat Biotechnol 22:1179–1183

Biesecker LG (2005) Mapping phenotypes to language: a proposal to organize and standardize the clinical descriptions of malformations. Clin Genet 68:320–326

Brown EG, Wood L, Wood S (1999) The medical dictionary for regulatory activities (MedDRA). Drug Saf 20:109–117

Bult CJ, Eppig JT, Kadin JA et al (2008) The Mouse Genome Database (MGD): mouse biology and model systems. Nucleic Acids Res 36(Database issue):D724–D728

Butte AJ (2008) Medicine. The ultimate model organism. Science 320:325–327

Butte AJ, Chen R (2006) Finding disease-related genomic experiments within an international repository: first steps in translational bioinformatics. AMIA Annu Symp Proc 2006:106–110

Butte AJ, Kohane IS (2006) Creation and implications of a phenome-genome network. Nat Biotechnol 24:55–62

Cantor MN, Sarkar IN, Bodenreider O et al (2005) Genestrace: phenomic knowledge discovery via structured terminology. Pac Symp Biocomput 2005:103–114

Chen DP, Weber SC, Constantinou PS et al (2008a) Novel integration of hospital electronic medical records and gene expression measurements to identify genetic markers of maturation. Pac Symp Biocomput 13:243–254

Chen ES, Hripcsak G, Xu H et al (2008b) Automated acquisition of disease drug knowledge from biomedical and clinical documents: an initial study. J Am Med Inform Assoc 15:87–98

Dudley J, Butte AJ (2008) Enabling integrative genomic analysis of high-impact human diseases through text mining. Pac Symp Biocomput 13:580–591

Eppig JT, Blake JA, Bult CJ et al (2007) The mouse genome database (MGD): new features facilitating a model system. Nucleic Acids Res 35(Database issue):D630–D637

Fan JW, Friedman C (2008) Word sense disambiguation via semantic type classification. AMIA Annu Symp Proc 2008:177–181

Freimer N, Sabatti C (2003) The human phenome project. Nat Genet 34:15–21

Hamosh A, Scott AF, Amberger J et al (2000) Online Mendelian Inheritance in man (OMIM). Hum Mutat 15:57–61

Hernandez-Boussard T, Woon M, Klein TE et al (2006) Integrating large-scale genotype and phenotype data. OMICS Winter 10:545–554

Hristovski D, Peterlin B, Mitchell JA et al (2005) Using literature-based discovery to identify disease candidate genes. Int J Med Inform 74:289–298

Hunter PJ, Borg TK (2003) Integration from proteins to organs: the Physiome Project. Nat Rev Mol Cell Biol 4:237–243

Juristica I (2007) Integrative computatinal biology. In: Juristica I, Wigle DA, Wong B (eds) Cancer informatics in the post genomic era: towards information-based medicine. Springer, New York, pp 129–145

Kahraman A, Avramov A, Nashev LG et al (2005) PhenomicDB: a multi-species genotype/phenotype database for comparative phenomics. Bioinformatics 21:418–420

Korbel JO, Doerks T, Jensen LJ et al (2005) Systematic association of genes to phenotypes by genome and literature mining. PLoS Biol 3:e134

Lander ES, Linton LM, Birren B et al (2001) Initial sequencing and analysis of the human genome. Nature 409:860–921

Lenffer J, Nicholas FW, Castle K et al (2006) OMIA (Online Mendelian Inheritance in Animals): an enhanced platform and integration into the Entrez search interface at NCBI. Nucleic Acids Res 34(Database issue):D599–D601

Lindberg DA, Humphreys BL, McCray AT (1993) The Unified Medical Language System. Methods Inf Med 32:281–291

Lussier YA, Liu Y (2007) Computational approaches to phenotyping: high-throughput phenomics. Proc Am Thorac Soc 4:18–25

Lussier YA, Sarkar IN, Cantor M (2002) An integrative model for *in-silico* clinical-genomics discovery science. Proc AMIA Symp 2002:469–473

Mendonca EA, Cimino JJ, Johnson SB (2001) Using narrative reports to support a digital library. Proc AMIA Symp 2001:458–462

Miller R, Masarie FE, Myers JD (1986) Quick medical reference (QMR) for diagnostic assistance. MD Comput 3:34–48

Organization WH. International Classification of Diseases for Oncology, 3rd ed (ICD-O-3). Available from: http://www.who.int/classifications/icd/adaptations/oncology/en/

Pare G, Trudel MC (2007) Knowledge barriers to PACS adoption and implementation in hospitals. Int J Med Inform 76:22–33

Perez-Iratxeta C, Wjst M, Bork P et al (2005) G2D: a tool for mining genes associated with disease. BMC Genet 6:45

Physionet. http://www.physionet.org/.

Rebhan M, Chalifa-Caspi V, Prilusky J et al (1997) GeneCards: integrating information about genes, proteins and diseases. Trends Genet 13:163

Rebhan M, Chalifa-Caspi V, Prilusky J et al (1998) GeneCards: a novel functional genomics compendium with automated data mining and query reformulation support. Bioinformatics 14:656–664

Rindflesch TC, Fiszman M (2003) The interaction of domain knowledge and linguistic structure in natural language processing: interpreting hypernymic propositions in biomedical text. J Biomed Inform 36:462–477

Rindflesch TC, Tanabe L, Weinstein JN et al (2000) EDGAR: extraction of drugs, genes and relations from the biomedical literature. Pac Symp Biocomput 5:517–528

Safran M, Solomon I, Shmueli O et al (2002) GeneCards 2002: towards a complete, object-oriented, human gene compendium. Bioinformatics 18:1542–1543

Sam L, Liu Y, Li J et al (2007) Discovery of protein interaction networks shared by diseases. Pac Symp Biocomput 12:76–87

Sam LT, Mendonca EA, Li J et al (2009) PhenoGO: an integrated resource for the multiscale mining of clinical and biological data. Bmc Bioinformatics 10:S8

Shortliffe EH, Sondik EJ (2006) The public health informatics infrastructure: anticipating its role in cancer. Cancer Causes Control 17:861–869

Smith B, Ashburner M, Rosse C et al (2007) The OBO Foundry: coordinated evolution of ontologies to support biomedical data integration. Nat Biotechnol 25:1251–1255

Spackman KA, Campbell KE, Cote RA (1997) SNOMED RT: a reference terminology for health care. Proc AMIA Annu Fall Symp 1997:640–644

Spasic I, Ananiadou S, McNaught J et al (2005) Text mining and ontologies in biomedicine: making sense of raw text. Brief Bioinform 6:239–251

Stevens R, Goble CA, Bechhofer S (2000) Ontology-based knowledge representation for bioinformatics. Brief Bioinform 1:398–414

The Medical Subject Headings (MeSH) National Library of Medicine. http://www.nlm.nih.gov/mesh/.

Venter JC, Adams MD, Myers EW et al (2001) The sequence of the human genome. Science 291:1304–1351

Wang X, Friedman C, Chused A et al (2008) Automated knowledge acquisition from clinical narrative reports. AMIA Annu Symp Proc 6:783–787

Watson JD (1968) The double helix; a personal account of the discovery of the structure of DNA, 1st edn. Atheneum, New York

Wheeler DL, Church DM, Edgar R et al (2004) Database resources of the National Center for Biotechnology Information: update. Nucleic Acids Res 32(Database issue):D35–D40

Wong B (2007) Informatics. In: Juristica I, Wigle DA, Wong B (eds) Cancer informatics in the post genomic era: towards information-based medicine. Springer, New York, pp 87–145

Xu H, Fan JW, Hripcsak G et al (2007) Gene symbol disambiguation using knowledge-based profiles. Bioinformatics 23:1015–1022

Chapter 12
Application of Bioinformatics in Cancer Research

Beatriz Stransky and Pedro Galante

Abstract Cancer is a disease determined by several genetic and epigenetic alterations. Due to technological advances in the omics disciplines, cancer research is going through a revolution. The technological advances that lead to the post-genome era have allowed molecular biologists to make meticulous studies on the DNA (genome), the mRNA (transcriptome) and the protein sequences (proteome). Initiatives that intend to describe cancer in a global dimension are providing an opportunity for investigators to have more useful data to analyze and integrate in novel ways. Despite the practical difficulties, a growing number of projects are being developed with the aim to integrate information about samples, protocols, and data from multiple sources. Cancer bioinformatics deals with the organization and analysis of the data so that important trends and patterns can be identified – the ultimate goal being the discovery of new therapeutic and/or diagnostic protocols for cancer. In this chapter, we will discuss some aspects of this revolution giving a special emphasis on Bioinformatics. Furthermore, we will discuss how the omics data is being analyzed and used to transform the way cancer patients are treated.

B. Stransky (✉)
Institute of Mathematics and Statistics, University of São Paulo, Rua do Matão, 1010, Cidade Universitária, São Paulo, CEP 05508-090, Brazil
e-mail: stransky@ime.usp.br

P. Galante
Computational Biology Group, Ludwig Institute for Cancer Research, São Paulo Branch, Rua João Julião 245-1° Andar – Paraíso, São Paulo, CEP 01323-903, Brazil
Decision Systems Group, Brigham and Women's Hospital, Harvard Medical School, Boston, MA, USA
e-mail: pgalante@ludwig.org.br

W.C.S. Cho (ed.), *An Omics Perspective on Cancer Research*,
DOI 10.1007/978-90-481-2675-0_12,

12.1 The Multidisciplinary Nature of Bioinformatics

The biomedical sciences are going through an important revolution that started a few decades ago. Technological innovations, such as high-throughput sequencing, have allowed genome scale measurements of many molecular species within the cell, tissue and body, leading to a massive expansion of the biomedical data. The challenge of converting all the information into a more useful knowledge is being overcome by the development of special computer programs and information technology-based techniques aimed to organize and analyze biological records. Bioinformatics and computational biology are multidisciplinary disciplines, which employ theories and applications from areas like computer science, mathematics, statistics, physics and engineering to solve biomedical problems and to improve the understanding of biological phenomena. Although the terms are frequently used as synonyms, they are different according to the NIH Biomedical Information Science and Technology Initiative (http://www.bisti.nih.gov/bistic2.cfm):

"*Bioinformatics*: Research, development, or application of computational tools and approaches for expanding the use of biological, medical, behavioral or health data, including those to acquire, store, organize, archive, analyze, or visualize such data".

"*Computational Biology*: The development and application of data-analytical and theoretical methods, mathematical modeling and computational simulation techniques to the study of biological, behavioral, and social systems".

Albeit distinct aspects of this multidisciplinary field define bioinformatics and computational biology, these areas have several common characteristics and purposes. Indeed, the fundamental objective is "…to enable the discovery of new biological insights as well as to create a global perspective from which unifying principles in biology can be discerned" (http://www.bisti.nih.gov/bistic2.cfm). In this chapter we will focus on the bioinformatics approaches to cancer, according to the NIH definition above.

As one of the first steps to organize and analyze biological data, bioinformaticians usually create databases, i.e. repositories of a large number of ordered and consistent data, usually associated with computer programs – to store all types of biological records, such as DNA or protein sequences. To handle all these information, specific software is continuously developed to update, query, and retrieve components of the data stored within the system. Nowadays, many of these databases are connected, allowing an integrated access and an easy recovery of detailed information from several sources using simple queries.

The International Nucleotide Sequence Database Collaboration (INSDC – www.insdc.org) has been developed for more than 18 years, covering the DNA DataBank of Japan (DDBJ), the European Molecular Biology Laboratory (EMBL), and GenBank at the National Center for Biotechnology Information (NCBI). This organization has a policy of free and unlimited access to all of its records. GenBank is the NIH's genetic sequence database located at NCBI (http://www.ncbi.nlm.nih.gov/Genbank/index.html). This comprehensive database contains an annotated collection of all publicly available DNA sequences for more than 260,000 identified organisms

(Benson et al. 2008). The EMBL Nucleotide Sequence Database, also known as EMBL-Bank (http://www.ebi.ac.uk/embl), constitutes Europe's primary nucleotide sequence resource. In Japan, the DDBJ (http://www.ddbj.nig.ac.jp) is the only DNA data bank that is officially certified to assemble DNA sequences. The main sources for DNA and RNA sequences are primarily submissions from individual researchers, from sequencing projects as batches of different types of sequences, including cDNA and genomic ones, and patent applications. The information exchanged and updated among these three repositories on a daily-basis ensures the incorporation of the most recent available sequence data and a worldwide coverage. This kind of initiative has led to many constructive projects and should proliferate in the biology community as data accumulates in a significant way.

As accomplished for INSDC, there has been a huge community effort to develop and improve databases and tools for all kinds of biological records –from sequences databases to metabolic pathways, proteomics, organelles, human diseases, plants and immunological databases, to cite some. A substantial effort has been made to make available molecular records to the scientific community in a reliable and appropriate manner (Wolfsberg et al. 2002; Kraj and McIndoe 2005; Harris 2008). A good reference is the Molecular Biology Database Collection, a public repository updated annually and published in the first issue of Nucleic Acid Research, which describes hundreds of databases every year (http://nar.oxfordjournals.org).

As these repositories become more complex, scientists start to use a gamma of technologies based on knowledge discovery and data mining to extract information from these databases. Knowledge Discovery in Database, or KDD, is a computational approach that consists basically in databases construction, i.e. data selection, pre-processing, transformation and dimensionality decrease (Barrera et al. 2004). This knowledge is used to search for regular patterns, association rules, temporal sequences or legitimate correlation between the records, something that is not normally recognized by the specialist. The expected product from KDD is a significant information system that can be applied by the decision-makers systems.

Database-mining methods, as part of the KDD approach, are also extremely useful to explore large amounts of data and basically consist of (a) data exploration, (b) pattern or model definition, and (c) validation of the model in other datasets. The application of models and algorithms to biological databases has produced valuable information related to pattern discovery in biological molecules, text mining in biomedical literature, data integration and probabilistic modeling of genome sequences (Fogel 2008; Haquin et al. 2008). The application of this kind of information has helped many biomedical research programs – from the hypotheses creation in a hypothesis-driven project to the design of large-scale experiments.

The integration of computer and experiment-based approaches is a significant challenge for the whole biomedical field. Although experimentalists are using computer-based approaches in a daily basis in some fields like gene expression and phylogenetics, the great majority of biologists are still far from using computer tools in an effective way. As illustrate in Fig. 12.1, the Physiome Project is a worldwide effort of several loosely connected research groups. It was created to define the

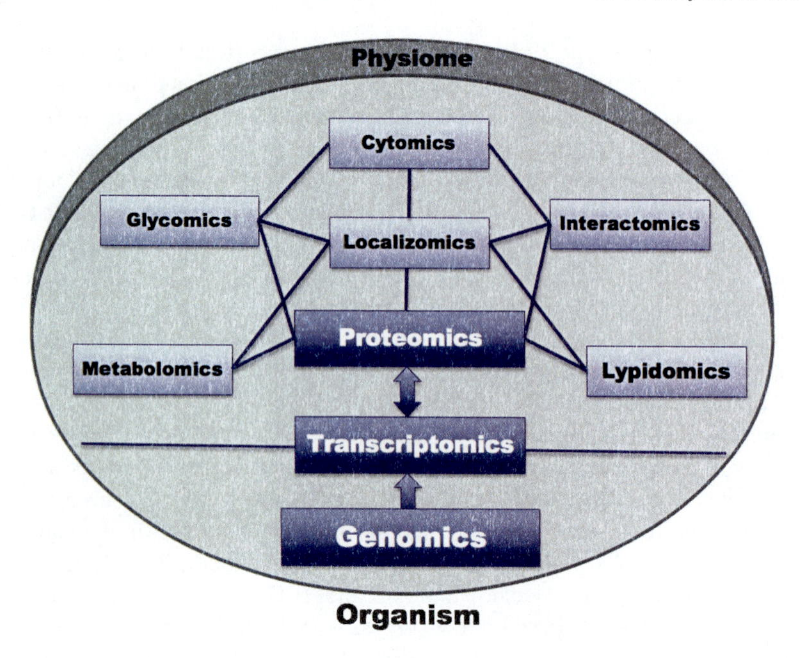

Fig. 12.1 Schematic representation of the relationship between different areas of biological organization in the physiome

physiome via the construction of databases and the development of integrated quantitative and descriptive modeling (http://nsr.bioeng.washington.edu). Biologists cannot afford to ignore such advances in this critical area. The development of biomedical science as a whole is dependent on a broader access to bioinformatics and computational biology.

12.2 Cancer Bioinformatics

Cancer is a disease determined by several genetic and epigenetic alterations (Balmain et al. 2003). In its simplest form, cancer is a genetic disease caused by alterations in the genome of a cell, ranging from point mutations to insertions, deletions and chromosomal translocations. These genetic changes can lead to an abnormal growth of cells and tissues that characterize the neoplastic phenotype of cancer. Although the molecular mechanisms governing cancer pathogenesis have been the focus of intense investigation over the last 50 years, including investigations that employed a number of basic molecular methods, the mechanisms underlying the development of human cancer still remain ill-defined (Nagl 2006). Despite the establishment of many molecular genetic and epigenetic alterations that lie at the root of cellular transformation, the complex processes that lead to the tumor phenotype are just beginning to be fully appreciated

(Rivenbark and Coleman 2007). Opportunely, currently studies on the genetics basis of cancer are undergoing a revolution.

The technological advances that lead to the post-genome era allowed the molecular biologists to make meticulous studies on the DNA (genome), the mRNA (transcriptome) and the protein sequences (proteome). Initiatives that intend to describe cancer in a global dimension have provided an opportunity to investigators to obtain more useful data to analyze and to integrate in novel ways. In spite of the practical difficulties, a growing number of projects are being developed with the aim to integrate hundreds of samples, studies, and data types combined from multiple sources.

The Cancer Genome Anatomy Project (CGAP) (http://cgap.nci.nih.gov), launched and maintained by the National Cancer Institute (NCI), has become one of the leading initiatives in cancer genetics. This project has produced more than three million expressed sequences tags (EST) from a large variety of tumor and normal samples, besides hundreds of libraries constructed using new methodologies like serial analysis of gene expression (SAGE) and massively parallel signature sequencing (MPSS). In Brazil, the FAPESP/LICR Human Cancer Genome Project (HCGP) initiative produced more than one million ESTs from prevalent tumors using a new methodology called Open Reading frame ESTs (ORESTES), biased toward the central part of the transcripts (Camargo et al. 2001). The expressed sequences generated by the CGAP and HCGP were incorporated into a database known as the International Database Cancer Gene Expression that constituted the foundation of the Human Cancer Index at the Institute of Genomic Research (http://www.tigr.org). The alliance between the two projects was established because both initiatives were working essentially with the same objective – the creation of a catalog of gene expression in cancer, and together they annotated and submitted millions of sequences from tumor and normal tissues to GenBank (Brentani et al. 2003). The proposal was to determine the distinctive gene expression patterns of normal, pre-cancer, and cancer cells, with the objective of improving detection, diagnosis, and treatment for the patient.

The Cancer Biomedical Informatics Grid (http://cabig.nci.nih.gov) is an ambitious and relatively new NCI-funded initiative that aims to create a cancer network, integrating information from four general categories: Interfaces, Vocabularies/Terminologies and Ontologies, Data Elements, and Information Models. The caBIG™ community – including researchers, physicians, and patients, represents more than 800 people from over 80 organizations on more than 70 projects ranging from analysis of gene expression data to clinical trials (Hanauer et al. 2007). caBig is a component of the Cancer Genome Atlas initiative funded and maintained by NCI (http://cancergenome.nih.gov). As stated, the goals of caBIG are: (1) to connect scientists and practitioners through a shareable and interoperable infrastructure; (2) to more easily share information by developing standard rules and a common language; and (3) to build or adapt tools for collecting, analyzing, integrating, and disseminating information associated with cancer research and care. The caBIG project seeks to create a collaborative information network to accelerate the discovery of new approaches and to improve patient outcomes.

Accordingly, cancer bioinformatics deals with the organization and analysis of the data in order that important trends and patterns can be identified – the ultimate goal is the discovery of new therapeutic and/or diagnostic protocols for cancer.

The first step to accomplish this objective is the search for a blueprint of genetic expression that characterizes specific cancer conditions. It is generally accepted that any biological state, physiological or not, is a representation of a differentiated set of gene expression patterns, which could not be characterized by the expression of a single gene (Nevins and Potti 2007). Therefore, in order to reveal the molecular marks that typify cancer initiation and progression, researchers carry out an extensive genome analysis, using for example gene expression microarrays, array comparative genomic hybridization (array CGH) and tissue microarrays. However, a significant number of alterations that would characterize specific cancer stages can take place after genome replication, during transcriptional, translational or posttranslational phases. Modifications such as gene amplification, alternative RNA splicing, phosphorylation, methylation and differences in protein stability and secretion are not envisaged by genome analysis. Proteomic analyses allow the identification and quantitative analysis of an entire protein set in biological samples (Posadas et al. 2005). The currently technologies are 2D polyacrylamide gel electrophoresis (2DE), isotope-coded affinity tags, matrix-assisted laser desorption ionization-mass spectrometry (MALDI-MS), liquid chromatography-MS/MS (LC-MS/MS), imaging MS, protein arrays and autoantibody expression techniques. To organize and analyze the considerable quantity of data produced by these high-throughput methods, specific computational software packages and databases have to be developed (Manning et al. 2007). The establishment of protein or genetic profiles, assisted by computational statistical analyses, has allowed the recognition of genetic signatures that could be valuable in the prognostic and development of new and individualized cancer therapies (Matharoo-Ball et al. 2007).

The application of bioinformatics approaches and clinical validation has been used to identify indicative profiles in many types of cancer. Kim et al. (2007) analyzed SAGE and EST data to produce a list of differentially expressed genes in lung cancers. A systematic examination of the annotated gene properties led to 20 genes, which were subjected to experimental validation using clinical specimens from lung cancer patients. Sjoblom et al. (2006) determined the sequence of well-annotated human protein-coding genes in colorectal and breast tumor. Analysis of 13,023 genes in 11 breast and 11 colorectal cancers revealed that individual tumors accumulated an average of approximately 90 mutant genes, but only a subset of them contributes to the neoplastic process. Using stringent criteria to delineate this subset, they identified 189 genes (average of 11 per tumor) that were mutated at significant frequency.

Therefore, statistical and bioinformatics tools can help to identify mutations with a role in tumorigenesis. The identification of molecular markers and profiles is being used in cancer classification and diagnosis as well as in envisaging clinical outcomes. The identification of specific genes, proteins and cellular pathways on which cancer cells depend is leading to the development of more effective therapeutic agents.

Designed to integrate data and results from multiple applications, platform projects such as GeneSpring Analysis Platform (http://www.chem.agilent.com/en-US/Pages/HomePage.aspx), or the open source and development software project, Bioconductor (http://www.bioconductor.org), are not only specifically developed to answer biological questions at the intersection of genomics, genetics, proteomics and biomarker screening, but also provide comprehensive statistical analysis, data mining and visualization tools. Furthermore, population-based studies of molecular and genetic variation may become the basis of individualized treatment. Some successful examples of therapeutic agents, which are now currently applied in clinics, are Gleevec, a kinase inhibitor for the treatment of some forms of adult and pediatric chronic myeloid leukemia (CML) and the monoclonal antibodies Rituxan (non-Hodgkin's lymphoma), Avastin (colorectal cancer and non-small-cell-lung cancer), and Herceptin (breast cancer) (Section 12.6.1). The present text is not intended to be an extensive overview of cancer bioinformatics. Rather, the aim is to highlight some of the main molecular/bioinformatics methodologies in cancer research and their clinical application.

12.3 Large-Scale Approach to the Study of Cancer

For decades, the traditional approach to the study of cancer was to select a few genes, genomic regions or proteins and then to compare their status in healthy *versus* cancer states. However, with the advent of technologies for large-scale data generation and analysis, the paradigm to study cancer is changing (Bonetta 2005). The use of genomics, transcriptomics, proteomics and bioinformatics has allowed the generation of a great number of new hypotheses to be tested and has stimulated a fast development of cancer research (Collins et al. 2003). For example, the use of these large-scale approaches is amplifying the numbers of genetic variants known to be associated with the risk of developing specific types of cancer, and integrating molecular signatures to predict cancer prognosis and treatment response. Figure 12.2 shows a schematic representation of how bioinformatics, together with genomics, transcriptomics and proteomics, are used to study cancer.

12.3.1 Genomics

Genomics is the large-scale study of the whole DNA sequence of an organism. Historically, Fred Sanger's group established genomics when they developed techniques to sequence, map and store the DNA sequence of the virus phi X174 (Sanger et al. 1977). The birth of Genomics, however, is more associated to the deciphering of the genome sequence of the bacteria *Haemophilus influenza* (Fleischmann et al. 1995). Today, genomics has become intrinsic to modern biological research, and two factors were essential to this: (a) the development of large-scale sequencing

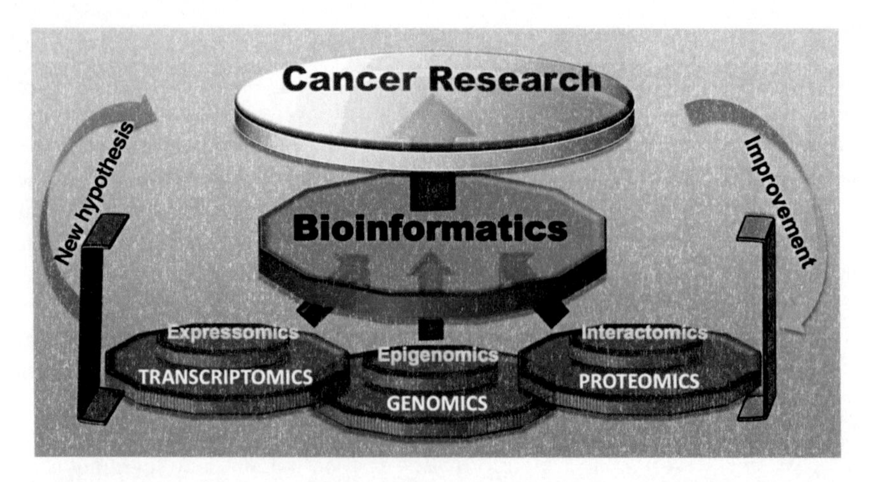

Fig. 12.2 Schematic representation of how bioinformatics, together with genomics, transcriptomics, proteomics, can be used to develop of cancer research

methods in the middle of the 1980s and (b) the development of computation methods to store, analyze and integrate these data (for a wide view of genomics, explore the UCSC (http://genome.ucsc.edu), NCBI (http://www.ncbi.nih.gov), and EBI (http://www.ebi.ac.uk) web sites).

The genomics approach has introduced an important new dimension into biomedical research and is one of the most relevant new areas is "cancer-genomics". This field integrates large-scale data generation and computational resources to study the structural changes in the genomes of a tumor tissue or cell line (Balmain et al. 2003). Below we describe two examples of how this genomic approach has revolutionized cancer research.

In a recent study, Campbell et al. (2008) identified and characterized, to the base-pair level, deletions, tandem duplications, inverted duplications, inversions and inter-chromosomal rearrangements in the genome of two lung cancer cell lines. Basically, the authors developed bioinformatics approaches to identify variations in the cancer genome using short sequences as input and the human public genome sequences as reference. Sjoblom et al. (2006) and Wood et al. (2007) looked for mutations in a set of breast and colon tumors sample. First, using bioinformatics tools, the authors selected 18,191 human genes available in the Reference Sequence database (http://www.ncbi.nlm.nih.gov/RefSeq). Next, they sequenced the protein-coding exons of these genes in all tumor samples, yielding ~800,000 possible mutations. They then used additional bioinformatics methods to remove artifacts, normal variants and synonymous substitutions to generate a more reliable set of somatic mutations occurring in those tumors. They found on average 80 mutated genes per each breast or colon tumor. More recently, the same leading group published a similar analysis for glioblastoma and pancreatic tumors (Jones et al. 2008; Parsons et al. 2008). In Table 12.1 there is a complete list of bioinformatics and genomics applications for cancer research.

Table 12.1 Description of genomics, transcriptomics, and proteomics bioinformatics techniques in cancer research. Some subcategories of genomics (epigenomics), transcriptomics (expressomics) and proteomics (interactomics) were also illustrated

Methodologies	What can be explored in cancer	Tools used
Genomics–bioinformatics	Mutations; polymorphisms; methylation changes; chromosomal amplification, deletion and rearrangement	DNA sequencing; next-generation sequencing; SNP arrays; comparative genomic hybridization; BLAST; FASTA; other specific software and packages; web-tools and public databases
Transcriptomics–bioinformatics	Definition of new exonic and intronic regions; gene expression; post-transcriptional modification; targets (genes) expressed preferentially in disease state	Microarray-based technologies; SAGE; MPSS; next-generation sequencing; PCR-based gene expression; BLAST; BLAT; sim4; other specific software or packages; web-tools and public databases
Proteomics–bioinformatics	Protein identification; protein level measurement; post-translational modification; protein–protein interaction; enzymatic activity	Immunohistochemistry; immunofluorescence; mass spectroscopy; 2D gels; protein microarray; specific software; web-tools and public databases

12.3.2 Transcriptomics

The complete set of transcripts produced by a genome at any time is called transcriptome, and, in analogy with the term genomics, transcriptomics is defined as the study of the transcriptome in a global way.

Unlike the genome, the transcriptome is extremely dynamic, varying not only among the different tissues in an organism but also between healthy and disease states, as is the case in cancer (Balmain et al. 2003). Based on this observation, many authors have studied the expression profiles of a large number of genes, and tried to identify patterns of gene expression in cancer (Nevins and Potti 2007). Below we describe two studies where transcriptomics was used to study genes expressed in cancer.

Rhodes et al. (2004) developed a computational protocol to find a meta-signature of gene expression in several types of cancer. These authors analyzed 40 published cancer microarray data sets with ~38 million gene expression measurements from more than 3,700 cancer samples. As a result, they obtained 67 genes over-expressed in more than 10 cancers relative to their normal tissues. Using almost the same strategy, many other works tried to identify signatures of gene expression in tumors (Greshock et al. 2007; Chanrion et al. 2008; Ivliev et al. 2008). Using a sequence-based approach, Sugarbaker et al. (2008) integrated a set of computational methods and next-generation sequencing to study the mesothelioma tumors from six patients. The author studied some characteristics of mesothelioma transcriptome, and identified 15 new non-synonymous mutations in this tumor type.

As pointed out for genomics, these reports are examples of how bioinformatics and transcriptomics can be integrated to produce fruitful results in the fight against cancer (Table 12.1).

12.3.3 Proteomics

Proteins are an essential part of organisms, participating in almost all-physiological and metabolic processes within a cell (Sharan et al. 2007). Also, in analogy with the term genomics and transcriptomics, the large-scale study of all expressed proteins in a cell or organism, as well as their modifications and interactions, is called proteomics.

Proteomics is often considered the next step in the study of biological systems, after genomics and transcriptomics (Lander et al. 2001). However, the study of proteome is much more complicated than genomics due to intrinsic features of proteins such as the variability of post-translational modifications. Moreover, the technologies available to study genomes in a large-scale manner are more powerful than the technologies for proteomes in spite of the advances in the mass spectrometry field.

Despite its methodological challenges, proteomics is growing rapidly and making important contributions to clinical diagnosis and disease management applied to cancer. Several works identified proteins that change in amount in breast, ovary, prostate, and esophagus cancer (Srinivas et al. 2002). For instance, tumor markers can be unambiguously identified in the blood of patients with ovarian cancers through proteomics approaches (Petricoin et al. 2002). Table 12.1 lists more examples of how proteomics can be used to study cancer.

12.4 Techniques of Large-Scale Analysis and Their Application in Cancer Research

As described previously, large-scale studies are essential for the development of modern biological research. However, these approaches – genomics, transcriptomics and proteomics – exist because experimental methodologies were developed and generated large-scale data, such as DNA sequencing, microarray (Schena et al. 1995; Lockhart et al. 1996), SAGE (Velculescu et al. 1995) and mass spectroscopy. These methodologies have also contributed to the development of bioinformatics. It is remarkable that almost all of these methodologies were mainly applied originally to the study of cancer. Below, we will describe some of these methodologies and how they are used in biological research.

12.4.1 Expressed Sequences Tags (ESTs)

Expressed sequence tags (ESTs) are short sequences, 500 base pairs on average, generated through a partial single sequencing run from full-length cDNAs (Adams et al. 1991).

This technology was first developed as an alternative to survey expressed genes, instead of the slow, small-scale and laborious "northern blot". In 1991, Adam et al. generated the first ESTs (609 sequences) and used them to study genes expressed in the central nervous system. After this seminal paper, many works have used EST to study several aspects of gene expression in diverse tissues and organisms, and the generation of ESTs grew exponentially. By now, the NCBI-based public database of EST, dbEST (Boguski et al. 1993), contains more than 54 millions sequences from 1,587 organisms (dbEST release 072508).

Essentially, the analysis of ESTs follows a well-defined bioinformatics pipeline: (a) in the first step, the sequences are trimmed and the low quality bases or contaminated regions are removed; (b) in the second step, the ESTs are clustered based on their similarities; (c) on the last step, the ESTs are annotated based on their overlapping, at sequence level, with known genes or known genomic regions. After these initial processing, the generated results can be used for gene discovery, gene expression analysis or studies of post-transcriptional variations, such as alternative splicing or alternative polyadenylation.

The relationship between EST and cancer research proved itself to be fruitful early. First, many works have used ESTs to study cancer (Nishiguchi et al. 1994; Papadopoulos et al. 1994; Watson and Fleming 1994). Second, two distinct cancer projects highlighted the importance of ESTs for the study of the transcriptome. The CGAP generated ~1.5 million ESTs from >10 tumor types (Strausberg et al. 2000). The HCGP generated ~1 million ESTs from >10 tumor types (Camargo et al. 2001). For many years, the number of ESTs from cancer was predominant in dbEST.

12.4.2 SAGE and MPSS

Serial analysis of gene expression and MPSS are two methodologies that quantify gene expression through the generation of short sequence tags (10 or 17 bp for SAGE; 13 or 20 bp for MPSS) adjacent to the most 3′ site of *Nla*III (SAGE) or *Dpn*II (MPSS) within polyadenylated transcripts (Velculescu et al. 1995; Brenner et al. 2000). The output of an experiment of SAGE and MPSS is a list of short sequences (tags), where the frequency of a given tag should be proportional to the abundance of the transcript from which the tag was derived.

Although SAGE and MPSS produce a similar output – a list of tags, the experimental protocols are completely different. For example, while SAGE uses the traditional cloning and DNA sequencing method, MPSS uses a novel cloning and a parallel sequencing protocol (proprietary) based on enzymatic digestion and hybridization. As a result, a SAGE output contains 100,000 tags on average while a MPSS output contains more than 1,000,000 tags (Brenner et al. 2000).

An essential step in the analyses of SAGE and MPSS data is the correct assignment/mapping of tags to genes. Basically, there are three strategies to perform this process:

(a) an annotation based on databases constructed using in-house computer approaches (Blackshaw et al. 2004; Silva et al. 2004); (b) an annotation, tag by tag, done on websites such as the SAGE Genie (Boon et al. 2002) or SAGEmap (Lash et al. 2000); and (c) a large-scale annotation using several reference datasets (Galante et al. 2007; Norambuena et al. 2007). Another essential step in SAGE analysis is the identification of tags differentially represented in each sample. For this purpose, there are several methods, such as those described in Baggerly et al. (2003), Thygesen and Zwinderman (2006); and Zuyderduyn (2007).

Since their inception, the use of SAGE and MPSS has grown dramatically (especially SAGE). Numerous publications have used this methodology for an analysis of global gene expression, particularly in cancer (Riggins and Strausberg 2001). One of the first comprehensive analyses of global gene expression in human cancer was performed using SAGE (Zhang et al. 1997). Three characteristics make SAGE and MPSS a powerful method to study gene expression in cancer: (a) SAGE and MPSS provide a description of the mRNA population without *a priori* selection of the genes to be studied allowing the discovery of new genes involved in the carcinogenesis for example; (b) the data obtained in one experiment can be directly compared with data generated from any other laboratory or with data available in public databases, allowing a large-scale comparison of genes expression; (c) the generated data, tag-frequencies, are in a digital format, allowing robust statistical analysis of gene expression.

12.4.3 Microarray

Since their conception in the mid-1990s (Fodor et al. 1991; Schena et al. 1995; Lockhart et al. 1996), the use of microarrays has spread rapidly throughout the research community. In recent years, microarray experiments have become the "must–have" item for many primary research articles in biology. Although there are many protocols and types of systems available for microarrays, the basic technique involves four main steps: (a) extraction of RNA from biological samples; (b) copy of RNA to cDNA, which includes the incorporation of either fluorescent nucleotides or a tag that is later stained with fluorescence; (c) hybridization of labeled RNAs (cDNAs) to the microarray chip; (d) scanning of the microarray chip under laser light and measurement of gene expression.

Today, the most common microarray platforms are "cDNA microarrays" (Schena et al. 1995) and "oligonucleotide microarrays" (Lockhart et al. 1996). The advantage of "oligonucleotide microarrays" lies on the fact that all probes are designed to have similar hybridization temperature and binding affinity. Despite the differences, both platforms are able to measure the gene expression of more than 10k genes per biological sample.

One critical step in the microarray experiments is the processing and analysis of results. Even though there are many software packages available (Gentleman et al. 2004), it is still difficult to find a single software framework to process the raw data (e.g. background correction and normalization) and execute the analysis (e.g. identify and present the genes differentially expressed). The best option is the development of a bioinformatics protocol (pipeline) that integrates several packages.

In addition to their obvious use in basic research, microarrays are broadly used to determine genes correlated with diseases, and, among these, cancer is the disease most commonly explored. For example, microarrays have been used to determine (a) genes differentially expressed in tumor *versus* healthy tissue, (b) genes correlated with tumor progression, and (c) genes that are able to distinguish cancer from normal state or even multiple subtypes of tumors accurately (Butte 2002).

12.4.4 Next-Generation Sequencing Technologies

Largely because of efforts necessary to sequence the human genome, the sequencing of DNA (and cDNA) has undergone a steady metamorphosis from a cottage industry into a large-scale production enterprise (Mardis, 2008a, b). During this process, the cost per reaction of DNA sequencing has fallen, while the number of bases sequenced per run has increased drastically. Now, we are within the next-generation (or second generation) sequencing technology, and the third generation (next-next-generation) is coming in the next few years (Mardis, 2008a, b). However, what does make the next-generation sequencing method cheaper and faster than the traditional sequencing methods? In contrast to the "Sanger" sequencing ("traditional method") that processes 96 sequence-reads at the same time, the next-generation sequencing processes millions of sequence-reads in parallel, generating a huge amount of sequences in a few hours. For example, Roche-454, Illumina-Solexa and ABI-SOLiD, generates 1.2 million reads (average read length of 400 bp), 40–50 million reads (read length of 36 bp) and ~100 million reads (read length of 25–35 bp) per run, respectively (Mardis, 2008a, b).

Next-generation sequencing has applications that are immediately relevant to the medical field. In cancer genetics, for example, these methods may facilitate the discovery of mutations, as well as, the improvement in the quantification of gene expression and the discovery/description of regulatory RNA molecules (non-coding RNA) whose myriad functions continue to be characterized in cancer. However, despite these enormous potential, next-generation sequencing methods are still offset by increased costs per run (at least $8,000) and difficulties on the bioinformatics front to handle the huge amount of data generated in each experiment (Mardis, 2008a, b).

12.4.5 Mass Spectrometry

Mass spectrometry is a powerful analytical technique that is used to identify unknown compounds, to quantify known compounds, and to elucidate the structure and chemical properties of molecules. Mass spectrometry (MS) is at the heart of virtually all proteomics experiments. Basically, the dominant MS workflow starts with a site-specific enzymatic digestion from proteins to peptides. Next, the peptides are volatized and the mass spectrometer generates the spectra for each sample. In the last step, the spectra are compared to peptide sequence from a database and the protein sequences are inferred.

The first step in a MS experiment, the enzymatic digestion, is done by a protease, such as trypsin. In order to volatize peptides, the two most common methods are matrix-assisted laser desorption/ionization (MALDI) and electrospray ionization (ESI). MALDI is used to volatize peptide mixture containing a small amount of peptide; ESI is used when the peptide mixture contains a large amount of peptides. The last step is a bioinformatics protocol whereby the spectra (observed peaks) are processed and compared to peptide sequences derived from a "virtual" digestion of protein sequences available in a protein database. Because protein identifications rely on matches with sequence databases, proteomics is currently restricted largely to those species for which comprehensive sequence databases are available.

A direct application of MS is the detection of protein or peptide peaks that differ in their mass/charge ratio in patients with cancer compared with healthy individuals. For example, using this strategy, Nakagawa et al. (2006) identified two polypeptides related to breast cancer, Hao et al. (2008) identified some polypeptides related to gastric cancer and Sun et al. (2008) identified 116 proteins that potentially could be used to distinguish between hepatocellular carcinoma and normal liver cells.

Even with the great potential of MS application to cancer research, the genetic variation among individuals (Nedelkov et al. 2005) and dynamic changes in the plasma proteome as a function of a multitude of factors (sex, age, health status, for example) are still an enormous barrier to this type of analysis (Ku et al. 2003).

12.5 The Integration of Omics Data

As pointed early, advances in genomics, transcriptomics and proteomics have provided biologists with a huge repertoire of data. However, most of this data comes from different platforms or is stored in databases where data integration is not easy or not even feasible. A specific area in bioinformatics has evolved to addresses this very important problem: the integromics (Venkatesh and Harlow 2002).

In integromics, algorithms and tools have been developed to make an inter- and intra-genomics, transcriptomics and proteomics data integration (Venkatesh and Harlow 2002). Theoretically, the methods in integromics can be divided in two subtypes: high-level and low-level primary data integration. The first sub-type involves the merging of presumed facts and conclusions obtained from the raw experimental data. The second type, the low-level primary data integration, includes the process of normalization between platforms, organization of data and data-mining techniques.

Examples of both sub-types of methods can be seen at NCBI (http://www.ncbi.nih.gov), UCSC (http://genomics.ucsc.edu) or EBI (http://www.ei.ac.uk) web sites.

An interesting example of data integration is the gene naming and gene product classification. Often a gene, discovered by different groups, has more than one name. For example, the gene *TP53* has five different names (Maglott et al. 2007). This is also true for gene products. To standardize gene names, among other goals, the Human Genome Organization (HUGO) project was created (http://www.hugo-international.org). HUGO assigns a unique symbol to each human gene and all newly discovered human genes should be submitted to HUGO to receive an official name. Several initiatives are trying to classify gene products. The most popular is the Gene Ontology (GO) consortium (Ashburner et al. 2000), which uses a well-controlled vocabulary to classify each gene product in terms of their biological processes, cellular components and molecular functions.

Despite these several difficulties, data integration is extremely important for many reasons, which range from technical necessities (like reinforcement of a signal and deletion of some platform-specific noises) to the development of areas that depends on many data to obtain effective results (like the development of molecular pharmacology of cancer) (Rhodes and Chinnaiyan 2005). In spite of the fact that many methods and protocols have been developed in integromics, there still remains much "omics" data to be integrated.

12.6 Clinical Bioinformatics

Clinical bioinformatics applies bioinformatics knowledge and techniques to help in the diagnosis, treatment, prevention and control of diseases, as well in the development of chemical, structural and biochemical methods for clinical research. In cancer research, bioinformatics tools are used to detect biomarkers in several kinds of cancers, during different stages – initiation, progression and advanced. Accordingly to NCI, "biomarkers are defined as cellular, biochemical, molecular (genetic and epigenetic) alterations by which a normal, abnormal, or simply biologic process can be recognized or monitored". Biomarkers can be measurable in biological media, such as in tissues, cells, or fluids and can be applied to evaluate early diagnosis, risk assessment, classification, and prognosis of cancer.

In 1999, the NCI Biomarker Developmental Laboratories (http://edrn.nci.nih.gov/about-edrn/scicomponents/bdl) was created to identify techniques for finding molecular, genetic, and biologic early warning signals of cancer. This division, together with Biomarker Reference Laboratories (formerly known as Biomarker Validation Laboratories), Clinical Epidemiology and Validation Centers (formerly known as Clinical and Epidemiologic Centers), and Data Management and Coordinating Center (DMCC), form the NCI's Early Detection Research Network (EDRN – http://edrn.nci.nih.gov/about-edrn) which has the responsibility for the development, evaluation, and validation of biomarkers for earlier cancer detection and risk assessment. Besides the classification and risk assessment, cancer biomarkers have the potential of being transformed into targets for vaccine development.

Clinical bioinformatics is a fundamental component of a personalized treatment for cancer. The association studies that provide the data necessary to establish the link between a genetic alteration or variation and a specific clinical outcome are heavily dependent on the bioinformatics methods discussed in this chapter. Furthermore, these bioinformatics methods are also important for the establishment of biobanks and their integration with clinical information.

12.6.1 Identification of Gene and Protein Targets to Drugs and Vaccine Development

Cancer is a genetic disorder, and as such, it can be characterized by the genetic alterations established during its initiation and progression. As previously discussed, gene and protein signatures generated by altered gene expression, somatic mutations and genomic instability open on the possibility of distinguishing normal from cancer cells and the use of this knowledge to improve the diagnostic, prognostic, and development of new drugs and therapies. During the last years, the development of bioinformatics technology has facilitated the identification of cancer markers leading to a better comprehension of the mechanisms underlying the tumorigenic process. Approaches aimed for target detection involve a high-throughput screening of genes and proteins and the quantification of their profiles and variants. The exploitation of bioinformatics methods in cancer research is leading to the detection of many targets that will need to be experimentally and clinically validated before a possible use in clinical practice.

In general, the process of drug discovery goes through five phases: identification of target, validation, high-throughput compound screening, lead optimization and clinical trials (Iorns et al. 2007). The first phase – target identification, aims to show a relationship between the alterations of gene or protein profile with a particular variety of cancer. Although this has led to identification of many markers, a challenge in the cancer drug discovery process is the second phase – target validation, which aims to establish the clinical importance of the target. In phase 3, high-throughput techniques are used to recognize compounds that could inhibit a validated protein target. It could happen in this phase that a particular compound might be able to block a target *in vitro*, but it does not present the basic chemical and physical characteristics of a pharmaceutics drug. The phase of lead optimization aims to enhance the drug efficiency against the selected target and to discard unspecific leads. Unfortunately, a common side effect of this enhanced efficient is an undesirable cross-recognition of unrelated proteins that may limit the clinical application. In the last phase, biomedical or health-related studies are conducted following a pre-defined protocol and the outcomes are measured and evaluated in clinical trials. During each of these phases, several factors such as – samples collection, processing and storage, patient classification, analytical incompleteness and limitations of bioinformatics tools and databases, may influence reproducibility and analysis, affecting the final outcome.

Despite the difficulties, omics analyses are speeding up the identification of many markers that potentially may be transformed in targets of therapeutic drugs (Strausberg et al. 2004). These targets can be classified in three main categories: (1) small-molecule inhibitors of oncogenic signals, (2) antibodies to surface components and intercellular communicating factors, and (3) molecularly defined vaccines. In the first category, the most evident targets are kinases, since their association to tumor development is clear. An example of a positive result of this drug development strategy, already in clinical use, is the small molecule, imatinib mesylate (Gleevec), which is used in the treatment of some forms of adult and pediatric chronic myelogenous leukemia (CML), and for the treatment of a rare form of cancer called gastrointestinal stromal tumor (http://www.cancer.gov/clinicaltrials/digestpage/gleevec). In the second and third categories, among the targets that could easily be recognized by therapeutic antibodies and vaccines are those with expression limited to one or more non-essential tissues or cells, irrespective of the disease state of the individual cells. In this manner, the damage caused by the immunotherapeutic intervention in organs such as breast and prostate, could be followed by clinical removal without causing a severe loss. Other successful examples already on clinical use are the monoclonal antibodies Rituxan, used to treat non-Hodgkin's lymphoma, Avastin, for colorectal cancer and non-small-cell-lung cancer and Herceptin, for breast cancer (Carter 2001; Carter and Senter 2008). Although approaches involving targeting of oncogene kinases and immunotherapeutic systems are under development and have already obtained some success, both methods are hindered by the fact that potential targets are found only in subsets of any tumor type. For example, colon cancer has a broad spectrum of low-frequency mutations in kinases, each of which could be a potential target. This molecular heterogeneity makes the development of universally applicable therapies a complicated and difficult enterprise (Carter and Senter 2008).

12.6.2 Individualized Treatment Based on Molecular and Genetic Variation

The completion of the human genome sequence opens the era of the functional genomic analysis and brings promises of the development of new disease therapies. However, regardless of our present knowledge of the human genome and a better understanding of the cancer genetic alterations, there is still a long way to go in developing widespread cancer treatments.

Traditionally, the treatment of cancer depends on the size and location of the tumor, the extent of the spread of the cancer and the person's underlying health. Although all these facts are important, the location of a tumor has been recognized as the basic rule in clinical oncology: the organ of origin should dictate the therapeutic approach to cancer therapy. The typical approach to primary cancer usually consists of surgical resection of the primary tumor, local irradiation according to the type of surgery and the extent of disease and adjuvant systemic therapy according to the tumor biology (http://www.cancer.net/portal/site/patient). However, the genomics research data obtained in recent years strongly indicates that each individual cancer – even from the same origin, presents an individual signature, composed of a set of different genetic alterations and might therefore require a collection of targeted agents. According to Lengauer et al. (2005)), "it is likely that the paradigm of 'one drug for one cancer' will eventually be replaced by multiple targeted drugs for multiple different genetic defects" – for one individual cancer, in one individual person.

In fact, the idea of individualized treatment is not new – it was already proposed decades ago (Scott 1969), but it's implementation is dependent on further scientific progress. The Individualized Cancer Chemotherapy (ICC) is a procedure that aims to maximize the efficacy of chemotherapy and to minimize its adverse effects. Based on the pathogenic status of a specific tumor and the knowledge of drug responses, clinicians can stipulate the best drugs combination in the therapy. This strategy can be designed based on the follow information: (a) the use of drug sensitivity tests that compare the anticancer activities of potential candidate drugs on surgically removed samples and those showing the best responses are selected for the use in succeeding treatments; (b) the use of tumorigenic markers as specific targets for drug antagonism and disruption; and (c) the integration of morphological information concerning primary tumor location and mode of spread to a rationally designed drug combination (Lu et al. 2006). Others approaches used in individualized treatment aim (a) to define pretreatment molecular markers that predict response and toxicity from a chemotherapeutic regimen for head and neck (Singh and Pfister 2008) and colorectal cancer (Bandres et al. 2007; Wilson et al. 2007), (b) to select the most cost-effective and least toxic chemotherapy for an individual (Lu et al. 2006) or (c) to identify reliable and validated marks that could select patients for specific treatments breast cancer patients (Pusztai et al. 2004; Dressman et al. 2007).

Although gene expression and protein profile of cancer are attractive technologies to fully characterized properties of a particular cancer, the development of successful predictors that can be applied in clinic treatments has proven to be difficult (Lengauer et al. 2005). In other words, whether the basic research seeks to identify a marker to be used as a predictor of response or resistance to chemotherapy, it remains to be seen what accuracy is need to a predictive marker to be clinically useful, i.e. whether a patient will respond to a particular drug or combination regime in question.

Regardless of the enormous difficulties imposed by the cancer complexity on treatments, the solution is to strengthen the association of basic and applied cancer research in areas such as germ-cell mutation analysis, genetic and gene expression profiles, targeted therapy, cancer stem cells, circulating cancer cells and single-nucleotide polymorphism. The increase of successful targeted therapies will significantly depend on the quality of the science that will lead to the drug discovery and development process, which is believed to refine future preventive and early intervention strategies on an individual basis.

12.7 Final Remarks

Cancer disease, as an epigenetic disorder, can be characterized by distinct genetic and protein profiles that are associated to its initiation and development. Several modifications, such as altered gene expression patterns, genomic instability and somatic mutations, can be used to discriminate cancer cells from normal ones, and this knowledge can be applied in the diagnosis support and development of therapies specifically directed to cancer cells. In this context, bioinformatics approaches have not only been an essential tool in the identification of biological markers but also are facilitating the understanding of the whole process of tumor development. The development of "omics" science involves the integration of a huge amount of data from many different sources. Data from high-throughput analysis of genes, gene variants, expression and proteomics are continually accumulating in public databanks and, as a corollary, bioinformatics technologies will also be continually developed to help the investigation of these datasets of complexes processes and networks. Therefore, bioinformatics analysis deals with data integration and such approach is providing a variety of information on cancer that begins to be used for clinical practices, both for diagnosis and treatment. The establishment of bioinformatics as a pillar of such enterprise is a natural consequence of its quantitative nature.

According to Pusztai and collaborators (Pusztai et al. 2004), before this practice can be adopted in the clinic, scientists and clinicians will need to overcome several challenges. The initial observations that genetic profiling of cancer can lead to the development of an individualized therapy will need to be confirmed, and the true accuracy of potential cancer targets will need to be determined. As stated by the authors, "…to draw an analogy between the clinical development of drugs and that of diagnostic predictive markers, the currently published results represent encouraging phase I/II clinical marker-discovery trial data. To demonstrate the clinical utility and evaluate the accuracy of these markers with a narrow confidence interval, a larger, randomized, phase III marker validation studies will be needed".

Despite all these difficulties, it is clear that, although the application of bioinformatics in cancer research is still in an early stage, it has already become an obligatory technology to assist and improve the development of cancer therapy in this post-genomic era. In this chapter, we aimed to provide a general overview of this exciting field. We envisage that the new technological advances will represent an even more dramatic contribution to our fight against cancer.

Acknowledgements The authors would like to acknowledge the financial support. BS is supported by São Paulo State Funding Agency (FAPESP). PAFG was supported in part by grant 5D43TW007015-02 from the Fogarty International center, NIH.

References

Adams MD, Kelley JM et al (1991) Complementary DNA sequencing: expressed sequence tags and human genome project. Science 252:1651–1656

Ashburner M, Ball CA et al (2000) Gene ontology: tool for the unification of biology. The Gene Ontology Consortium. Nat Genet 25:25–29

Baggerly KA, Deng L et al (2003) Differential expression in SAGE: accounting for normal between-library variation. Bioinformatics 19:1477–1483

Balmain A, Gray J et al (2003) The genetics and genomics of cancer. Nat Genet 33:238–244

Bandres E, Zarate R et al (2007) Pharmacogenomics in colorectal cancer: the first step for individualized-therapy. World J Gastroenterol 13:5888–5901

Barrera J, Cesar RM Jr et al (2004) An environment for knowledge discovery in biology. Comput Biol Med 34:427–447

Benson DA, Karsch-Mizrachi I et al (2008) GenBank. Nucleic Acids Res 36:D25–30

Blackshaw S, Harpavat S et al (2004) Genomic analysis of mouse retinal development. PLoS Biol 2:1411–1431

Boguski MS, Lowe TM et al (1993) dbEST – database for "expressed sequence tags". Nat Genet 4:332–333

Bonetta L (2005) Going on a cancer gene hunt. Cell 123:735–737

Boon K, Osorio EC et al (2002) An anatomy of normal and malignant gene expression. Proc Natl Acad Sci USA 99:11287–11292

Brenner S, Johnson M et al (2000) Gene expression analysis by massively parallel signature sequencing (MPSS) on microbead arrays. Nat Biotechnol 18:630–634

Brentani H, Caballero OL et al (2003) The generation and utilization of a cancer-oriented representation of the human transcriptome by using expressed sequence tags. Proc Natl Acad Sci USA 100:13418–13423

Butte A (2002) The use and analysis of microarray data. Nat Rev Drug Discov 1:951–960

Camargo AA, Samaia HP et al (2001) The contribution of 700,000 ORF sequence tags to the definition of the human transcriptome. Proc Natl Acad Sci USA 98:12103–12108

Campbell PJ, Stephens PJ et al (2008) Identification of somatically acquired rearrangements in cancer using genome-wide massively parallel paired-end sequencing. Nat Genet 40:722–729

Carter P (2001) Improving the efficacy of antibody-based cancer therapies. Nat Rev Cancer 1:118–129

Carter PJ, Senter PD (2008) Antibody-drug conjugates for cancer therapy. Cancer J 14:154–169

Chanrion M, Negre V et al (2008) A gene expression signature that can predict the recurrence of tamoxifen-treated primary breast cancer. Clin Cancer Res 14:1744–1752

Collins FS, Green ED et al (2003) A vision for the future of genomics research. Nature 422:835–847

Dressman HK, Berchuck A et al (2007) An integrated genomic-based approach to individualized treatment of patients with advanced-stage ovarian cancer. J Clin Oncol 25:517–525

Fleischmann RD, Adams MD et al (1995) Whole-genome random sequencing and assembly of Haemophilus influenzae Rd. Science 269:496–512

Fodor SP, Read JL et al (1991) Light-directed, spatially addressable parallel chemical synthesis. Science 251:767–773

Fogel GB (2008) Computational intelligence approaches for pattern discovery in biological systems. Brief Bioinform 9:307–316

Galante PA, Trimarchi J et al (2007) Automatic correspondence of tags and genes (ACTG): a tool for the analysis of SAGE, MPSS and SBS data. Bioinformatics 23:903–905

Gentleman RC, Carey VJ et al (2004) Bioconductor: open software development for computational biology and bioinformatics. Genome Biol 5:R80

Greshock J, Nathanson K et al (2007) Cancer cell lines as genetic models of their parent histology: analyses based on array comparative genomic hybridization. Cancer Res 67:3594–3600

Hanauer DA, Rhodes DR et al (2007) Bioinformatics approaches in the study of cancer. Curr Mol Med 7:133–141

Hao Y, Yu Y et al (2008) IPO-38 Is identified as a novel serum biomarker of gastric cancer based on clinical proteomics technology. J Proteome Res 7:3668–3677
Haquin S, Oeuillet E et al (2008) Data management in structural genomics: an overview. Methods Mol Biol 426:49–79
Harris MA (2008) Developing an ontology. Methods Mol Biol 452:111–124
Iorns E, Lord CJ et al (2007) Utilizing RNA interference to enhance cancer drug discovery. Nat Rev Drug Discov 6:556–568
Ivliev AE, t'Hoen PA et al (2008) Microarray retriever: a web-based tool for searching and large scale retrieval of public microarray data. Nucleic Acids Res 36:W327–W331
Jones S, Chen WD et al (2008) Comparative lesion sequencing provides insights into tumor evolution. Proc Natl Acad Sci USA 105:4283–4288
Kim B, Lee HJ et al (2007) Clinical validity of the lung cancer biomarkers identified by bioinformatics analysis of public expression data. Cancer Res 67:7431–7438
Kraj P, McIndoe RA (2005) caBIONet–A.NET wrapper to access and process genomic data stored at the National Cancer Institute's Center for Bioinformatics databases. Bioinformatics 21:3456–3458
Ku JH, Kim ME et al (2003) Influence of age, anthropometry, and hepatic and renal function on serum prostate-specific antigen levels in healthy middle-age men. Urology 61:132–136
Lander ES, Linton LM et al (2001) Initial sequencing and analysis of the human genome. Nature 409:860–921
Lash AE, Tolstoshev CM et al (2000) SAGEmap: a public gene expression resource. Genome Res 10:1051–1060
Lengauer C, Diaz LA Jr et al (2005) Cancer drug discovery through collaboration. Nat Rev Drug Discov 4:375–380
Lockhart DJ, Dong H et al (1996) Expression monitoring by hybridization to high-density oligonucleotide arrays. Nat Biotechnol 14:1675–1680
Lu Y, Chen XL et al (2006) Individualized cancer chemotherapy integrating drug sensitivity tests, pathological profile analysis and computational coordination – an effective strategy to improve clinical treatment. Med Hypotheses 66:45–51
Maglott D, Ostell J et al (2007) Entrez gene: gene-centered information at NCBI. Nucleic Acids Res 35:D26–31
Manning AT, Garvin JT et al (2007) Molecular profiling techniques and bioinformatics in cancer research. Eur J Surg Oncol 33:255–265
Mardis ER (2008a) Next-generation DNA sequencing methods. Annu Rev Genomics Hum Genet 9:387–402
Mardis ER (2008b) The impact of next-generation sequencing technology on genetics. Trends Genet 24:133–141
Matharoo-Ball B, Ball G et al (2007) Clinical proteomics: discovery of cancer biomarkers using mass spectrometry and bioinformatics approaches – a prostate cancer perspective. Vaccine 25:B110–121
Nagl S (2006) Cancer Bioinformatics: from therapy design to treatment. John Willey & Sons Ltd, West Sussex.
Nakagawa T, Huang SK et al (2006) Proteomic profiling of primary breast cancer predicts axillary lymph node metastasis. Cancer Res 66:11825–11830
Nedelkov D, Kiernan UA et al (2005) Investigating diversity in human plasma proteins. Proc Natl Acad Sci USA 102:10852–10857
Nevins JR, Potti A (2007) Mining gene expression profiles: expression signatures as cancer phenotypes. Nat Rev Genet 8:601–609
Nishiguchi S, Joh T et al (1994) A survey of genes expressed in undifferentiated mouse embryonal carcinoma F9 cells: characterization of low-abundance mRNAs. J Biochem 116:128–139
Norambuena T, Malig R et al (2007) SAGExplore: a web server for unambiguous tag mapping in serial analysis of gene expression oriented to gene discovery and annotation. Nucleic Acids Res 35:W163–168

Papadopoulos N, Nicolaides NC et al (1994) Mutation of a mutL homolog in hereditary colon cancer. Science 263:1625–1629

Parsons DW, Jones S et al (2008) An integrated genomic analysis of human glioblastoma multiforme. Science 321:1807–1812

Petricoin EF, Ardekani AM et al (2002) Use of proteomic patterns in serum to identify ovarian cancer. Lancet 359:572–577

Posadas EM, Simpkins F et al (2005) Proteomic analysis for the early detection and rational treatment of cancer – realistic hope? Ann Oncol 16:16–22

Pusztai L, Rouzier R et al (2004) Individualized chemotherapy treatment for breast cancer: Is it necessary? Is it feasible? Drug Resist Updat 7:325–331

Rhodes DR, Chinnaiyan AM (2005) Integrative analysis of the cancer transcriptome. Nat Genet 37:S31–37

Rhodes DR, Yu J et al (2004) Large-scale meta-analysis of cancer microarray data identifies common transcriptional profiles of neoplastic transformation and progression. Proc Natl Acad Sci USA 101:9309–9314

Riggins GJ, Strausberg RL (2001) Genome and genetic resources from the Cancer Genome Anatomy Project. Hum Mol Genet 10:663–667

Rivenbark AG, Coleman WB (2007) Dissecting the molecular mechanisms of cancer through bioinformatics-based experimental approaches. J Cell Biochem 101:1074–1086

Sanger F, Air GM et al (1977) Nucleotide sequence of bacteriophage phi X174 DNA. Nature 265:687–695

Schena M, Shalon D et al (1995) Quantitative monitoring of gene expression patterns with a complementary DNA microarray. Science 270:467–470

Scott RB (1969) Accurate cervical diagnostic studies: a necessity for individualized treatment of cancer of the uterine cervix. Obstet Gynecol Surv 24:985–992

Sharan R, Ulitsky I et al (2007) Network-based prediction of protein function. Mol Syst Biol 3:88

Silva AP, De Souza JE et al (2004) The impact of SNPs on the interpretation of SAGE and MPSS experimental data. Nucleic Acids Res 32:6104–6110

Singh B, Pfister DG (2008) Individualized treatment selection in patients with head and neck cancer: do molecular markers meet the challenge? J Clin Oncol 26:3114–3116

Sjoblom T, Jones S et al (2006) The consensus coding sequences of human breast and colorectal cancers. Science 314:268–274

Srinivas PR, Verma M et al (2002) Proteomics for cancer biomarker discovery. Clin Chem 48:1160–1169

Strausberg RL, Buetow KH et al (2000) The cancer genome anatomy project: building an annotated gene index. Trends Genet 16:103–106

Strausberg RL, Simpson AJ et al (2004) Oncogenomics and the development of new cancer therapies. Nature 429:469–474

Sugarbaker DJ, Richards WG et al (2008) Transcriptome sequencing of malignant pleural mesothelioma tumors. Proc Natl Acad Sci USA 105:3521–3526

Sun Y, Mi W et al (2008) Quantitative proteomic signature of liver cancer cells: tissue transglutaminase 2 could be a novel protein candidate of human hepatocellular carcinoma. J Proteome Res 7:3847–3859

Thygesen HH, Zwinderman AH (2006) Modeling SAGE data with a truncated gamma-Poisson model. BMC Bioinformatics 7:157

Velculescu VE, Zhang L et al (1995) Serial analysis of gene expression. Science 270:484–487

Venkatesh TV, Harlow HB (2002) Integromics: challenges in data integration. Genome Biol 3:reports 4027.1–4027.3

Watson MA, Fleming TP (1994) Isolation of differentially expressed sequence tags from human breast cancer. Cancer Res 54:4598–4602

Wilson PM, Ladner RD et al (2007) Exploring alternative individualized treatment strategies in colorectal cancer. Clin Colorectal Cancer 7:S28–36

Wolfsberg TG, Wetterstrand KA et al (2002) A user's guide to the human genome. Nat Genet 32:1–79

Wood LD, Parsons DW et al (2007) The genomic landscapes of human breast and colorectal cancers. Science 318:1108–1113

Zhang L, Zhou W et al (1997) Gene expression profiles in normal and cancer cells. Science 276:1268–1272

Zuyderduyn SD (2007) Statistical analysis and significance testing of serial analysis of gene expression data using a Poisson mixture model. BMC Bioinformatics 8:282

Chapter 13
Translational Medicine: Application of Omics for Drug Target Discovery and Validation

Xuewu Zhang, Wei Wang, Kaijun Xiao, and Lei Shi

Abstract Drug research and development is a long, expensive and risky process. The novel omics technologies (genomics, transcriptomics, proteomics and metabonomics) and systems biology have brought unprecedented abilities to screen cells at the gene, transcript, protein, metabolite and their interaction network level in searching of novel drug targets, elucidating the primary mechanism-of-action of a drug, understanding side effects in unanticipated off-target interaction, validating existing drug candidates and finding new potential therapeutic applications for an established drug, hence to facilitate the translation from bench to bedside. This chapter provides an overview of recent applications of various omics technologies and systems biology to drug development.

13.1 Introduction

Translational medicine is to translate basic science achievements into clinical practice, particularly facilitating the transition "from bench to bedside" in the drug discovery and development process. Today in the pharmaceutical industry, higher investments in R&D are providing lower than anticipated returns, the process of discovering new medicines is long (10–15 years), expensive ($0.8–$1.5 billion), and risky (10% success following first dose in humans) (Austin and Babiss 2006). To improve such a process, appropriate biomarkers need to be developed and validated, facilitating the transition of a compound into a drug, i.e. the transition from the test tube and animal experiments to application in humans.

A biomarker is a biologically derived molecule in the body, which is indicative of normal biologic processes, pathogenic processes, or pharmacologic responses to a therapeutic or nutritional intervention. Different functional categories of biomarkers

X. Zhang (✉), W. Wang, K. Xiao, and L. Shi
College of Light Industry and Food Sciences, South China University of Technology, Guangzhou, China
e-mail: snow_dance@sina.com

W.C.S. Cho (ed.), *An Omics Perspective on Cancer Research*,
DOI 10.1007/978-90-481-2675-0_13, © Springer Science+Business Media B.V. 2010

are involved in different phases of drug development. For example, in the phase of drug discovery, disease-related biomarkers are applied to monitor disease causality, progression and susceptibility; in the phase of pre-clinical toxicology studies, the safety and efficacy-related biomarkers are used to evaluate the risk/benefit ratio; in the phase of clinical development, the previously discovered biomarkers need to be qualified and validated for human use. In general, the identification and validation of useful biomarkers will allow researchers to make early go/no go decisions, hence to decrease the cost of late phase testing of an ineffective agent. Moreover, biomarkers can be used as surrogate for treatment approval by regulatory agencies, and to decrease the lag between product development and marketability. This chapter presents an overview of omics technologies applied to drug target discovery and validation.

13.2 Genomics in Drug Target Discovery and Validation

Genomics is to investigate DNA sequences and DNA changes like DNA rearrangements, DNA copy numbers, single nucleotide polymorphism (SNP), DNA methylation, etc. Currently, two technologies are developed to simultaneously profile hundreds of thousand of SNPs in a single assay, one is the randomly ordered bead arrays from Ilumina (http://www.illumina.com/) and 454 Life Sciences (http://www.454.com/), another is the photolithographic-based *in situ* synthesized gene chips from Affymetrix (http://www.affymetrix.com/). Both platforms are suitable for whole-genome genotyping. For DNA methylation analysis, there are three commonly used approaches for high-throughput array-based DNA methylation analysis: (1) MeDIP (methylated DNA immunoprecipitation); (2) HELP (*Hpa* II tiny fragment enrichment by ligation-mediated PCR); (3) fractionation by McrBC, an enzyme that cuts most methylated DNA. Recently, to overcome significant limitations to these methods (bias toward CpG islands in MeDIP, relatively incomplete coverage in HELP, and location imprecision in McrBC), Irizarry et al. (2008) developed a comprehensive high-throughput array for relative methylation (CHARM). Unlike other approaches, CHARM has a substantial advantage that it is highly quantitative. In general, DNA biomarkers can be applied to establish the link among genetic variations, disease, environmental influences and treatments. Specially, DNA methylation and gene polymorphisms like SNP play an important role in drug discovery and development.

Gene polymorphisms can occur at the level of proteins directly involved in drug action and drug metabolizing enzymes or transporters, leading to alterations in drug efficacy and toxicity. The identification of functional polymorphisms in patients undergoing chemotherapy may help the clinician prescribe the optimal drug combination and predict the drug response to these prescriptions with more accuracy. Especially, predicting drug toxicity is very important and should be determined in routine before drug administration, for example, three frequent SNPs, Lys751 Gln in the xeroderma pigmentosum complementation group D (*XPD, ERCC2*) gene, Asp1104His in the xeroderma pigmentosum complementation group G (*XPG, ERCC5*) gene and Ile105Val in the glutathione S-transferase P1 (*GSTP1*) gene, were found to be related to the cytotoxicity of some anticancer drugs (Le Morvan et al. 2006). Recently, Liu et al. (2008) performed a genome-wide association scan for obesity by examining approximately 500,000 SNPs in a sample of 1,000 unrelated US Caucasians. They found that multiple SNPs in a newly identified gene, *CTNNBL1*, were associated with body mass index (BMI) and fat mass, among them the most significant SNP is rs6013029, suggesting a novel mechanism for the development of obesity, hence providing a new drug target for anti-obesity drug development.

DNA methylation of the promoter region of genes and associated gene silencing has been recognized to play a very important role in tumorigenesis. DNA methylation biomarkers can be detected in the presence of large amounts of background DNA with high sensitivity, they have been applied to molecular diagnostic tests for routine clinical use (Lesche and Eckhardt 2007). Specially, it has been suggested that DNA methylation biomarkers can be applied to monitor drug response. For example, the disappearance of *RASSF1* methylation in serum has been observed to correlate with response of breast cancer treatment to tamoxifen (Fiegl et al. 2005), the status of the enzyme o-6-methylguanine-DNA methyltransferase (MGMT) methylation is the best independent predictor of response to BCNU (carmustine) and temozolomide in gliomas (Hegi et al. 2005). Recently, Shen et al. (2007) showed that different promoter hypermethylation profiles can effectively predict the sensitivity of NCI-60 cancer cell lines to a library of 30,000 drugs, this highlights the potential of DNA methylation biomarkers in drug development. In fact, the FDA approved the first inhibitor of DNA methylation, Azacytidine, for the treatment of myelodysplastic syndromes (Issa and Kantarjian 2005), and new drugs targeting DNA methyltransferases have been introduced into clinical trials for the treatment of advanced prostate cancers (Nelson et al. 2007).

13.3 Transcriptomics in Drug Target Discovery and Validation

Transcriptomics is to investigate gene expression patterns based on the relative estimation of messenger RNA (mRNA) copy number under a given condition for a given organism. Currently, the most widely used tool for transcriptomics is DNA microarrays, which allows to measure the expression level of thousands of genes, or even entire genomes, simultaneously. Oligonucleotide microarrays and cDNA microarrays are two major formats of DNA microarrays commercially available. A typical DNA microarray experiment consists of the following steps: (1) DNA sequences are immobilized on a solid surface; (2) mRNA extraction from a sample (cells, tissues, or organs) and labeled with fluorescence dyes; (3) Labeled nuclei acids are then hybridized to the DNA sequences; (4) using an appropriate scanning device (fluorescence scanner or CCD camera) to detect signal; and (5) data analysis by pattern recognition technologies and bioinformatics tools. Gene expression microarray technology is benefiting all phases of drug discovery and development.

For target identification and validation, comparison of gene expression profiles in the normal and disease tissue will enhance the understanding of disease pathology and identify potential therapeutic intervention points. In the context of cancer and drug development, application of gene expression analysis to cellular processes such as cell cycle and signal transduction may identify genes involved in tumorigenesis that may be potential drug targets. Numerous reports have demonstrated the potential power of gene expression profiling in normal and pathological tissues for the identification and validation of biomarkers and new molecular targets (Zhang 2007; Zhang et al. 2007a). In particular, Rhodes et al. (2004) employed large-scale meta-analysis of cancer microarrays and identified some common cancer biomarkers. For example, *TOP2A* is present in 18 cancers *versus* normal signatures, representing 10 types of cancer. Similarly, seven gene pairs were identified as common biomarkers for four types of cancer (colon, melanoma, ovarian and esophageal cancers) (Basil et al. 2006). These biomarkers could be used for cancer diagnosis and could also be considered as potential drug targets.

Furthermore, gene expression microarrays can be used to profile the pharmacological effects of lead compounds on a genome-wide basis, to investigate the molecular mechanism of action of drugs in clinical trials, to identify genes and expression patterns related to toxicity, drug sensitivity or resistance, and to predict which patient is most likely to benefit from which particular drug. For example, Staunton et al. (2001) used the expression profile of the NCI 60 human tumor cell line panel to predict sensitivity or resistance to 232 compounds. This study not only reveals information on factors governing drug resistance/sensitivity but also provides information on the potential target. Zembutsu et al. (2002) employed genome-wide cDNA microarray screening to correlate gene expression profiles with sensitivity of 85 human cancer xenografts to anticancer drugs. Some interesting associations between gene expression and anticancer drug sensitivity were observed, for example, increased topoisomerase II expression and increased doxorubicin resistance, a negative correlation between thymidylate synthetase expression and 5-FU sensitivity, and also a negative correlation between aldehyde dehydrogenase 1 and camptothecin sensitivity. Recently, whole-genome gene expression profiling has become a promising approach for defining responsiveness to treatment. A good example is dedicated to breast cancer. In a number of breast cancer studies whole-genome gene expression have been used to profile breast samples of patients treated with Tamoxifen or Tamoxifen plus chemotherapy before surgery, the resulting profiles are scored to assess the risk of recurrence and the need of adjuvant therapy, and have been associated with chemotherapy response (Paik et al. 2006).

13.4 Proteomics in Drug Target Discovery and Validation

Proteomics involves the large-scale identification and functional characterization of all the expressed proteins in a given cell or tissue, including all protein isoforms and modifications. Typically, the workflow for the proteomics analysis essentially consists of sample preparation, protein separation, and protein identification. The frequently used tools for proteomic investigations include two-dimensional gel electrophoresis (2D-gel) and mass spectrometry (MS). A simple and affordable approach is that proteins are separated with 2D-gel and quantified by direct staining with visible or fluorescent detection dyes (i.e. Coomassie blue or SYPRO Ruby), then the protein spots of interest are excised, digested, and the resulting peptides are identified by MS. However, there are some technical limitations to this method: (1) the narrowed window of proteins separation, subject to their pH and abundances, i.e. inability to detect low-abundant proteins and proteins with extreme properties (very small, very large, very hydrophobic and very acidic or basic proteins); (2) the lack of reproducibility in the separation step (gel-to-gel variations), in the enzymatic digestion, and in the ionization process; (3) the lack of automation, the labor-intensive, time-consuming and costly properties. Instead of the gel approaches, an exciting and powerful tool is MS-based proteomics technology, which has been widely used for biomarker discovery and early diagnosis of human cancers (Zhang et al. 2007b). The mostly used MS instruments for proteomics experiments are electrospray ionization (ESI)-MS, matrix-assisted laser desorption ionization time-of-flight (MALDI-TOF)-MS and its variant surface-enhanced laser desorption/ionization (SELDI)-TOF-MS. The superior power of MS in the proteomic analysis can be further enhanced when MS is combined with a separation technique (e.g. gas chromatography (GC), liquid chromatography (LC) or capillary electrophoresis (CE)) or when tandem MS (MS/MS) is used.

The recent developments in proteomic technologies have brought them with ability to comparatively screen large numbers of proteins within clinically distinct samples. The differential expression of proteins detected in normal *versus* disease samples can be used to characterize which biological pathways are involved, which can later be targeted with drugs. Many interesting results have been obtained by application of MS-based proteomics technologies to biomarker discovery and validation. For example, with SELDI–TOF-MS analysis, Cho et al. (2004) identified two isoforms of serum amyloid A protein as useful biomarkers to monitor relapse of nasopharyngeal carcinoma (NPC), which are correlated with relapse and a sharp fall with response to salvage chemotherapy. Further examination was conducted to identify other cancer targets, inter-α-trypsin inhibitor precursor and platelet factor-4 that were associated with active disease or chemoresponse in NPC patients treated with chemotherapy respectively. These disease- and treatment-associated serum biomarkers might serve for the diagnosis and chemotherapy monitoring of NPC patients (Cho et al. 2007). Zhang et al. (2004) identified and validated three biomarkers by SELDI-TOF-MS analysis of the serum proteome of patients with early-stage ovarian cancer, apolipoprotein A1, a truncated form of transthyretin and a cleavage fragment of inter-α-trypsin inhibitor heavy chain H4. By applying a proteomics technology, Keay et al. (2004) identified *CKAP4* as a receptor of an antiproliferative factor (APF) and a possible druggable target to treat patients suffering the adverse effects of interstitial cystitis. Huang et al. (2006a) used differential gel electrophoresis (DIGE) and MALDI-TOF/TOF-MS to screen biomarker candidates in serum samples obtained from 39 patients with breast cancer and 35 controls.

They revealed proapolipoprotein A-I, transferrin, and hemoglobin as up-regulated and apolipoprotein A-I, apolipoprotein C-III, and haptoglobin alpha 2 as down-regulated in patients, and routine clinical immunochemical reactions were used to validate transferrin as potential biomarker. Holly et al. (2006) analyzed urinary proteins from septic rats with acute renal failure (ARF) by DIGE and MALDI-MS and identified meprin-1-alpha as a potential biomarker and drug target. Dear et al. (2007) employed a comparative proteomics to a clinically relevant mouse model of sepsis, identified a number of novel proteins that changed in abundance and facilitated the discovery of new therapeutic target cyclophilin receptor CD147.

Recognizing specific protein changes in response to drug administration in humans has the potential for significant utility in clinical research. Proteomics can be used to investigate the molecular mechanism of action of drugs in clinical trials and to predict drug response. Lee et al. (2005) used a proteomic analysis including 2D-gel and MALDI-TOF-MS to investigate the anti-cancer mechanism of paclitaxel against cervical carcinoma cells and identified several cellular proteins that are responsive to paclitaxel treatment in HeLa cells. This study demonstrates the power of proteomic profiling with functional analysis using RNAi technology for the discovery of novel molecular targets and a better understanding of the actions of paclitaxel at the molecular level in cervical carcinoma cells. Patil et al. (2007) used a MS-based proteomics technique to measure changes in proteins related to drug administration in the plasma and cerebrospinal fluid (CSF) proteomes of 11 subjects given atomoxetine. They detected statistically significant changes in the CSF protein pattern after drug treatment, suggesting that identification of changes in the CSF proteome associated with the administration of centrally active drugs is feasible, and may be valuable in the development of new drugs.

13.5 Metabonomics in Drug Target Discovery and Validation

Metabonomics is to investigate the fingerprint of biochemical perturbations caused by disease, drugs, and toxins (Goodacre 2007). It is not to be confused with "metabolomics", which is defined as the comprehensive analysis of all metabolites generated in a given biological system, focusing on the measurements of metabolite concentrations and secretions in cells and tissues (Fiehn 2002). Typically, two main analytical platforms are used in metabonomics experiments: nuclear magnetic resonance (NMR) and MS technologies. Both platforms are very complementary, NMR allows the identification and quantification of small polar molecules, MS enables the profiling of larger non-polar molecules. The frequently used MS instruments include MS in combination with some chromatography technologies such as GC, LC, CE and UPLC (a new HPLC systems using sub-2 μm packing columns combined with high operating pressures). Further recent development in MS-based metabonomics is the chip-based nanoelectrospray MS developed at Novartis, which can reduce matrix effects, measurement time (no chromatographic separation needed), and improve sensitivity.

In principle, there are some particular advantages to using metabolites as biomarkers, especially in the context of translational research (Keun and Athersuch 2007). First, the number of metabolites is a few thousand, *versus* tens of thousands of genes and up to hundreds of thousands of proteins. Second, metabolite measurements are usually accessible via non- or minimally-invasive biological fluids like urine, rather than tissue. Third, metabolites remain the same chemical entity irrespective of their origin, while genes and gene products are subject to sequence variation, post-translational modifications, etc. These features make bench-to-bedside translation much easier without increasing patient morbidity or cost. However, another distinguished feature of metabonome from the genome and proteome is the major, direct and sustained input of key exogenous sources such as diet, life styles, smoking, physical activities, and exposure to xenobiotics, etc. This makes metabonomic profiling very sensitive to many environmental factors and brings metabonomics investigations a significant challenge to discriminate between exogenous causes and endogenous effects.

Currently, there are more and more examples of metabonomics technologies being successfully used to drug development process. For example, Chen et al. (2006) combined desorption electrospray ionization mass spectrometry (DESI-MS) and NMR for differential metabonomics on urine samples from diseased (lung cancer) and healthy mice and successfully identified over 80 different metabolites under the condition of no sample preparation. Dieterle et al. (2006) used NMR urinalysis to rank compounds based on toxicity, showing the potential of metabonomics in lead selection and optimization. Al-Saffar et al. (2006) employed NMR technology and xenograft models to examine the effects of MN58b treatment on human colon and breast cancer cell lines. They showed that the choline kinase inhibitor MN58b reduced levels of total choline, phosphocholine and total phosphomonoesters both *in vitro* and *in vivo*, demonstrating the potential value of applying metabonomics to biomarker discovery and the identification of novel targets in a clinical translational environment. Clayton et al. (2006) used a combination of pre-dose metabolite profiling and chemometrics to model and predict the responses of individual subjects with paracetamol (acetaminophen) administered to rats. They showed pre-dose prediction of an aspect of the urinary drug metabolite profile and an association between pre-dose urinary composition and the extent of liver damage sustained after paracetamol administration. This suggests that pretreatment metabolic profiles can be predictive of response to drug exposure. Soga et al. (2006) applied a metabolome differential display method based on CE-TOF-MS to profile liver metabolites following acetaminophen-induced hepatotoxicity. They globally detected 1,859 peaks in mouse liver extracts and specifically found that serum ophthalmate is a sensitive indicator of hepatic GSH depletion, and may be a new biomarker for oxidative stress.

13.6 Systems Biology in Drug Target Discovery and Validation

In general, human diseases are enormously complex and involve simultaneous pathologies of multiple organ systems, hence a comprehensive analysis of such a complex system is highly required to identify molecular biomarkers of different toxic endpoints during drug development process. Systems biology, the simultaneous measurement of genomic, proteomic, and metabonomic parameters in a given system under defined conditions (Davidov et al. 2003), offer exciting opportunities for drug research and development.

There is an increasing trend in systems biology applied in the drug-discovery setting. For example, Ruepp et al. (2002) used microarray expression and proteomics to evaluate acetaminophen toxicity in mice liver. Heijne et al. (2003) employed a combined transcriptomics and proteomics approach to investigate hepatotoxicity induced in rats by bromobenzene administration. The results revealed a modest overlap in results from proteomics and transcriptomics, suggesting that transcriptomics and proteomics technologies are complementary to each other and provide new possibilities in molecular toxicology. Coen et al. (2004) applied transcriptomics and metabonomics to identify biochemical changes arising from hepatotoxicity in mice dosed with acetaminophen. They found that an increased rate of hepatic glycolysis was consistent with the altered levels of gene expression relating to lipid and energy metabolism in liver, showing that these two technology platforms together offer a complementary view into cellular responses to toxic processes. Recently, Li et al. (2006) applied an integrative omics approach (genomics, transcriptomics and proteomics) to identify potential biomarkers for the diagnosis and therapy of lung cancer. First, 183 up-regulated genes were identified by genomic and transcriptomic methods, Second, 42 over-expressed proteins were identified by 2D-gel and MS, then four genes (*PRDX1, EEF1A2, CALR and KCIP-1*) were found to be correlated with elevated protein expression by the comparison between the 183 genes and 42 proteins. The further validation experiments by Southern, Northern, and Western blotting demonstrated that the amplification of *EEF1A2* and *KCIP-1* was associated with elevated protein expression, strongly suggesting that the two genes could be potential biomarkers for the diagnosis and therapy of lung cancer. Craig et al. (2006) utilized an integrated systems approach to understanding the toxic mechanisms of the histamine antagonist methapyrilene administered in rats by combining proteomics, metabonomics by H-1 NMR spectroscopy and genomics by microarray gene expression profiling. They found that the changes occurred in signal transduction and metabolic pathways during methapyrilene hepatotoxicity are reflected in both gene/protein expression patterns and metabolites. Based on these data, the authors suggest a new cytochrome P450 target for the specific periportal activation of methapyrilene.

However, it is noted that it is not easy to establish a direct link between genes and/or proteins and metabolites: multiple mRNAs could be formed from one gene; multiple proteins from one mRNA; multiple metabolites can be formed from one enzyme; and the same metabolite can participate in many different pathways. This complicates the interpretation and integration of the various omics data. In fact, some studies have now shown that there is a poor correlation between the amount of a protein and its transcript's abundance. For example, Conradas et al. (2005) conducted a combined proteome and microarray investigation of inorganic phosphate-induced pre-osteoblast cells. Comparison of the mRNA microarray data with the 24-h quantitative proteomic data resulted in a generally weak overall correlation. Post-transcriptional processing events, temporal differences in mRNA and protein expression, or other factors may be the potential reasons for this lack of correlation.

13.7 Emerging Applications in Clinical Practice and Perspectives

Over the past few years, the rapid development of whole genome profiling technologies has greatly expanded our knowledge of cancer development and progression. Currently, numerous commercialized multigene assays have entered the expanding landscape of cancer diagnostics. In particular, several multigene assays are now available for breast cancer (Ross 2008), for example, ProExBr (www.bd.com/tripath/labs/mo_oncology.asp), Mammostrat (www.appplied-genomics.com), eXagenBC (www.exagen.com), oncotype DX (www.genomichealth.com/oncotype/default.aspx), Invasiveness gene signature (www.oncomed.com), MammaPrint (www.agendia.com), Two-gene ratio H/I (www.aviaradx.com), Rotterdam signature (www.veridex.com), Cytochrome P450CYP2D6 (www.roche.com/med_backgr-ampli.htm), etc. Among them, the two tests that have achieved the most advanced commercial success are oncotype DX and MammaPrint, each test has some advantages and disadvantages. The 21-gene oncotype DX use widely available starting material like formalin-fixed paraffin-embedded (FFPE), while the 70-gene MammaPrint cannot currently be performed on FFPE tissues and requires either fresh-frozen tumor samples or tissues collected into an RNA preservative solution. The MammaPrint test has a wider patient eligibility than oncotype DX by including ER-positive and ER-negative and younger patients. The oncotype DX can be served as both a prognostic test and predictive test for certain hormonal and chemotherapeutic agents, and the MammaPrint is validated as a prognostic test only and has not been formally tested as a predictive test for specific therapy regimes. In terms of US FDA status, the MammaPrint assay has received 510(k) clearance, whereas oncotype DX has been exempt.

With the further development of omics technologies, more and more biochip-based assays (genes chips, protein chips or metabolite chips) will be moved from bench to bedside for applications of diagnosis, prognosis and response to therapy. In particular, nanotechnology raises new possibilities in diagnosis and treatment of human cancers. Lee et al. (2007) employed the NanoChip Molecular Biology Workstation to validate a CYP2D6 genotyping assay, CYP2D6 is a highly polymorphic phase I enzyme that metabolizes 20–25% of clinically used drugs. Corradi et al. (2008) combined a multiplex RT-PCR approach with the electronic hybridization and fluorescent detection on the Nanogen NanoChip Molecular Biology Workstation to screen for the most common fusion gene transcripts in human leukemia, providing a multi-purpose platform for relevant comprehensive diagnostics of hemato-oncology patients.

However, regardless of the final assay platform, all these tests are to use profiling technology for the discovery of the test's gene, mRNA, protein or metabolite biomarkers. One critical issue is that these biomarkers must be fully validated on large and independent patient samples, it remains to be seen whether these new assays will hold up over time as more patients are tested. On the other hand, it should be noted that these new tests can easily be misused, including applying the test in the wrong clinical setting and ending up with misleading reassurance about test-driven decisions.

Finally, it must be pointed out that the future direction on cancer drug discovery and therapy should be pathway-oriented. Very recently, three landmark genome scans of cancer have affirmed the complexity of genetic changes in solid tumors, they all share a core group of perturbed pathways, although each tumor in each patient is different, it appears that cancer is really a pathway disease (Chin et al. 2008; Jones et al. 2008; Parsons et al. 2008). These new results point towards a future where we should change the way in thinking about cancer from gene-centric to pathway-centric. It may be a good choice that the drug development aims at entire signaling pathways rather than just one mutation at a time, and the determination of therapy regimes by looking at genetic changes in patients, as part of a personalized medicine, could be changed to look directly at the core pathway. To achieve these goals, systems biology offers exciting opportunity. For example, Nikolsky et al. (2005) used network analysis tool to compare the effects of different drug treatment (4-hydroxytamoxifen and estrogen) on the MCF-7 breast cancer cell line, they identified topological motifs in expression network and provided a novel type of biomarker for drug exposure. Huang et al. (2006b) applied the pathway-mapping approach to the quantitative evaluation of anticancer drug action, they proposed that the way to improve the efficacy of anticancer drugs is to screen for the compounds that selectively target the pathways which exhibit reduced coherence in cancer.

13.8 Conclusions

Novel omics technologies are showing the potential value in the drug development process (Fig. 13.1): target identification and validation; lead discovery and optimization; preclinical efficacy and safety assessment; mode of action and mechanism;

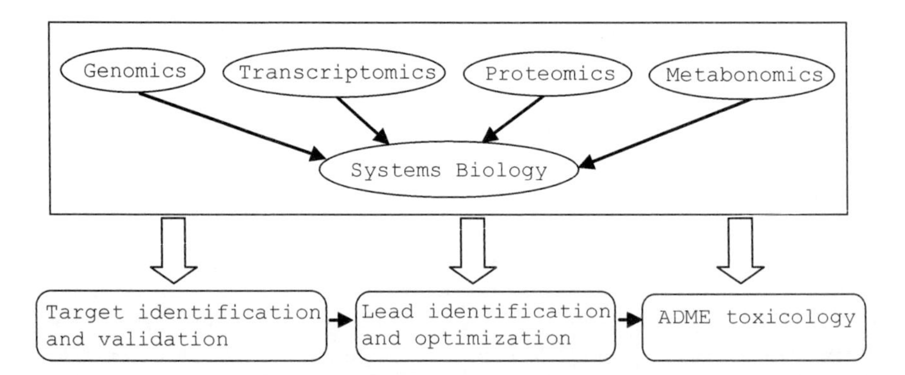

Fig. 13.1 Flow scheme of the contributions of omics technologies (genomics, transcriptomics, proteomics and metabonomics) and systems biology to the drug discovery processes. ADME: absorption, distribution, metabolism and excretion

patient stratification; and clinical pharmacological monitoring. In particular, systems biology, more than the simple merge of various omics technologies (genomics, transcriptomics, proteomics, and metabonomics), will greatly facilitate the understanding of the complex interaction network between drugs and molecules in biological systems and accelerate the discovery and validation of drug target as well as the translation from bench to bedside, hence to achieve the goal of personalized medicine and healthcare, although many significant challenges still exist ahead.

Acknowledgements This work was partly supported by Guangdong Scientific Development Grant 2006B13001003 and 2007A020100001-4.

References

Al-Saffar NM, Troy H, de Molina AR et al (2006) Noninvasive magnetic resonance spectroscopic pharmacodynamic markers of the choline kinase inhibitor MN58b in human carcinoma models. Cancer Res 66:427–434

Austin MJF, Babiss L (2006) Commentary: where and how could biomarkers be used in 2016. AAPS J 8:E185–189

Basil CF, Zhao YD, Zavaglia K et al (2006) Common cancer biomarkers. Cancer Res 66: 2953–2961

Chen H, Pan Z, Talaty N et al (2006) Combining desorption electrospray ionization mass spectrometry and nuclear magnetic resonance for differential metabolomics without sample preparation. Rapid Commun Mass Spectrom 20:1577–1584

Chin L, Meyerson M, Aldape K et al (2008) Comprehensive genomic characterization defines human glioblastoma genes and core pathways. Nature 455:1061–1068

Cho WC, Yip TT, Yip C et al (2004) Identification of serum amyloid a protein as a potentially useful biomarker to monitor relapse of nasopharyngeal cancer by serum proteomic profiling. Clin Cancer Res 10:43–52

Cho WC, Yip TT, Ngan RK et al (2007) ProteinChip array profiling for identification of disease- and chemotherapy-associated biomarkers of nasopharyngeal carcinoma. Clin Chem 53:241–250

Clayton TA, Lindon JC, Cloarec O et al (2006) Pharmaco-metabonomic phenotyping and personalized drug treatment. Nature 440:1073–1077

Coen M, Ruepp SU, Lindon JC et al (2004) Integrated application of transcriptomics and metabonomics yields new insight into the toxicity due to paracetamol in the mouse. J Pharm Biomed Anal 35:93–105

Conradas KA, Yi M, Simpson KA et al (2005) A combined proteome and microarray investigation of inorganic phosphate-induced pre-osteoblast cells. Mol Cell Proteomics 4:1284–1296

Corradi B, Fazio G, Palmi C et al (2008) Efficient detection of leukemia-related fusion transcripts by multiplex PCR applied on a microelectronic platform. Leukemia 22:294–302

Craig A, Sidaway J, Holmes E et al (2006) Systems toxicology: Integrated genomic, proteomic and metabonomic analysis of methapyrilene induced hepatotoxicity in the rat. J Proteome Res 5:1586–1601

Davidov EJ, Holland JM, Marple EW, Naylor S (2003) Advancing drug discovery through systems biology. Drug Discov Today 8:175–183

Dear JW, Leelahavanichkul A, Aponte A et al (2007) Liver proteomics for therapeutic drug discovery: inhibition of the cyclophilin receptor CD147 attenuates sepsis-induced acute renal failure. Crit Care Med 35:2319–2328

Dieterle F, Schlotterbeck GT, Ross A et al (2006) Application of metabonomics in a compound ranking study in early drug development revealing drug-induced excretion of choline into urine. Chem Res Toxicol 19:1175–1181

Fiegl H, Millinger S, Mueller-Holzner E et al (2005) Circulating tumor-specific DNA: a marker for monitoring efficacy of adjuvant therapy in cancer patients. Cancer Res 65:1141–1145

Fiehn O (2002) Metabolomics – the link between genotypes and phenotypes. Plant Mol Biol 48:155–171

Goodacre R (2007) Metabolomics of a superorganism. J Nutrit 137:S259–266

Hegi ME, Diserens AC, Gorlia T et al (2005) MGMT gene silencing and benefit from temozolomide in glioblastoma. N Engl J Med 352:997–1003

Heijne WH, Stierum RH, Slijper M et al (2003) Toxicogenomics of bromobenzene hepatotoxicity: a combined transcriptomics and proteomics approach. Biochem Pharmacol 65:857–875

Holly MK, Dear JW, Hu X et al (2006) Biomarker and drug–target discovery using proteomics in a new rat model of sepsis-induced acute renal failure. Kidney Int 70:496–506

Huang HL, Stasyk T, Morandell S et al (2006a) Biomarker discovery in breast cancer serum using 2-D differential gel electrophoresis/MALDI-TOF/TOF and data validation by routine clinical assays. Electrophoresis 27:1641–1650

Huang RL, Wallqvist A, Covell DG (2006b) Targeting changes in cancer: assessing pathway stability by comparing pathway gene expression coherence levels in tumor and normal tissues. Mol Cancer Ther 5:2417–2427

Irizarry RA, Ladd-Acosta C, Carvalho B et al (2008) Comprehensive high-throughput arrays for relative methylation (CHARM). Genome Res 18:780–790

Issa JP, Kantarjian H (2005) Azacitidine. Nat Rev Drug Discov:S6-7

Jones S, Zhang XS, Parsons DW et al (2008) Core signaling pathways in human pancreatic cancers revealed by global genomic analyses. Science 321:1801–1806

Keay SK, Szekely Z, Conrads TP et al (2004) An antiproliferative factor from interstitial cystitis patients is a frizzled 8 protein-related sialoglycopeptide. Proc Natl Acad Sci USA 101:11803–11808

Keun HC, Athersuch TJ (2007) Application of metabonomics in drug development. Pharmacogenomics 8:731–741

Le Morvan V, Bellott R, Moisan F et al (2006) Relationships between genetic polymorphisms and anticancer drug cytotoxicity vis-a-vis the NCI-60 panel. Pharmacogenomics 7:843–852

Lee KH, Yim EK, Kim CJ et al (2005) Proteomic analysis of anti-cancer effects by paclitaxel treatment in cervical cancer cells. Gynecol Oncol 98:45–53

Lee HK, Lewis LD, Tsongalis GJ et al (2007) Validation of a CYP2D6 genotyping panel on the NanoChip molecular biology workstation. Clin Chem 53:823–828

Lesche R, Eckhardt F (2007) DNA methylation markers: A versatile diagnostic tool for routine clinical use. Curr Opin Mol Ther 9:222–230

Li R, Wang H, Bekele BN et al (2006) Identification of putative oncogenes in lung adenocarcinoma by a comprehensive functional genomic approach. Oncogene 25:2628–2635

Liu YJ, Liu XG, Wang L et al (2008) Genome-wide association scanc identified CTNNBL1 as a novel gene for obesity. Human Mol Genetics 17:1803–1813

Nelson WG, Yegnasubramanian S, Agoston AT et al (2007) Abnormal DNA methylation, epigenetics, and prostate cancer. Front Biosci 12:4254–4266

Nikolsky Y, Ekins S, Nikolskaya T et al (2005) A novel method for generation of signature networks as biomarkers from complex high throughput data. Toxicol Lett 158:20–29

Paik S, Tang G, Shak S et al (2006) Gene expression and benefit of chemotherapy in women with nodenegative, estrogen receptor-positive breast cancer. J Clin Oncol 24:1–10

Parsons DW, Jones S, Zhang XS et al (2008) An integrated genomic analysis of human glioblastoma Multiforme. Science 321:1807–1812

Patil ST, Higgs RE, Brandt JE et al (2007) Identifying pharmacodynamic protein markers of centrally active drugs in humans: a pilot study in a novel clinical model. J Proteome Res 3:955–966

Rhodes DR, Yu JJ, Shanker K et al (2004) Large-scale meta-analysis of cancer microarray data identifies common transcriptional profiles of neoplastic transformation and progression. Proc Natl Acad Sci USA 101:9309–9314
Ross JS (2008) Multigene predictors in early-stage breast cancer: moving in or moving out? Expert Rev Mol Diagn 8:129–135
Ruepp SU, Tonge RP, Shaw J et al (2002) Genomics and proteomics analysis of acetaminophen toxicity in mouse liver. Toxicol Sci 65:135–150
Shen L, Kondo Y, Ahmed S et al (2007) Drug sensitivity prediction by CpG island methylation profile in the NCI-60 cancer cell line panel. Cancer Res 67:11335–11343
Soga T, Baran R, Suematsu M et al (2006) Differential metabolomics reveals ophthalmic acid as an oxidative stress biomarker indicating hepatic glutathione consumption. J Biol Chem 281:16768–16776
Staunton JE, Slonim DK, Coller HA et al (2001) Chemosensitivity prediction by transcriptional profiling. Proc Natl Acad Sci USA 98:10787–10792
Zembutsu H, Ohnishi Y, Tsunoda T et al (2002) Genome-wide cDNA microarray screening to correlate gene expression profiles with sensitivity of 85 human cancer xenografts to anticancer drugs. Cancer Res 62:518–527
Zhang XW (2007) Biomarker validation: movement towards personalized medicine. Expert Rev Mol Diagn 7:469–471
Zhang Z, Bast RC, Yu YH et al (2004) Three biomarkers identified from serum proteomic analysis for the detection of early stage ovarian cancer. Cancer Res 64:5882–5890
Zhang XW, Li L, Wei D et al (2007a) Moving cancer diagnostics from bench to bedside. Trends Biotechnol 25:166–173
Zhang XW, Wei D, Yap YL et al (2007b) Mass spectrometry-based 'omics' technologies in cancer diagnostics. Mass Spectrom Rev 26:403–431

Chapter 14
Integration of Omics Data for Cancer Research

Luis Martín, Alberto Anguita, Víctor Maojo, and José Crespo

Abstract The development of high-throughput techniques for analyzing cell components has provided vast amounts of data in recent years. This development of gene-sequencing methods was followed by advances in techniques for analyzing and managing data from transcriptomes, proteomes, and other omics data. The so-called omics revolution has led to the development of numerous databases describing specific cell components. A recent study suggests that cell behavior cannot be modeled by analyzing its constituents separately, but rather calls for an integrative approach (Barabási and Oltvai 2004). Thus, specialized techniques are being developed to integrate omics information. To enable new research avenues that can take advantage of and apply this information to new therapies – e.g. in cancer research – methods must be designed that provide a seamless integration of these new databases with classical clinical data.

The problem of database integration has been studied at length over the last 15 years, with special emphasis in the post-genomic era on publicly available online data. In the field of genomic medicine, the integration of phenotype and genotype data is of special interest for the prevention of patient intolerance to specific drugs and for defining personalized therapies. Patients' individual characteristics will play a fundamental role in future treatment design. These characteristics include, of course, genetic profiles.

To address the issues surrounding omics data integration, this chapter is organized as follows. Section 14.1 describes the role of data integration in cancer research. Section 14.2 analyzes omics data integration problems and techniques. Section 14.3 introduces a range of international efforts in database integration. Finally, Section 14.4 presents future trends in omics data integration for cancer research.

L. Martín (✉), A. Anguita, V. Maojo, and J. Crespo
Biomedical Informatics Group, Universidad Politécnica de Madrid, Madrid, Spain
e-mail: lmartin@infomed.dia.fi.upm.es

W.C.S. Cho (ed.), *An Omics Perspective on Cancer Research*,
DOI 10.1007/978-90-481-2675-0_14, © Springer Science+Business Media B.V. 2010

14.1 The Role of Data Integration in Cancer Research

The joint analysis of phenotypic and omics data can drastically improve the ability to diagnose, identify, characterize and finally treat cancer. These different types of data are normally stored in disparate heterogeneous databases. Thus, accessing, retrieving, interpreting and evaluating these data are time-consuming and, often, tricky processes within clinical environments. There is a need to present this heterogeneous information in a readable, easy-to-understand form (Collins and McKusick 2001). However, attempts at integrating these data come up against all sorts of problems. Most of these difficulties are related to semantic and syntactic heterogeneities, security, and accessing disparate and remote locations. These difficulties can be overcome by using technologies to bridge different types of heterogeneity and data federation.

The applications of heterogeneous data integration in medicine are manifold. Some of the most interesting uses for this chapter are related to genomic medicine. This area covers different uses of genotypic information to enhance diagnosis and treatments. The use of integrated genetic and phenotypic information will lead to enhanced diagnosis and treatment (Maojo and Tsiknakis 2007) – i.e. personalized application of a cancer treatment such as chemotherapy based on the patient's genetic profile could avoid cases of intolerance. If results from tests such as biopsies were subjected to analysis at a molecular level, and these results were stored and made accessible to other organizations for seamless integration with other kinds of data, individualized molecular profiles could be built to improve the performance of a therapy in a particular case (Nature 2004).

There are a range of possibilities for integrating data for cancer treatment. Cross-institution data integration refers mainly to the exchange of similar data with the aim of completing a data set – i.e. laboratory tests taken at different times and places. Even though these data are of the same type, issues like how they were gathered and stored, together with the privacy and proprietary technologies used, must be addressed. Another case of integration is classical heterogeneous data integration. This type of integration involves different types of data sources – e.g. relational phenotypic data sources, images or genetic data. Another case is biological multi-level data integration. This involves integrating data of different levels of granularity – i.e. from genetic or molecular data to macroscopic images of a tumor, for example. A particular type of multi-level data integration is omics data integration. The next section focuses on this issue in more detail.

14.1.1 Types of Omics Data

Omics is a term used informally to refer to any type of biological data gathered using the high-throughput techniques, like genome-sequencing, developed in recent years. Omics data can be divided into three main categories, depending on what type of cellular network description they provide (Joyce and Palsson 2006). These are (a) component data, which describes a specific part of a cell, (b) interaction data, which stores the relationship between molecular elements inside a cell, and (c) functional-states, which specify the global behavior of cells. For each of these categories, there exist specialized fields of research. The most important of these fields are briefly described below.

The component data field is divided as follows:

a. Genomics, which refers to the complete sequencing of an organism's genomes. Numerous genomics efforts have been carried out over the last few years on different species (like the Human Genome Project resulting in the 2001 publication of a draft sequence of the human genome).
b. Transcriptomics, which analyzes transcriptome data – the set of all mRNA molecules inside a cell or group of cells. The goal is to study gene expressions in order to uncover active cell components. This information can help to provide an understanding of many biological processes, such as carcinogenesis, by studying which genes are active in which situations.
c. Proteomics, which deals with the study of proteins. The ultimate goal is to assess the cellular levels of each protein encoded in the genome.
d. Metabolomics, which studies the cell metabolome – i.e. the complete set of metabolites. Metabolites are "the end products of cellular regulatory processes" (Fiehn 2002), and studying metabolite levels can give insight into the responses of biological systems to genetic or environmental changes. The related field of metabonomics researches the metabolic response to external perturbations.
e. Glycomics, which attempts to study the glycome – the set of all glycan molecules. Glycan molecules can be found acting as an interface between cells to coordinate biological processes (Iozzo 2001). Their study can hopefully lead to the development of new drugs (Shriver et al. 2004).
f. Lipidomics, which studies lipids and interacting factors (Wenk 2005). Numerous studies have demonstrated the relationship between lipid metabolic enzyme disruptions and diverse diseases, including cancer.
g. Localizomics, which aims to discover the sub-cellular locations of proteins. This information can provide important insight into the specific role of individual proteins.
h. Epigenomics, dealing with the study of heritable changes not coded in the DNA sequence (epigenetics) on a large-scale.
i. Immunomics, which studies immune networks and pathways, making use of transcriptomics, genomics and proteomics.
j. Cytomics, which refers to the study of the molecular architecture of cell-systems. Different bioinformatics techniques are employed to this end – e.g. confocal and laser scanning microscopy.

There are two separate specialized fields of research for interactions data (also called interactomics):

a. Protein-DNA interactome refers to interactions between proteins and DNA. These interactions are the basis of the cell's genetic regulatory network.

b. Protein-protein interactions, which are crucial for numerous cellular processes. Their analysis is essential for correctly understanding biology as an integrated system (Cusick et al. 2008).

Finally, functional-states data are classified as:

a. Fluxomics, which aims to analyze the metabolic fluxes that occur within a cell, for example in metabolic or signaling pathways (Wiechert et al. 2007).
b. Phenomics, which refers to the analysis of phenotypes – i.e. the observable expression of the genotype in a given environment.

Figure 14.1 summarizes the description of fields in each of the three categories.

14.1.2 Need for Integration of Omics Data

Describing a complex system through its inner components usually provides an easy-to-understand representation of the global entity. However, important properties can be missed by analyzing the elements separately. In general systems theory this phenomenon is known as emergence. Hurricanes, which are formed due to a positive feedback between wind, humidity, water evaporation and the Coriolis force, are a standard example of emergent entities (Abbott 2004). Biological organisms are not a case apart. As Burgun and Bodenreider (2008) pointed out, the evolution of biomedical research from traditional clinical and biological practices towards the omics

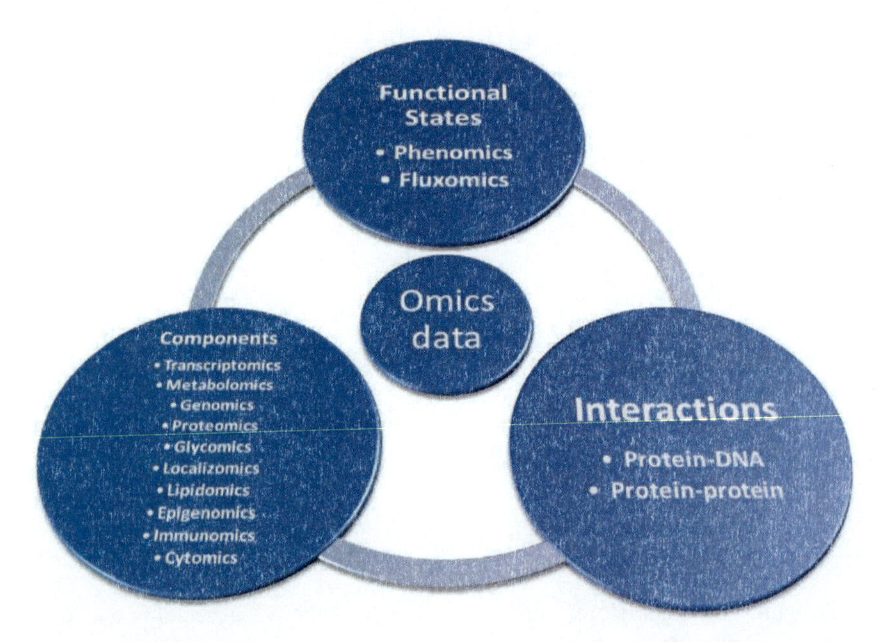

Fig. 14.1 The three types of omics data

era calls for the integration of disparate data sources, bridging phenotypic data with omics knowledge to model the molecular basis of diseases, and developing support systems for analyzing research data. The various types of omics data above describe different elements of biological organisms – such as cells. But to be able to understand the inner processes taking place inside such organisms, an integrated approach is required. There are many examples where omics data integration provides new knowledge that can result in new research paths or lead to the discovery of new treatments. The integration of transcriptomics and metabolomics is driving studies aiming to assess the impact of genetic modifications on crops. Kristensen et al. inserted high-flux pathways into *Arabidopsis thaliana*, modifying the inner mechanisms for synthesis and storage of high quantities of cyanogenic glucoside (Kristensen et al. 2005). The study concluded that such modifications could be performed without noticeably affecting the transcriptome and metabolome. Hirai et al. (2004) studied the impact of nutrient starvation in *A. thaliana*. To do this, he conducted a series of analyses on transcriptome and metabolome data, including Fourier transform-ion cyclotron MS experiments. The results showed the presence of general responses to sulfur and nitrogen deficits. The integration of metabolomics and transcriptomics is used by the same author to elucidate gene-to-gene and metabolite-to-gene networks, providing strategies for discovering new gene functions. These strategies included batch learning and self-organizing mappings of integrated metabolomics and transcriptomics data (Hirai et al. 2005). Coen et al. (2004) integrated metabonomics and transcriptomics to study cellular responses to toxic processes. The study involved analyzing the liver tissue of mice injected with different doses of paracetamol. Extraction and integration of the data confirmed the relationship between paracetamol toxicity and global energy metabolism failure. Other toxicology-related works (Heijne et al. 2003) employed a combination of transcriptomics and proteomics to study the cellular mechanisms of toxicity. This study revealed that transcriptomics and proteomics are complementary and opened up new opportunities for knowledge discovery in molecular toxicology.

Evolution studies benefit from the integration of omics data as well. Genomics, transcriptomics and proteomics have been utilized in different studies to evaluate how variations in gene and protein patterns in the brain relate to differences in humans and chimpanzees (Enard et al. 2002; Khaitovich et al. 2004, 2005). In this work, microarrays and human protein expression patterns were studied, showing a great many changes in gene expression, mainly related to the human brain, between closely related mammals. Genomics and transcriptomics integration has launched other studies to assess the effect of changes in transcriptional regulatory networks on yeast evolution (Ihmels et al. 2005). The same type of integration served to study the evolutionary dynamics of transcriptional regulatory networks in different yeast species (Tanay et al. 2005). These studies prove that the annotation of uncharacterized data with well-known concepts enables the integration of genomic and phenotypic data, facilitating the reconstruction of regulatory network evolution. Clinical studies have also taken advantage of the integration of diverse omics data.

Cancer research is already benefiting from this strategy. One example is the identification of biomarkers associated with diseases. Petrik et al. (2006) investigated methods for using genomics, proteomics and metabolomics data from brain tumor tissues to identify biomarker signatures. Results indicate better diagnosis accuracy with biomarker signatures compared to individual biomarkers. Ippolito et al. (2005) presented a study of feature identification related to poor human neuroendocrine cancer outcomes based on integrated genomics and metabolomics. The results of this study show different features of poor prognosis in this type of cancer. Other studies, like the one presented by Sohal et al. (2008), demonstrate the viability of integrating genomic data collected at different laboratories and public databases, such as the NCBI's Gene Expression Omnibus (GEO). In this work, they were able to circumvent experimental variability and discovered common gene expression signature characteristics of cells involved in leukemia processes.

The inherent complexity of biological systems makes the task of integrating omics a far from simple one. The Munich Systems Biology Forum recognizes two different dimensions of omics integration (MSBF). One carries out integration across levels of structure and scale, starting from components and ending with the large-scale organization of the system. The other integrates data across process phases, linking the insights from different omics. Combining these two approaches would uncover the spatio-temporal characteristics of biological systems. This, however, raises a series of challenges: (a) the discovery of new methods and techniques for extracting accurate data from biological tissue, (b) the development of new modeling and computational approaches that can handle extremely large amounts of data, and (c) establishment of interdisciplinary collaborations among experimentalists and modelers. By advancing in these fields, biologists hope to achieve a better understanding of the global behavior of biological processes and, eventually, to apply this new knowledge to new treatments for common diseases, like cancer.

14.2 The Problem of Data Integration in Cancer Research

The problem of heterogeneous data integration has been the object of a plethora of studies over the last 20 years. The main goal of database integration is to offer end users seamless, homogeneous access to a set of data combined in a single view (Lenzerini 2002). The problems to be overcome to integrate heterogeneous data are varied. They can be categorized as follows: (1) technological issues, (2) instance level heterogeneities, (3) schema level heterogeneities, and (4) legal, ethical issues. Technological issues arise when different software platforms, database management systems, languages or interfaces are used to access different information repositories. Instance level heterogeneities are related to the heterogeneity present at the data level, while schema level heterogeneity refers mainly to the structure of and the relations among the data. In an environment, such as genomic medicine, where privacy is important, legal and ethical issues in relation to the usage of sensitive data should also be taken into consideration. To solve this type of problems, proper middleware software layers should be built, providing homogenous access to the upper layers dealing with other kinds of heterogeneities. These middleware software components are called *wrappers*. Wrapping services leverage communication when different interfaces are used, but they often include syntactic heterogeneities processing at the schema level. This is the case of the Data Access Services in the ACGT platform (Martín et al. 2007).

14.2.1 Database Integration Approaches

A number of software components normally deal with instance and schema level heterogeneities. The behavior and architecture of the components depend on the selected database integration approach. Database integration approaches can be divided into three categories: (1) information linkage, (2) data translation and (3) query translation approaches (Sujanski 2001). Information linkage is a simple method used to integrate public sources available on the Internet using cross references. This requires *ad hoc* built software working with specific data repositories. Some examples of online public sources whose information contents can be integrated using these techniques are MEDLINE and OMIM. By contrast, data translation approaches are based on the actual storage of data in a centralized repository, usually with its own schema. Data from the actual data sources are translated and homogenized in order to fit the requirements of this repository, which offers unified access to final users. The most representative example of an integrated data translation repository is a data warehouse (Kimball 1996). On the other hand, query translation approaches do not make any previous transformation or translation of the actual data. In this case, mediation software is responsible for translating the queries and retrieving partial results from the actual databases.

14.2.1.1 Centralized Approaches

Centralized database integration approaches rely on the transformation and storage of data in a centralized repository. This implies the creation and maintenance of a new database, with its own database management system and interfaces. The most representative example of this type of approach is a data warehouse. A data warehouse has an architecture comprising several layers. Data in one layer are derived from data in the lower layer (Jarke et al. 1998). The bottom layer of a data warehouse contains the *operational databases*. Operational databases are usually heterogeneous data repositories that were not built with the intention of being integrated. The core layer is the so-called *global data warehouse*. Integrated data from the underlying databases are stored at this level. This integrated repository is updated periodically, maintaining a historical record of past states. On top of the architecture are situated the *local warehouses*, containing data specifically aggregated for different purposes, such as decision making or historical analysis. Clearly, centralized approaches require a huge amount of space allocation to store both the global and the local warehouses. Another drawback is that the data that the warehouses contain may not be completely up to date – i.e. updating operational databases will not necessarily update warehouses. In contrast, queries are answered quickly, and the use of highly aggregated data enables complex analysis and faster decision making.

14.2.1.2 Query Translation Approaches

Query translation approaches are not based on an *a priori* transformation of data. The sources of information are in fact an active part of the system, and are accessed each time a user launches a query. This vision contrasts with the idea of transforming and homogenizing data for the construction of a centralized integrated repository. In query translation, data are transformed and homogenized, but this happens every time a query is made. This type of approach has one main advantage: retrieved information is always up to date. In addition, there is no need with query translation to allocate space for the centralized repository. The functioning is more control based than data driven – i.e. there exists mediation software responsible for translating the queries and homogenizing data. However, there are some drawbacks associated with this type of approach. First of all, the single view displayed to the users – i.e. the so-called *virtual schema* – is not real. This means that this is not the real database schema, but a view built for the purpose of giving users the feeling that they are accessing a single database. Depending on how elements in this schema link to items in the actual databases, the translation of the query and the data may not be straightforward.

There are two main ways of building a query translation system: (1) global as view (Cali et al. 2001), and (2) local as view (Levy et al. 1996). In global as view, the global schema is built using predefined views of the real databases. In this type of approach, the links between the virtual schema and the data sources are specified early on. This way, query translation is direct and there are no problems of consistency with the results. However, global as view-based systems are not very adaptable to a dynamic environment – i.e. if any kind of change occurs in the original data sources, the whole model needs to be changed. Conversely, single descriptions of the individual sources are built in local as view, usually using elements from a common homogenization framework – e.g. an ontology.

14.2.1.3 Levels of Heterogeneity: Instance *versus* Schema

It is worth establishing a clear difference between two types of common semantic heterogeneities, namely the instance and schema levels. Instance heterogeneities refer mainly to the problems encountered when integrating the actual data contained in the sources. Some examples of instance level heterogeneities are synonymy – e.g. different identifiers used to refer to the same protein – homonymy – e.g. same identifier with different meanings – scale or granularity. Each of these types of heterogeneity requires a set of transformations that need to be designed and implemented *ad hoc*. Instance level heterogeneities are also hard to detect, since each type of inconsistency has its own peculiarities. Consequently, a method designed to detect the heterogeneity for one cannot usually be used for the others. There are not many relevant works, like Ontodataclean (Pérez-Rey et al. 2006), studying how to leverage this problem. Solving heterogeneities at this level is the biggest challenge in modern database integration.

Conversely, schema level heterogeneities are semantic problems present in data relations and database structures. A database normally has a defined structure of relations between its stored data. The description of this structure is called the database *schema*. When integrating heterogeneous sources of information, the database integration system has to deal with heterogeneities in the way these schemas are defined in different databases. A popular way to overcome this kind of problems is to use a common framework to homogeneously describe the data domain and relations. Database schemas are transformed to match the common framework specifications. The most accepted type of framework used to address this issue is ontology (Gruber 1993). Some examples of works where ontologies have been used to integrate databases are Ontofusion (Pérez-Rey et al. 2005), ACGT (Tsiknakis et al. 2006) and TAMBIS (Baker et al. 1998).

14.2.1.4 Public Database Integration

There are a great many publicly available databases containing different kinds of biological data (Galperin 2008). These sources contain a lot of heterogeneous information regarding genes, proteins and diseases, etc. Each database is usually populated and maintained by the institution that created it, and access is normally offered via Internet. Such web interfaces are based on text queries – i.e. like popular search engines such as Google or Yahoo. Such queries can include natural language information or specific technical descriptors. Others require HTML forms converted to XML documents. The main issue when trying to integrate public data sources is that they are unstructured. There is no definite schema giving a view of how data are related in the database. In actual fact, data are not related at all in many cases and are simply stored in the bodies of text documents.

The most common way to integrate public data sources is to use wrappers to carry out the information retrieval process. Each web-based interface needs to have its own wrapper designed to extract information matching the interface requirements of the service. These wrappers solve the syntactic problems, and offer a homogeneous interface to the upper layers of the database integration system. The main problem with this type of wrappers is that they need to be built completely *ad hoc* for each database. If a change takes place in the web service interface, the wrapper should be re-designed and re-implemented. It is even harder to integrate this kind of non-structured sources with classical relational databases such as patients' hospital records. The Ontofusion system is an example of a solution for this type of integration (Alonso-Calvo et al. 2007).

14.2.2 Techniques for Integrating Omics Data

Given the acknowledged importance of integrated omics for research on biological systems, a big effort has gone into advancing this topic lately. Research has focused on addressing the challenges presented in Section 1.2. Numerous algorithms and a lot of software have been developed to enable researchers to perform more effective analyses and create more accurate models. In regard to interdisciplinary collaborations, several standards have been drafted. This has facilitated research by enabling data sharing across different work groups. Probably one of the most fruitful developments is the Systems Biology Markup Language (SBML), a machine-readable format that is useful for representing models of biochemical reaction networks. SBML was created as a common language enabling information sharing among developers, modelers and researchers. A great many applications have been developed in conformity with this standard, making SBML the de facto language for biological systems modeling. Another widespread language for modeling biological systems is CellML (CellML). This is similar to SBML, although it has a broader scope. Although CellML was originally designed to describe biological models, it can actually store any mathematical model. CellML facilitates modeling by enabling component-based design of new models, as well as by using encapsulation techniques and import mechanisms (Lloyd et al. 2004). These features allow users to build models based on previously tested models.

Other standardization efforts involve creating common semantics for modeling biological systems. The Systems Biology Ontology (SBO) 'project (SBO: systems biology ontology 2008) aims at building controlled vocabularies and ontologies for systems biology-related problems (in the context of computational modeling). The purpose is to build the standard semantics of these areas of research complementarily to standard syntaxes like SBML of CellML.

14.2.3 Omics Integration Algorithms

The latest research has made it clear that most biological characteristics cannot be attributed to single molecules. Rather biological systems activity is determined by complex interactions between the numerous components of a cell, such as DNA, RNA, proteins and diverse molecules (Barabási and Oltvai 2004; Albert 2005). Models that simply describe the behavior of individual elements cannot cope with the extreme complexity of biological systems, and more sophisticated models are required. Networks have been widely used for this purpose, causing the development of numerous algorithms for biological network analysis. One example is the SANDY algorithm (Luscombe et al. 2004). SANDY integrates transcriptional regulatory information and gene expression data in a wide variety of conditions. It employs network analysis techniques to uncover the dynamics of biological networks on a genomic scale. Another example is the GRAM algorithm

(GRAM). This algorithm combines genomics and transcriptomics data in order to discover gene modules – i.e. sets of genes co-expressed and regulated by the same set of transcription factors (TFs). This information can lead to the discovery and understanding of the network representing the general regulation of gene expression in a cell. The gene modules are found by establishing thresholds to the *p*-value binding genes with TFs, using randomization tests to reduce unwarranted assumptions. Variants on the network-based approach are employed to analyze different biological aspects. For example, developmental genetic regulatory networks integrate genomic and phenomic data to give insight into development phenomenology (Davidson et al. 2003). These models are vital for analyzing and comprehending of how genetic regulatory networks work, as they can lead to DNA-specific predictions.

Another approach for describing biological data is to use matrices to relate genes with a set of conditions (Madeira and Oliveira 2004). Biclustering algorithms are used to extract biclusters – i.e. subsets of genes sharing similar conditions – from such matrices. Although biclustering was introduced in the mid-1970s, it was not until the year 2000 that they were first used to analyze biological data (Cheng and Church 2000). Since then, many biclustering algorithms have been developed for biological data analysis. For example, the SAMBA algorithm calculates high quality clusters in polynomial time by combining graph theory and statistical data analysis (Tanay et al. 2002). The algorithm is applied to genomics and DNA-protein interactions to discover new biological associations. It was tested with data from carcinogenic cells, achieving better results than earlier approaches.

The growth of techniques for integrating omics data has produced a parallel growth in tools and resources for researchers to visualize data and create increasingly accurate biological models. Many of these tools support the most widespread standards, such as SBML, which facilitates data and results sharing. (A full list of tools supporting SBML can be found at http://sbml.org/SBML_Software_Guide/SBML_Software_Summary) Tools for modeling, visualizing and simulating cellular behavior and biological pathways are quite common – e.g. CellDesigner, SmartCell or Cell Illustrator. Other resources make biological data available for researchers to analyze with the available tools. Panther Pathway (PANTHER) features over 165 pathways (primarily signaling) that can be browsed on the web (via CellDesigner). PathArt (PathArt) contains a database of biomolecular interactions, with tools for searching, analyzing and visualizing data. It uses curated data from the literature and public databases for more than 2,100 signaling and metabolic pathways.

14.3 Examples of Omics Integration: International Efforts

Numerous international efforts at developing integration platforms are now under way. The goal is to help researchers to seamlessly access the vast amounts of data available today.

14.3.1 ACGT

ACGT (Advancing Clinico-Genomic Trials on Cancer) is an R&D project funded by the European Commission under the e-Health program. The main aim of ACGT is to develop an ontology-driven infrastructure comprising a set of semantic grid services enabling the execution of analytical workflows in the context of multi-centric, post-genomic clinical trials. This project focuses on providing an open source platform, allowing the cancer research community to integrate clinical and genomic data at different levels. The data integration process is carried out by the Semantic Mediation Layer together with the Data Access Services. Semantic and syntactic problems are dealt with separately, providing seamless domain-independent data integration. The project uses data from two existing clinical trials on cancer, namely TOP, regarding breast cancer, and SIOP, dealing with nephroblastoma. These trials are used in the requirements acquisition and testing processes. More information about ACGT can be found in (Tsiknakis et al. 2006)

14.3.2 caBIG

caBIG (cancer Biomedical Informatics Grid) is an information network supported by the US National Cancer Institute with the main aims of providing an infrastructure to connect scientists and practitioners, developing standard rules and a common language to lever information exchange, and building and adapting tools for the collection, analysis, integration and dissemination of cancer-related data. One of the tools that caBIG is providing for data integration is semCDI (Shironoshita et al. 2008), a query formulation model that uses the semantic metadata provided in caBIG to build queries and integrate heterogeneous data. This model has been tested by querying integrated data about human proteins using caBIO (Kraj and McIndoe 2005), GeneConnect (geneConnect) and Pathway Interaction Database (PID) data sources. The use of the caBIG semantic metadata allows the construction of high-level semantically rich queries for these integrated repositories. More information about caBIG can be found in (Langella et al. 2007).

14.3.3 HeC

The European Commission funded Health e-Child (HeC) project aiming at developing an integrated healthcare platform for European pediatrics, based on Grid-based services aimed at manipulating and sharing heterogeneous biomedical data (Freund et al. 2006). In the HeC methodology for vertical knowledge integration, patient information is integrated according to disease models, instead of using public biomedical databases. Patient data collected in the hospitals are annotated using a common model called Integrated Disease Knowledge Model (IDKM). One IDKM is built for each particular disease applying a methodology that uses well-known biomedical ontologies and public databases. This methodology identifies six different levels of granularity in the information, namely population, individual, organ, tissue, cellular and molecular. Elements gathered from the different ontologies are classified into these categories. Relations among concepts in different models are also identified. More information about HeC can be found in (Branson et al. 2008).

14.3.4 ONTOFUSION

ONTOFUSION is a database integration system. It was designed to provide unified access to multiple, heterogeneous biological and medical sources that are publicly available over the Internet (Alonso-Calvo et al. 2007), but can be used to access private patient databases as well. In ONTOFUSION, a wrapper is built for each one of the data sources to grant homogeneous access. Each data source is mapped to a global ontology in order to semantically homogenize the language of the concepts and relations. The mapping process produces a virtual repository, a homogenized view of a single database. With the ONTOFUSION approach, these repositories can be automatically combined into abstract integrated virtual repositories, offering a unique view of a set of integrated databases. ONTOFUSION offers a web-based end-user interface that helps users with the query formulation process. The tool has been tested by integrating data from multiple information sources of different kinds, such as Ensembl, SwissProt, OMIM, Prosite, SNP, PDB, ENZYME, LocusLink, and InterPRO (Pérez-Rey et al. 2005).

14.3.5 BIRN

The Biomedical Informatics Research Network (BIRN) is an NCRR/NIH sponsored project for creating and maintaining a virtual community of partners who share information across a data management, integration and analysis infrastructure (BIRN Web Site). Each partner contributes a database of its specific domain to the project. A semantic mediator and a series of analysis tools have been developed to offer clinicians the necessary infrastructure to conduct brand new experiments.

The integration module in BIRN is based on a mediator/wrapper design. The wrappers provide an SQL interface to all underlying data, relieving the mediator from syntactic heterogeneities. The actual semantic mediator, named Metropolis-II, employs external ontologies as global schemas and uses a global-as-view approach. The mediator includes functionality for underlying repositories to export both data and functions, providing a highly flexible architecture. By adopting this approach not only databases but also computational resources can be integrated (Astakhov et al. 2005).

14.3.6 SIG

In the field of standardization, the HL7 Clinical Genomics special interest group (SIG) is working on the development of standards to enhance the exchange of clinical and genomic data between different institutions (HL7). The core of this standardization effort is the 'Genotype' model that encapsulates a range of genomic data types. The SIG has been working on breast cancer with different institutions involved in cancer research, like Massachusetts General Hospital. This initiative is led by the IBM Haifa Group.

14.4 Future of Data Integration in Genomics Medicine

Biological model generation using omics data still has a long way to go before it is a mature area. A lot of progress has to be made on omics data analysis, processing and integration techniques before we get a proper understanding of the biological processes that take place inside our organism. Nevertheless, the field of omics data integration is expected to provide important advances in a wide range of clinical areas, including cancer research.

Accurate biological modeling will signify a great advance in many areas of medicine. Using such models, medical researchers could, for example, explain the cause of numerous diseases, or establish a person's predisposition to suffer from a specific disease. Drug design applications are expected to benefit from omics integration as well (Nikolsky et al. 2005). Even now, such disparate fields as nutrigenomics or toxicogenomics are beginning to take advantage of the results of integrating heterogeneous omics data (Stierum et al. 2005; Corthésy-Theulaz et al. 2005). Additionally, clinicians will be able to foresee the effect of a treatment on individual patients. Therapies will be designed specifically for each person based on her/his genomic signature, resulting in greatly improved healthcare.

14.4.1 Personalized Genomic Medicine

It has been proven that genomic similarity across random persons lies at around 99.5%, compared to 99.9% as was generally believed (Levy et al. 2007). Mason et al. (2007) reviews the types of variations in the human genome, and divides them into five categories, namely: (a) large-scale, (b) rampant small-scale, (c) rogue agents, (d) transcribed based and (e) epigenomics based. It is known that genomic variations can be the reason for patients' positive predisposition to specific diseases, or the level of responsiveness to a drug – e.g. inherited genetic markers have been recently tested for predicting clinical outcomes in cancer patients (Rebbeck 2006). There are already initiatives to establish infrastructures for researchers to test their hypotheses and uncover the genomic relation to disease predisposition. The Personal Genome Project (PGP) at Harvard University collects genetic samples from volunteers for publication in the interests of scientific research. The goal is to provide data for eventually achieving the personalized genomic medicine, developing tools for the design of patient-specific therapies.

In addition, there are efforts at developing comprehensible systems to enable clinical and omics data access and interpretation. A relevant example is the Mayo Clinic/IBM Computational Biology Collaboration (de Groen et al. 2003), led by Professor de Groen. This consortium has created a data warehouse containing clinical and genomic data from millions of patients. The database management system is enhanced by a powerful search engine, able to handle a large variety of data types and query forms. This engine is complemented by an *ad hoc* designed end-user interface.

To be able to implement real personalized genomic medicine, all omics information that researchers have at their disposal has to be contextualized. New technologies have yet to emerge before this can be achieved. Bhowmick et al. (2003) described the challenges that lie ahead of this enterprise. Of these, we should stress the need to develop common standards for storing omics information that can be adopted by different researchers – SBML and CellML are a first step in this direction – and to seamlessly integrate heterogeneous databases and properly combine omics data techniques so that researchers are able to access rich data repositories in a uniform manner.

Personalized genomic medicine is expected to change health care and improve our lives. It remains, however, to be seen how drastic this transformation will be. Currently, two different opinions coexist (Billings et al. 2005): on the one side, some predict that genomic medicine will imply a radical healthcare revolution, making current models obsolete. Others, however, foresee a more gradual application of new genomic technologies in the marketplace. In any case, personalized genomic medicine will have important implications for cancer research. It has been proven that, just like many common diseases, a great number of cancers are caused by interactions between genetic and environmental factors (Rubinstein and Roy 2005). Personalized genomic medicine can help to uncover such interactions, enabling precise predictions of whether a person is predisposed to suffer from a specific cancer, or her/his expected response to a treatment. Effective therapies could be designed against cancer, greatly enhancing our ability to fight this disease.

References

Abbott R (2004) Emergence, entities, entropy, and binding forces. The Agent 2004 Conference on Social Dynamics: Interaction, Reflexivity and Emergence, Argonne National Labs and

University of Chicago. http://abbott.calstatela.edu/PapersAndTalks/abbott_agent_2004.pdf, Accessed 7 November 2008

Abbott R, "Emergence, Entities, Entropy, and Binding Forces," The Agent 2004 Conference on: Social Dynamics: Interaction, Reflexivity, and Emergence, Argonne National Labs and University of Chicago, October 2004

Albert R (2005) Scale-free networks in cell biology. J Cell Sci 118:4947–4957

Alonso-Calvo R, Maojo V, Billhardt H et al (2007) An agent- and ontology-based system for integrating public gene, protein, and disease databases. J Biomed Inform 40:17–29

Astakhov V, Gupta A, Santini S et al (2005) Data integration in the Biomedical Informatics Research Network (BIRN), Data integration in the life sciences, 1st edn. Springer, Berlin

Baker PG, Brass A, Bechhofer S et al (1998) TAMBIS: Transparent access to multiple bioinformatics information sources. An overview. In: Proceedings of the Sixth International Conference of Intelligent Systems for Molecular Biology (ISMB98), Montreal.

Barabási AL, Oltvai ZN (2004) Network biology: understanding the cell's functional organization. Nat Rev Genet 5:101–113

Bhowmick SS, Singh DT, Laud A (2003) Data management in metaboloinformatics: issues and challenges. LNCS 2736:392–402

Billings PR, Carlson RJ, Carlson J et al (2005) Ready for genomic medicine? Perspectives of health care decision makers. Arch Intern Med 165:1917–1919

Biomedical Informatics Research Network. http://www.nbirn.net/index.shtm. Accessed 7 November 2008

Branson A, Hauer T, McClatchey R et al (2008) A data model for integrating heterogeneous medical data in the health-e-child project. Stud Health Technol Inform 138:13–23

Burgun A, Bodenreider O (2008) Accessing and integrating data and knowledge for biomedical research. Yearbook of medical informatics, pp 91–101

Cali A, De Giacomo G, Lenzerini M (2001) Models for information integration: turning local-as-view into global-as-view. In: Proceedings of International Workshop on Foundations of Models for Information Integration (10th Workshop in the series foundations of models and languages for data and objects), Viterbo.

The CellML web page. http://www.cellml.org/index_html. Accessed 7 November 2008

Cheng Y, Church GM (2000) Biclustering of expression data. Proc Int Conf Intell Syst Mol Biol 8:93–103

Coen M, Ruepp SU, Lindon JC et al (2004) Integrated application of transcriptomics and metabonomics yields new insight into the toxicity due to paracetamol in the mouse. J Pharm Biomed Anal 35:93–105

Collins FS, McKusick VA (2001) Implications of the Human Genome Project for medical science. JAMA 285:540–544

Corthésy-Theulaz I, den Dunnen JT, Ferré P et al (2005) Nutrigenomics: the impact of biomics technology on nutrition research. Ann Nutr Metab 49:355–365

Cusick ME, Klitgord N, Vidal M et al (2008) Interactome: gateway into systems biology. Hum Mol Genet 14:R171–181

Davidson EH, McClay DR, Hood L (2003) Regulatory gene networks and the properties of the developmental process. Proc Natl Acad Sci USA 100:1475–1480

de Groen PC, Dettinger R, Johnson P (2003) Mayo Clinic/IBM computational biology collaboration: a simple user interface for complex queries. In: Universal access in HCI – volume 4 of the proceedings of human–computer interaction (HCI) international, pp 1083–1087

Enard W, Khaitovich P, Klose J et al (2002) Intra- and interspecific variation in primate gene expression patterns. Science 296:340–343

Fiehn O (2002) Metabolomics – the link between genotypes and phenotypes. Plant Mol Biol 48:155–171

Freund J, Comaniciu D, Ioannis Y et al (2006) Health-e-Child Consortium. Health-e-child: an integrated biomedical platform for grid-based paediatric applications. Stud Health Technol Inform 120:259–270

Galperin MY (2008) The molecular biology database collection: 2008 update. Nucleic Acids Res 36:D2–4

geneConnect. https://cabig.nci.nih.gov/tools/GeneConnect. Accessed 7 November 2008.
The GRAM Algorithm. http://www.psrg.lcs.mit.edu/Networks/alg/GRAM.pdf. Accessed 7 November 2008
Gruber TR (1993) A translation approach to portable ontologies. Knowl Acquis 5:199–220
Heijne WH, Stierum RH, Slijper M et al (2003) Toxicogenomics of bromobenzene hepatotoxicity: a combined transcriptomics and proteomics approach. Biochem Pharmacol 65:857–875
Hirai MY, Yano M, Goodenowe DB et al (2004) Integration of transcriptomics and metabolomics for understanding of global responses to nutritional stresses in Arabidopsis thaliana. Proc Natl Acad Sci USA 101:10205–10210
Hirai MY, Klein M, Fujikawa Y et al (2005) Elucidation of gene-to-gene and metabolite-to-gene networks in arabidopsis by integration of metabolomics and transcriptomics. J Biol Chem 280:25590–25595
Clinical Genomics special interest group. http://www.haifa.ibm.com/projects/software/cgl7/specifications.html. Accessed 7 November 2008
Ihmels J, Bergmann S, Gerami-Nejad M et al (2005) Rewiring of the yeast transcriptional network through the evolution of motif usage. Science 309:938–940
Iozzo RV (2001) Heparan sulfate proteoglycans: intricate molecules with intriguing functions. J Clin Invest 108:165–167
Ippolito JE, Xu J, Jain S et al (2005) An integrated functional genomics and metabolomics approach for defining poor prognosis in human neuroendocrine cancers. Proc Natl Acad Sci USA 102:9901–9906
Jarke M, Jeusfeld M A, Quix C et al (1998) Architecture and quality in data warehouses. In: Pernici B, Thanos C (eds) Proceedings of the 10th international conference on advanced information systems engineering (08–12 June 1998). Lecture notes in computer science, volume 1413. Springer, London, pp 93–113
Joyce AR, Palsson B (2006) The model organism as a system: integrating 'omics' data sets. Nat Rev Mol Cell Biol 7:198–210
Khaitovich P, Muetzel B, She X et al (2004) Regional patterns of gene expression in human and chimpanzee brains. Genome Res 14:1462–1473
Khaitovich P, Hellmann I, Enard W et al (2005) Parallel patterns of evolution in the genomes and transcriptomes of humans and chimpanzees. Science 309:1850–1854
Kimball R (1996) The data warehouse toolkit: practical techniques for building dimensional data warehouses. New York: John Wiley
Kraj P, McIndoe RA (2005) caBIONet-A.NET wrapper to access and process genomic data stored at the National Cancer Institute's Center for Bioinformatics databases. Bioinformatics 21:3456–3458
Kristensen C, Morant M, Olsen CE et al (2005) Metabolic engineering of dhurrin in transgenic Arabidopsis plants with marginal inadvertent effects on the metabolome and transcriptome. Proc Natl Acad Sci USA 102:1779–1784
Langella SA, Oster S, Hastings S et al (2007) The Cancer Biomedical Informatics Grid (caBIG) Security infrastructure. In: AMIA annual symposium proceedings, pp 433–437
Lenzerini M (2002) Data integration: a theoretical perspective. In: Proceedings of the twenty-first ACM SIGMOD-SIGACT-SIGART symposium on principles of database systems. PODS '02 ACM, New York, pp 233–246
Levy AY, Rajaraman A, Ordille JJ (1996) Querying heterogeneous information sources using source descriptions. In: Proceedings of the twenty-second international conference on very large databases. Mumbai, India, pp 251–262
Levy S, Sutton G, Ng PC et al (2007) The diploid genome sequence of an individual human. PLoS Biol 5:e254
Lloyd CM, Halstead MD, Nielsen PF (2004) CellML: its future, present and past. Prog Biophys Mol Biol 85:433–450
Luscombe NM, Babu MM, Yu H et al (2004) Genomic analysis of regulatory network dynamics reveals large topological changes. Nature 431:308–312
Madeira SC, Oliveira AL (2004) Biclustering algorithms for biological data analysis: a survey. IEEE/ACM Trans Comput Biol Bioinform 1:24–45
Maojo V, Tsiknakis M (2007) Biomedical informatics and healthGRIDs: a European perspective. IEEE Eng Med Biol Mag 26:34–41

Martín L, Bonsma E, Anguita A et al (2007) Data Access and Management in ACGT: Tools to solve syntactic and semantic heterogeneities between clinical and image databases, in First International Workshop on Conceptual Modelling for Life Sciences Applications (CMLSA 2007): 4802 (LNCS) / pp 24-335-9 Nov 2007, Auckland (New Zealand)

Mason CE, Seringhaus MR, Sattler de Sousa e Brito C (2007) Personalized Genomic Medicine with a Patchwork, Partially Owned Genome. Yale J Biol Med 80:145–151

What is systems biology? The Munich systems biology forum. http://www.msbf.mpg.de/ho_sys_ch.html. Accessed 7 November 2008

Editorial (2004) Making data dreams come true. Nature 428:239

Nikolsky Y, Nikolskaya T, Bugrim A (2005) Biological networks and analysis of experimental data in drug discovery. Drug Discov Today 10:653–662

PANTHER classification system. http://www.pantherdb.org/pathway/. Accessed 7 November 2008

PathArt database. http://bioinformatics.unc.edu/software/pathart/index.htm. Accessed 7 November 2008

Pérez-Rey D, Maojo V, García-Remesal M et al (2005) ONTOFUSION: ontology-based integration of genomic and clinical databases. Comput Biol Med 36:712–730

Pérez-Rey D, Anguita A, Crespo J (2006) OntoDataClean: ontology-based integration and preprocessing of distributed data. Lecture Notes Comput Sci 4345:262–272

Petrik V, Loosemore A, Howe FA et al (2006) OMICS and brain tumor biomarkers. Br J Neurosurg 20:275–280

Personal genome project. http://www.personalgenomes.org/. Accessed 7 November 2008

PID. http://pid.nci.nih.gov/. Accessed 7 November 2008

Rebbeck TR (2006) Inherited genetic markers and cancer outcomes: personalized medicine in the postgenome era. J Clin Oncol 24:1972–1974

Rubinstein WS, Roy HK (2005) Practicing medicine at the front lines of the genomic revolution. Arch Intern Med 165:1815–1817

Russ Abbott, Emergence, Entities, Entropy and Binding Forces, In Proceedings of "The Agent 2004 Conference on: Social Dynamics: Interaction, Reflexivity and Emergence", Chicago, 2004

The systems biology markup language. http://sbml.org/Main_Page. Accessed 7 November 2008

SBO: systems biology ontology. http://www.ebi.ac.uk/sbo/. Accessed 7 November 2008

Shironoshita EP, Jean-Mary YR, Bradley RM et al (2008) semCDI: a query formulation for semantic data integration in caBIG. J Am Med Inform Assoc 15:559–568

Shriver Z, Raguram S, Sasisekharan R (2004) Glycomics: a pathway to a class of new and improved therapeutics. Nature Rev Drug Discov 3:863–873

Sohal D, Yeatts A, Ye K et al (2008) Meta-analysis of microarray studies reveals a novel hematopoietic progenitor cell signature and demonstrates feasibility of inter-platform data integration. PLoS ONE 3:e2965

Stierum R, Heijne W, Kienhuis A et al (2005) Toxicogenomics concepts and applications to study hepatic effects of food additives and chemicals. Toxicol Appl Pharmacol 207:179–188

Sujanski W (2001) Heterogeneous database integration in biomedicine. J Biomed Inform 34:285–298

Tanay A, Sharan R, Shamir R (2002) Discovering statistically significant biclusters in gene expression data. Bioinformatics 18:S136–144

Tanay A, Regev A, Shamir R (2005) Conservation and evolvability in regulatory networks: the evolution of ribosomal regulation in yeast. Proc Natl Acad Sci USA 102:7203–7208

Tsiknakis M, Kafetzopoulos D, Potamias G et al (2006) Building a European biomedical grid on cancer: the ACGT Integrated Project. Stud Health Technol Inform 120:247–258

Wenk MR (2005) The emerging field of lipidomics. Nat Rev Drug Discov 4:594–610

Wiechert W, Schweissgut O, Takanaga H et al (2007) Fluxomics: mass spectrometry *versus* quantitative imaging. Curr Opin Plant Biol 10:323–330

Index

P

Q

R

S

T

U

Y